Statistical Methods in Environmental Epidemiology

Statistical Methods in Environmental Epidemiology

Duncan C. Thomas

OXFORD
UNIVERSITY PRESS

Great Clarendon Street, Oxford OX2 6DP

Oxford University Press is a department of the University of Oxford.
It furthers the University's objective of excellence in research, scholarship,
and education by publishing worldwide in

Oxford New York

Auckland Cape Town Dar es Salaam Hong Kong Karachi
Kuala Lumpur Madrid Melbourne Mexico City Nairobi
New Delhi Shanghai Taipei Toronto

With offices in

Argentina Austria Brazil Chile Czech Republic France Greece
Guatemala Hungary Italy Japan Poland Portugal Singapore
South Korea Switzerland Thailand Turkey Ukraine Vietnam

Oxford is a registered trade mark of Oxford University Press
in the UK and in certain other countries

Published in the United States
by Oxford University Press Inc., New York

British Library Cataloguing in Publication Data
Data available

Library of Congress Cataloging in Publication Data
Data available

Typeset by Newgen Imaging Systems (P) Ltd., Chennai, India
Printed in Great Britain
by
CPI Antony Rowe, Chippenham, Wiltshire

ISBN 978–0–19–923289–5 (hbk)
      978–0–19–923290–1 (pbk)

10 9 8 7 6 5 4 3 2 1

To my father, a role model for my career
Any my mother, who taught me to write

Contents

Preface

Environmental epidemiology is arguably the basic science upon which government regulatory agencies rely for setting standards to protect the public from environmental and occupational hazards. While other disciplines such as toxicology are obviously also relevant, it is generally agreed that direct observation of risks in humans is the best basis for setting public health policy. Recent years have seen a shift in emphasis in epidemiology toward molecular and genetic epidemiology, with a focus on using epidemiologic methods to understand the fundamental mechanisms of disease or to incorporate basic science concepts and measurements of genetic modifiers and biomarkers into the conduct of epidemiologic studies. While laudable, such efforts should not denigrate the value of traditional epidemiologic studies of exposure–response relationships in humans that will ultimately provide the basis for environmental safety standards. This book represents an attempt to lure young investigators back into a field that may have lost some of its cachet of late in this recent enthusiasm for molecular epidemiology.

This book was conceived during a vacation in the Galapagos Islands, where the inspirational spirit of Charles Darwin is strong. I began to appreciate that Darwin was as interested in the environment as a driving force of evolution as he was in genetics. The first few chapters of the book were drafted in longhand there, between conversations with our outstanding naturalist guide, and mailed back to my long-suffering assistant, Stephanie Cypert-Martinez, who did a remarkable job of deciphering my scribbles. The rest of the book was drafted back at the University of Southern California during the fall semester, while I was struggling to keep a week ahead of my students during my new course on Environmental Biostatistics. I am grateful the USC for awarding me a sabbatical leave the following year, so that I could focus on completing the manuscript, while visiting a number of units that are major research centers in environmental epidemiology. In particular, I would like to thank my hosts at the Beijing Medical University and Peking University Department of Environmental Engineering, the Radiation Effects Research Foundation, the Columbia University Bangladesh Arsenic Study, the Bhopal Memorial Hospital and Research Institute, INSERM Unit 574 on Cancer and the Environment in Villejuif Paris, and the numerous researchers at each institution who shared their experiences, read portions of the book, and provided helpful comments. I am also grateful for the extensive comments I received from

Bryan Langholz, Mark Little, Nino Künzli, Don Pierce, Jonathan Samet, Dan Stram, Jonathan Samey, Erich Wichmann, and several anonymous reviewers. And of course to my wife, Nina, who again supported me throughout this process and kept my spirits up.

The book is intended primarily as a textbook for an advanced graduate course in epidemiology and biostatistics. As such, it provides only a brief overview in Chapters 2–4 of some of the basic design and analysis techniques that one might find in any of the standard epidemiology textbooks. A general familiarity with this material from basic courses in epidemiology and biostatistics is assumed, and these chapters are provided only so the book will be somewhat self-contained. The remainder of the book aims to cover a range of methods that are in some sense unique to *environmental* epidemiology. While there have been many textbooks covering one or more of these more specialized topics, when I began this project there appeared to be none that attempted to cover such a broad range of topics in a unified manner. Perhaps the closest to achieving this goal is the recent book by Baker and Nieuwenhuisen, "*Environmental Epidemiology: Study Methods and Application*" (Oxford University Press, 2008). While providing an excellent overview of epidemiologic study design principles and basic statistical methods, illustrated with a broad range of applications in environmental epidemiology, the present text aims to take the treatment of statistical methods to a higher level of sophistication. Much of this advanced material is my attempt to explain to myself some of the new literature I had to read to be able write the book; I hope that established epidemiologic and statistical researchers in the field will also find some of this material helpful as a reference book.

As in my previous book on *Statistical Methods in Genetic Epidemiology* (Oxford University Press, 2004), where I used the story of breast cancer genetics to illustrate the statistical methods, here I have tried to illustrate the various methodological challenges with examples from the environmental epidemiology literature. As no one story proved sufficient to illustrate the full range of problems, I have had to rely on two that I have been involved with for many years—ionizing radiation and air pollution. My aim is not, however, to provide a systematic review of either of these massive literatures or a quantitative risk assessment, only to pick and choose amongst them to illustrate methodologic points.

A tale of two exposures

Throughout history, mankind has experienced numerous and diverse environmental challenges. Some life-threatening problems have doubtless influenced the course of human evolution. Others, less severe, have nevertheless afflicted us with disease or affected us in other ways. Some are natural: climate, the micro-cosmos that surrounds us; others manmade: the myriad forms of pollution of the air, water, soil. Some are highly localized (toxic waste disposal sites), some regional (air pollution), some truly global in extent (climate change, depletion of the ozone layer). And new environmental problems are continually appearing, like fallout from nuclear weapons testing or novel chemicals produced by technological progress.

The Greeks believed that all matter was composed of four "elements": air, water, earth, and fire. We now recognize a more complex hierarchy of molecules, formed from 92 naturally occurring chemical elements (atoms), themselves composed of protons, neutrons, electrons, and a veritable zoo of subatomic particles and ultimately the truly elementary particles known as quarks. Nonetheless, the Greek's concept provides a useful metaphor for the scope of environmental epidemiology: Air, and the various pollutants we have contaminated it with; Earth, the source of all our nourishment, the nutritional choices we make and the pesticides we have used to boost its productivity; Water, the oceans from which life evolved, the lakes and rivers that transport the contaminants from earth and air; and Fire, or more generally energy, solar, ionizing radiation, and the entire electromagnetic spectrum.

This book is concerned with the study of the effects of the environment on human health—the field of environmental epidemiology—specifically, approaches to study design and methods of statistical analysis. If epidemiology is broadly the study of the distribution and determinants of disease in human populations, environmental epidemiology is more specifically focused on environmental factors in disease. The environment can be considered broadly as all risk factors other than genetics—including both exogenous and endogenous factors (e.g., nutritional status, hormones)—but here we will be focused on the *external* environment, particularly the changes brought about by human activity. The "nature vs. nurture" controversy is, of course passé, as we now recognize that most diseases that concern us today are the result of complex interactions between genes and environment. Here our field overlaps with that of genetic epidemiology (the subject of my previous book (Thomas 2004)), but no treatment of environment epidemiology would be complete without some discussion of gene–environment interactions.

The physical environment is also intimately tied to the social environment. Social, economic, and racial factors also affect both the distribution of environmental exposures and the risk of disease. Poor people are more likely to live near major sources of pollution and their host resistance to environmental insults more likely to be compromised. Here, we are not primarily concerned with social risk factors directly, but we cannot ignore their potential confounding and modifying effects on environmental risk factors.

There are at least three distinguishing features of environmental epidemiology from epidemiology more generally. First, most environmental risk factors—at least man-made factors—are modifiable. Thus, once we understand their effects on human health, interventions and regulations may be possible to ameliorate the problem. Environmental epidemiology thus has profound implications for public health policy, and can even be thought of as the basic science of risk assessment and risk management.

Second, environmental factors are spatially distributed, varying locally, regionally, and globally. These spatial correlations pose unique challenges to study design and analysis. But individuals also differ in their exposures due to behavioral and host factors—even individuals living at the same location. This heterogeneity poses both a challenge and an opportunity.

Third, environmental exposures vary temporally. Here we are not particularly concerned with the major infectious diseases of the past, although infectious disease epidemiologists continue to be confronted with new agents or strains of old ones or resurgence of those previously under control. But within the course of a person's lifetime, exposures come and go or vary in intensity. Most chronic diseases are due to the accumulated effects of a lifetime history of exposure. Again, this poses both a challenge and an opportunity. The epidemiologist must carefully assess this entire history, but comparisons over time can shed light on exposure–response relationships, or more generally what we will call exposure–time–response relationships.

These three distinguishing features combine in interesting ways to pose unique methodological challenges that are the subject of this book. Rather than systematically orienting the reader to the organization of the chapters that follow, I begin with a capsule summary of the two "tales" that will be used throughout to illustrate the various methodological problems and some of the creative solutions that have been found. First, ionizing radiation, then air pollution—these two examples provide rich fodder and illustrate quite different points. To preserve the narrative flow, we defer providing citations to the original literature until later chapters, where these studies are discussed.

The ionizing radiation tale

In 1945, the United States dropped atomic bombs on Hiroshima and Nagasaki, with devastating consequences. Approximately 100,000 people are estimated to have died from the immediate effects of the blast. Though they dwarf the numbers of deaths attributable to the radiation released by the bombs, these are not the primary concerns of this tale. Their causes are obvious and need no sophisticated epidemiologic study to reveal.

The radiation effects are early and late. The former comprise the various well-known symptoms of "acute radiation sickness" in varying degrees of severity depending upon dose. (Various units are used in different context to describe radiation exposures, but here we will be concerned with measures of energy deposition known as the Gray (Gy) and of "biologically effective tissue dose" known as the *Sievert* (Sv), the now standard international units replacing the previously used *rad* (1 Gy = 100 rad) and *rem* (1 Sv = 100 rem) respectively. Individuals closer than about 0.75 km from the hypocenters typically received whole body doses in excess of 6 Gy (Preston et al. 2004), generally a lethal dose (even if they survived the immediate blast effects) and would probably have died within a week or so. Their causes of death are also obvious and not directly a part of our tale, as they were thus not at risk of the late effects that concern us here. Individuals not quite so close to the hypocenters, but still receiving substantial doses (say >1 Gy, at about 1.25 km), may have experienced milder symptoms of acute radiation sickness, including hair loss (epilation) and chromosomal aberrations. Most such individuals will have survived the immediate blast and acute radiation effects, and gone on to be at risk of the late effects. We shall see how their experience of these symptoms can serve as a form of "biological dosimetry," to be combined with the "physical dosimetry" based on location and shielding. We can also ask whether such symptoms can provide evidence of genetic susceptibility to the late effects of radiation. Beyond about 2.5 km, doses were generally less than 0.01 Gy.

The primary late effect we will discuss is cancer, although the reader should be aware that many other endpoints have been studied and a few of those definitively associated with radiation, notably cataracts and, among those 12–15 weeks of gestation at the time of bombing, severe mental retardation. Most, but not all, cancer sites have been found to be radiation related, but with important differences in the slope of the dose–response relation and modifying effects of age, latency, and other risk factors. One cancer—leukemia—stands out as being uniquely radiosensitive and arising much earlier than the others. Of the remainder, numbers of cases are sometimes inadequate for detailed analysis, so for our purposes, it suffices to consider the broad category of all solid tumors combined.

In the immediate aftermath of the bombing, medical research on these late effects was, needless to say, not the highest priority. Some years later the joint U.S.–Japan Atomic Bomb Casualty Commission was established, later renamed the Radiation Effects Research Foundation (RERF). Despite concerns expressed by some about using the survivors as human guinea pigs, the overwhelming sentiment has been that the lessons that could be learned from this experience should not go to waste, and the voluntary participation of the survivors in research that has been conducted under the highest ethical standards has been exemplary. The full collaboration of scientific researchers from two former enemy counties (and many others) has been a model of multidisciplinary, international cooperation. As a result of this research (and a larger body of data from other exposure contexts), we know more about the health effects of radiation than virtually any other environmental exposure.

In 1950, the commission established two overlapping cohort studies (Chapter 2), the Life Span Study (LSS) and the Adult Health Study (AHS). The LSS comprised approximately 100,000 survivors of all ages who would be followed passively for mortality until the complete extinction of the cohort. At present, approximately half remain alive. The AHS is a smaller group of about 20,000 individuals who agreed to participate in periodic clinical examinations and provide biological specimens. For the purpose of this tale, we focus on the LSS cohort.

Radiation doses for the vast majority of the LSS cohort were estimated in 1965 and have been refined several times since then. These estimates are based on physical modeling of the spatial distribution of gamma and neutron radiation from the bomb, combined with information about the subjects' locations and shielding by buildings of the time of bombing. Despite the enormous efforts over decades of research that have gone into developing this dosimetry system, the reader should appreciate that these dose estimates are imprecise—a recurring theme in environmental epidemiology. We will return to the influence of these uncertainties shortly.

Identification of cohort members and passive mortality follow-up has been facilitated by the Japanese system of family registration known as *koseki*. Each nuclear family is registered in their town of residence. When an offspring marries and starts a new family, this event is recorded at their new place of residence, together with pointers to the records for their families of origin. The obvious privacy concerns were carefully addressed so that these records could be used for research purposes. In principle, these also provide a potential gold-mine for genetic research, but as of this writing, the greater sensitivity of such use has precluded the establishment of a family study. Instead, each member of the cohort is treated as an unrelated individual.

RERF statisticians have been at the forefront of developing sophisticated methods of analysis of dose–time–response relationships, the details of which will be a major theme of subsequent chapters. The major approach has been to prepare on extensive cross-tabulation of cause-specific deaths and person-years at risk by dose, age at exposure, attained age, sex, city, and other factors. These tabulations have been made available in public use data sets that have become one of the classic databases for both substantive and statistical methods research. Models can then be fitted to these data using the technique of Poison regression (Chapter 4). A variety of model forms have been explored, but most entail a general form for the baseline risk—the risk that would have occurred in the absence of exposure—either multiplied by or added to a model for the excess risk due to radiation. The latter is generally taken to be a linear (for solid cancers) or linear-quadratic (for leukemia) function of radiation dose, with slope coefficients that could depend upon the various modifying factors. Nonparametric dose–response analysis (Chapter 6) has shown no significant evidence of departure from linearity at doses as low as 0.1 Gy, although the possibility of a threshold below which there is no effect can never be excluded statistically. (Indeed, some have even suggested the possibility of a protective effect of low doses of radiation ("hormesis") using on evolutionary argument based on natural selection in the presence of background radiation, but there is no evidence to support this conjecture in the LSS data.) Nevertheless, radiobiological theory about the interactions of radiation with DNA (Chapter 13) strongly supports the linear-no-threshold hypothesis.

All cancers have some latent period between exposure and increases in risk. That latent period is the shortest for leukemia. Since the LSS data do not begin until 1950, they provide no information about risk less than five years after exposure, but some dose-related excess is already apparent 5–9 years after. The risk rises to a peak about 10–14 years after (the timing depending upon age of exposure) and then begins to decline, although some excess remains to this day, 50 years later. For most solid tumors, no excess is seen for at least 10 years, and the absolute excess rate continues to rise with age, although the relative rate (relative to the natural rise in rates of cancer rates with age) starts to decline after about 20 years (depending on cancer site and other factors). These estimated dose–response relationships—together with similar data from other occupational, environmental, and medically exposed populations—provide the scientific basis for radiation standards, as well as for resolving tort litigation and compensation claims, as discussed below. Exposure–time–response relationships can also shed light on basic mechanisms of carcinogenesis, like the multistage theory (Chapter 13).

Earlier, we mentioned that dose estimates had an inherent uncertainty. Considerable efforts have been devoted to quantifying the likely extent of these uncertainties and allowing for them in the statistical analysis. As well known in other areas of epidemiology, the general effect of random errors (here "nondifferential," not influenced by disease status, as in recall bias in case-control questionnaire data) is to dilute dose–response relationships, biasing the relative risk towards the null. Nevertheless, there are several important qualifications to this general principle (see Chapter 11), and, in particular the types of errors expected from the RERF dosimetry system are less subject to this bias than those based, say, on film badge dosimetry because the relationship between true and estimated doses differ by design of the respective dosimetry systems. In brief, the effect depends upon whether the measured doses are distributed around the true doses, as would be expected for a well-calibrated dosimeter ("classical error") or whether individuals' true doses are distributed around some assigned dose, as would be expected based on a dose reconstruction system based on external predictors like distance and shielding ("Berkson error"). The general rule of thumb about bias towards the null strictly applies only to classical error. Under appropriate assumptions, Berkson error does not produce a bias towards the null, although power will be reduced and confidence limits widened.

Earlier, we mentioned the idea of using early effects as a form of biological dosimetry. The basic idea is that for two individuals with the same assigned dose based on physical dosimetry, one manifesting early radiation symptoms is likely to have had a higher true dose than one who did not. Of course, it is also possible that there is real variation between individuals in their sensitivity to both early and late effects of radiation, perhaps due to genetic factors. The observation that the slope of the dose–response for cancer is steeper for those with epilation or chromosomal anomalies is, in principle, compatible with either hypothesis. However, these data are compatible with an average uncertainty of 35% without invoking any variability in sensitivity to early and late effects.

Nevertheless, there remains great interest in whether any host factors affect cancer radiosensitivity. Because detailed risk factor information and biological samples are not available on all LSS numbers, investigation of such questions requires some form of hybrid study design, as discussed in Chapter 5. For example, a nested case-control study of breast cancer was conducted, matching each breast cancer case in the LSS cohort with a randomly selected cohort member who was the same age at exposure and received the same dose. These members were then interviewed to learn about such established breast cancer risk factors as menstrual and reproductive histories. A similar design showed that for lung cancer, radiation and tobacco smoking interacted more nearly additively then multiplicatively. (As we shall see below, this interaction is quite different for uranium miners exposed to radiation.) Even more sophisticated designs are possible, exploiting information already available on the entire cohort to select controls in a manner that would considerably improve the power of the design, such as the "counter-matching" strategy discussed below.

Without biological specimens, it would still be possible to obtain indirect evidence of variability in genetic susceptibility using a family-based study to see if specific cancers aggregate in families, and whether the extent of such familial clustering varies with dose. The possibility of doing such a study using the *koseki* system is currently being explored. Various assays of biological samples from the clinical examination cohort have established that there is indeed variability in radiosensitivity as a function of an individual's genotype of various DNA repair and cell cycle control genes. Similar studies in medical irradiation cohorts are discussed below and more extensively in Chapter 12.

Before leaving this tale, it is worth considering some unique methodological challenges posed by other forms of radiation exposure: occupational, environmental, and medical. Several of these stories will be told in greater detail in subsequent chapters, so here we touch only on a few highlights, beginning with the uranium miners

Uranium for the production of nuclear weapons during World War II came primarily from foreign sources, notably the Belgian Congo and Czechoslovakia. Following the war, uranium mining began in earnest in the four-state Colorado Plateau (Colorado, Utah, New Mexico, and Arizona) of the United States, while on the other side of the Iron Curtain, several hundred thousand miners were employed in East German uranium mines. In total, perhaps about 10,000 U.S. miners were involved in what was in its early days, a largely unregulated industry. Uranium undergoes a series of radioactive decays, leading to radon gas, which accumulates in high concentrations in uranium mines (and also to a lesser extent in homes in certain geological areas). Radon is still radioactive and undergoes further disintegrations, through a sequence of "radon daughters," releasing a series of alpha particles (charged helium nuclei) that deliver most of the dose to the individuals exposed to radon. Unlike gamma rays and neutrons released by the atomic bombs that penetrate deeply, alpha particles cannot penetrate even the thickness of the skin. Inhaled, however, they can deliver a substantial dose to the lung epithelium, leading to lung cancer. Ultimately hundreds, if not thousands, of uranium miners died of this disease in what can only be characterized as a public health travesty. (See the report of the President's Advisory Committee on Human Radiation Experiments for an account of how knowledgeable government officials failed to require ventilation that could have prevented this epidemic, despite knowledge of the risks already available at the time.)

Belatedly, the U.S. Public Health Service launched a cohort study of 3,545 uranium miners in 1960 that is still on-going. Like the atomic bomb cohort study, a major challenge in this study was dosimetry, including efforts to characterize the uncertainties in dose estimates. There are two important differences, however. First, rather than using predictions based on location and shielding, actual measurements of radon or radon daughter concentrations were available. Second, unlike the instantaneous dose delivered by the atomic bombs, uranium miners received low dose-rate exposures over their entire working lifetimes.

The first difference is important because the measurement error structure is more like the classical error situation of other occupational settings like nuclear workers, where doses are based on film badges, but because the available measurements were sparse (some mines with few, if any, measurements for several years), dose estimation required an elaborate system of interpolation between nearby mines or over time. Furthermore, these uncertainties are correlated between individuals who worked in the same mines at the same time. We will discuss the effects of the shared uncertainties in Chapter 11.

The second difference raises both conceptual and analytical problems. Whereas the A-bomb data could be neatly summarized by a cross-tabulation of individuals by dose and other factors, this is not possible for the uranium miners because their lung cancer risks depend on the entire history of exposure. Furthermore, since the effect of each increment of exposure is modified by such factors as latency, age, and dose rate, no simple summary of an individual's exposure can tell the whole story. We will explore methods for modeling the effects of such extended exposure histories in Chapter 6.

From a biological perspective, differences in the nature of sparsely ionizing, penetrating forms of radiation (low Linear Energy Transfer, LET) like gamma rays and densely ionizing (high LET) radiation like alpha particles can have different implications for their interactions with DNA and subsequent carcinogenesis. Low LET radiation produces very narrow ionization tracks that are more easily repaired then the double-strand breaks typically produced by high-LET radiation. Since the probability of any DNA damage, single-stranded or double-stranded, is proportional to dose, the probably of two independent events leading to a double-strand break should be proportional to the square of dose. However, dose rate is also relevant, as a high dose-rate exposure is more likely to yield two independent breaks close enough in time that the first break will not have been repaired before the second occurs. Conventionally, it has been believed that high-LET and low dose-rate low-LET radiation would produce linear dose response relationships, whereas high dose-rate low-LET radiation would produce linear-quadratic dose–response relations. More recently, it has been noted that high-LET radiation can produce sublinear dose–response relations (steeper slopes at lower doses), a phenomenon that is difficult to reconcile with classical microdosimetry. One possible explanation is "bystander" effects, in which cells not directly hit by a radiation track, can be affected by damage to neighboring cells. The implications for formal models of carcinogenesis are discussed in Chapter 13.

In 1990, the U.S. Congress enacted the Radiation Exposure Compensation Act, which provided criteria for compensating uranium miners and residents downwind of the Nevada Test Site. (A separate act provided compensation for "Atomic Veterans" involved in nuclear weapons tests.) These criteria were loosely based on a synthesis of epidemiologic evidence from the world literature, incarnated in a set of "Radioepidemiologic Tables" published by NIH in 1985 and recently updated in the form of a computer program. These tables and program provide estimates of the "probability of causation" (PC) and uncertainty bounds on these probabilities that a cancer that occurred to an individual with a particular exposure history (and other possible causes or modifying factors) was caused by that exposure. (Note that the PC is a very different quantity from the risk that disease will occur in someone with a particular history, which forms the basis for risk management.) Under the principle of "balance of probabilities" that governs tort litigation, a claimant with a PC greater than 50% should be entitled to compensation. In practice, government agencies and courts have interpreted these criteria flexibly, often giving the claimant the benefit of the doubt—a principle formally enshrined in the legislation for the Veterans Administration, for example. While these public policy issues are beyond the scope of the volume, there are fundamental scientific issues about the estimability of PCs and the implications of their uncertainties that are discussed further in Chapter 16.

Although their lung cancer risks were high, the number of miners exposed to radon daughter products is small in comparison with the general population exposed to low-dose radon in their homes. The comparison of risks from occupational and residential settings raises yet other challenges, such as the extrapolation from high dose to low dose (Chapter 15), and methods of studying low-dose risks directly. Many case-control studies of residential radon exposure have been done, typically obtaining residence histories and attempting to measure radon concentrations at multiple locations over extended periods in as many of these homes as possible. Risk estimates derived from such studies have been highly variable, but generally compatible with those obtained by extrapolation from the uranium miner cohorts. In contrast, ecological correlation studies relating lung cancer rates at the county level to estimated average radon concentrations from sample surveys have often yielded negative correlations, even after adjustment for available population smoking and other demographic data. The reasons for this apparent paradox involve the "ecologic fallacy," namely that associations at the population level may not reflect those among individuals, due to confounding by individual and/or "contextual" variables, as explored in detail in Chapter 10.

As a consequence of nuclear weapons production and testing, as well as nuclear reactor accidents at Three Mile Island and more seriously at Chernobyl (Chapter 17), radionuclides have been widely dispersed throughout the environment, with the heaviest exposures to humans downwind of these facilities. Various epidemiologic studies have been conducted to assess whether any such exposures have had demonstrable health effects. These studies will be discussed in subsequent chapters, but here it is worth emphasizing another unique element of such studies, the major effort typically involved in "dose-reconstruction" using techniques of environmental pathway analysis and spatial statistics (Chapters 9 and 11). Pathway analysis entails reconstruction of the magnitude of the releases from the source (weapons test, industrial stack, waste disposal site, etc.), its transport by air or water, deposition on the ground, uptake in foodstuffs, distribution to humans, and ultimate fate in target organs. This is done for each radionuclide for each release—a mammoth undertaking, requiring much reliance on expert judgment as well as data and careful analysis of uncertainties of each step. Spatial analysis entails use of sophisticated statistical models for the correlations in measurements as a function of the distance between them to interpolate exposures at locations for which no measurements are available. Modern Geographic Information Systems (GIS) technologies have revolutionized the compilation and integration of spatially distributed data of different types (points, lines, areas) on different scales, and facilitate their analysis by spatial statistics methods. These methods will be illustrated in the following tale about air pollution, and have broad applicability to many types of environmental, socio-demographic, and medical data.

The final example of radiation exposure illustrates yet other unique methodological challenges. Unlike the previous examples, medical irradiation—diagnostic or therapeutic—is intended to improve health, but nevertheless also has the potential to cause harm. Needless to say, the risk/benefit balance is clearly in favor of its use, but the adverse effects are a valuable source of information about late radiation effects. A great advantage of medical irradiation studies is that dose is often known with much greater precision than is possible in environmental or occupational settings. Furthermore, because of the medical setting, much more detail may be available about other risk factors, long-term follow-up may be easier, and biological specimens may be readily available. This is particularly valuable for studies of gene–environment interactions and molecular mechanisms.

As an example, consider the question of whether DNA repair pathways—particularly those involved in repair of double-strand breaks— modify the effect of ionizing radiation. Various lines of investigation suggest that heterozygous mutations in the gene *ATM* (for Mutated in Ataxia Telangiectasia (A-T)) confer increased risk of cancer generally, and breast cancer in particular, and greater sensitivity to radiation.

It is known, for example, that *ATM* "homozygotes" (who develop the recessive A-T disease) are exquisitely sensitive to radiation and that their parents (obligate carriers of a single mutation, or "heterozygotes") are at increased risk of cancer. In order to investigate this hypothesis further, a study of second breast cancer in survivors of a first cancer has been launched. The advantage of this design is that by virtue of having had a first cancer, the study population is more likely to carry mutations of etiologic significance (including *ATM*) and that about half of such women will have had high dose radiation for treatment of the first cancer, which can yield exposure of 1–3 Gy to the contralateral breast at risk of a second cancer. This study illustrates a number of methodological challenges, including a novel study design for improving power by exploiting radiotherapy information in a multistage sampling fashion (Chapter 5), allowance for variation in dose across the contralateral breast rather than treating it as a single homogeneous organ as in most other studies (Chapter 6) and analysis of gene–radiation interactions (Chapter 12) involving ATM and other genes in an integrated model for DNA repair pathways (Chapter 13).

The air pollution tale

Rich as the radiation tale is, it is not sufficient to illustrate all of the methodological challenges that can face the environmental epidemiologist. For one thing, the endpoint of primary interest is cancer, a long-latency condition, which does not well illustrate problems of studying short-term effects. For another, risks are often high and unequivocal, allowing detailed study of the fine structure of dose–time–response relationships and modification by other factors, whereas much of environmental epidemiology is concerned with assessing very small risks—risks that may be at the limit of detection by epidemiologic methods. Nevertheless, even very small risks from a ubiquitous exposure like air pollution, when multiplied by a large population, can impose a large public health burden. Third, radiation doses can often be estimated relatively well at the individual level, whereas environmental epidemiology is often concerned with exposures that are ubiquitous or broadly distributed geographically. Thus, similarly situated individuals are likely to have very similar exposures, but comparisons between people living far away are likely to be strongly confounded. For these and other reasons, we need other examples to illustrate some of the problems to be addressed here, so we turn to another rich literature, that on air pollution.

That air pollution can have serious health effects—including death—has not been in serious dispute since the experience of the London Fog episode of 1952, in which 3,000 died within a week of the peak of pollution and another 12,000 over the following three months. (In an earlier episode in Donora, Pennsylvania, an even higher proportion of the population died in the first week than in London, although the total numbers were much smaller.) What is in question is whether present day regulations are sufficient to keep these health effects to an acceptable level, what the full range of health effects is, and whether there are subgroups of the population at especially high risk.

As with the radiation tale, we begin by distinguishing acute and chronic effects, but here we must consider both, rather than dismissing the acute effects as unambiguous. Amongst the acute effects are increased mortality, hospitalizations, absenteeism, asthma exacerbations, and other respiratory and cardiovascular symptoms immediately following an air pollution episode. Unlike the London Fog, however, such increases from modern air pollution levels tend to be very small—of the order of a few percent above background levels—and dwarfed by the normal range of variation. Chronic effects, on the other hand, include reduced lung function, increased risk of asthma, cardiovascular disease, and death due to long-term exposure to air pollution. An important question is whether these chronic effects represent an accumulation of insults from the acute effects or whether the acute effects are reversible and of no long-term significance. Alternatively, one could ask whether short-term associations between increases in mortality and air pollution levels immediately preceding represent additional deaths that would not have occurred otherwise or merely represent the advancement of the time of death by a few days of individuals who were about to die anyway. In short, a major challenge is to reconcile the estimates of excess risk of death or life shortening from acute and chronic effects studies.

Studies of acute and chronic effects require quite different designs. Broadly speaking, acute effects have been investigated using time-series correlation studies (Chapter 8), in which daily fluctuations in population mortality or hospitalization rates are correlated with daily variations in air pollution levels for the same area, after a suitable lag and adjusted for temporal variations in potentially confounding factors like weather. For endpoints for which population data are not routinely available, such as asthma exacerbations, one might conduct a panel study, in which, for example, a one or more cohorts of asthmatics are observed repeatedly over a relatively short period of time.

Such temporal comparisons are largely useless for studying chronic effects. For these, the most informative comparisons are spatial, relating morbidity or mortality rates in different geographic areas to air pollution levels in these areas, controlling for differences in potential confounding factors. The fundamental difficulty with such "ecologic correlation" studies is that associations across groups may not accurately reflect associations across individuals within groups. This so-called "ecologic fallacy" is discussed in depth in Chapter 10. The only way to overcome this problem is to study individuals, not aggregate data, so that the joint distribution of exposure and potential confounders can be assessed.

This leads us to consider various hybrid designs that combine elements of all these types of comparisons. One example is the multi-city time-series design, aimed at studying both the acute effects of air pollution overall, as well as spatial patterns of differences in effects between cities. Examples include the National Morbidity and Morbidity Air Pollution Study (NMMAPS), a comprehensive study of acute effects of air pollution in the 100 largest U.S. cities, and similar studies in Europe (APHEA).

For chronic effects, an example is the Southern California Children's Health Study (CHS). The CHS was designed to exploit three levels of comparisons: between communities ("ecologic" or aggregate); between individuals within communities (individual); and between times within individuals (temporal). Twelve communities with very different levels and types of air pollution (ozone, particulates, nitrogen dioxide, and acid vapors) were selected that were otherwise as demographically similar as possible. Within each community, cohorts of 4th, 7th, and 10th grade school children were enrolled, totaling initially about 300 children per community, 3,600 in total, from 45 different schools. Subsequent expansion of the study added three more communities, increased the total number of subjects to 12,000, and reduced the age of enrollment to kindergarten. Each child completed an extensive baseline questionnaire and lung function measurement, and was then re-examined annually thereafter until high school graduation. Another component monitored daily school absences, similar to the time series studies discussed above, except at the individual rather than aggregate level. Exposure assessment was initially based on continuous monitoring (hourly or with two-week integrated sampling depending upon the pollutant) at central sites established within each community. In addition, each child provided data on their usual time-activity patterns and household characteristics (e.g., air conditioning, indoor sources such as gas stoves and smokers in the household). These exposure data informed between-community comparisons of health effects (ecologic correlations), while controlling for confounders at the individual level. Later, as a result of the growing recognition of the importance of traffic as a source of local variation in air pollution levels, emphasis shifted to within-community comparisons. Measurements of local exposures at a sample of locations (homes and schools) within communities were launched to support the development of individual exposure models based on GIS-based traffic assessment, atmospheric dispersion modeling, and spatial statistics (see Chapters 9 and 11). A hierarchical random effects statistical model was used to assess air pollution effects at the aggregate, individual, and temporal levels (Chapter 7). A major advantage of a longitudinal design is that comparisons can be made *within* individuals—that is rates of change—so that each person serves as their own control, rendering comparisons at the individual and aggregate levels more resistant to confounding by personal factors.

The different air pollutants derive from multiple sources and interact with each other in a complex system of atmospheric chemical reactions. Teasing out the health effects of the different pollutants and their sources is an important goal of air pollution epidemiology, both to understand the biological basis of these effects and for setting air pollution regulations. An intervention aimed at the wrong source could have little benefit or even prove counter-productive. Some sources are regional, some local, some individual (ignoring for the moment global influences like climate change), and epidemiologic studies can take advantage of these different levels of spatial resolution to help tease apart these effects. Furthermore, some are natural while some are man-made, and again the two interact. For example, industrial emissions and motor vehicles in Southern California contribute various oxides of nitrogen (NO_x) amongst other chemicals that, under the influence of strong sunlight, are transformed to ozone (O_3) giving the Los Angeles region the highest ozone levels in the country. Ozone being a powerful oxidant, these reactions are reversible at the local level, so that fresh nitrous oxide (NO) emissions from motor vehicles react with O_3 to form NO_2, thereby paradoxically reducing the levels of O_3 near major roadways. Thus, depending upon whether NO_x or O_3 has a greater health impact, one might see either a positive or negative association with distance from major highways! Other chemical reactions transform the various primary pollutants into particles which further aggregate as they age in their journey across the Los Angeles basin toward the surrounding mountains, driven by the prevailing the Western winds. Both the physical and chemical properties of the resulting particles could affect their biological activity—their depth of inhalation, absorption into the blood stream, the body's local and systemic responses. The difficulty is that the various pollutants tend to be highly correlated with each other, so that separating their effects is difficult. Fortunately, this correlation structure can be rather different at different levels of spatial and temporal resolution, offering some hope to the epidemiologist clever enough to design studies to exploit multiple levels of comparisons. For example, comparisons between communities can shed light on regional pollutant effects, whereas comparisons within communities based on traffic exposure or household characteristics shed light on local sources and indoor sources respectively.

These various exposure sources are also subject to different sources of confounding and effect modification. Temporal comparisons have the great advantage of being completely immune to confounding by personal factors (except possibly by time-dependent factors, but these are unlikely to have much influence at the population level unless these temporal variations are synchronized across individuals, say by infectious diseases episodes). However, they are very sensitive to confounding by temporal factors like weather that affect the whole population more or less uniformly. Fortunately, weather data are routinely available and could be readily incorporated into time/series analysis. Flexible multivariate models may be needed to adequately adjust for the complex dependences: for example, mortality can be increased by periods of both hot and cold temperatures, at various lags. To overcome temporal confounding by unmeasured risk factors (or incomplete adjustment for measured factors), time-series methods or filtering are used. One must also take care not to over-adjust: wind speed, for example, can have a direct effect on air pollution levels but may not have any direct effect on health (conditional on air pollution); thus, adjustment for wind speed could incorrectly explain away a real causal effect of air pollution.

"Ecological" comparisons, on the other hand, are completely immune to temporal confounding but exquisitely sensitive to socio-demographic and other risk factors. In some cases, it may be possible to control for such factors at the aggregate level using routinely available population data, such as census data on income, race, and education, but such analyses are still subject to the ecological fallacy. Better would be to collect individual data on potential confounders for adjustment, even if the final analysis is done at the aggregate level. In Chapters 5 and 10, we will consider multistage sampling designs that combine population level data with individual data on a manageable subsample. The CHS is an example of such a study, which might be better characterized as "semi-individual," in that outcomes and confounders are assessed at the individual level, even though the basic comparison of exposure effects is performed at the aggregate level, thereby overcoming most of the problems with purely ecologic studies.

Except for demographic factors like age, sex, and race, for which population outcome data may be tabulated routinely, study of modifying factors generally requires collection of individual data. This might include pre-existing disease (asthma, heart disease), other risk factors (smoking, obesity, nutrition), genetics (family history, candidate genes), determinants of personal exposure or dose (time spent outside, physical activity, household ventilation), or biomarkers of intermediate biological processes (exhaled NO as a marker of inflammation).

Considering the complexity of exposure—multiple pollutants, different scales of spatial and temporal resolution, multiple sources, and determinants of personal exposure—it is not surprising that the CHS required a complex exposure assessment protocol. This was briefly summarized earlier in terms of its core components: central site monitoring, samples for inter-community variation, GIS-based traffic density, personal time-activity and household characteristics. Tying these disparate sources of information together requires extensive model building (Chapter 9), the aim being to assign to each study subject an estimate of their personal exposure, based on data that are available for the entire cohort, without requiring actual measurements for each person, which would have been a monumental and hopelessly expensive task. In addition, we aim to have an estimate of the uncertainty of each exposure assignments that could be incorporated into exposure–response models that account for measurement error (Chapter 11).

Some specific aims require more extensive data collection than is feasible for the entire cohort. For these, various nested substudies can be done. For example, for studying asthma, extensive phenotyping efforts are needed to confirm the cases' diagnosis and characterize their subtypes (and to document the absence of disease in controls), and to gather more extensive risk factor information, particularly about early-life or *in utero* exposures. For this purpose, a counter-matched case-control design was used, selecting for each newly diagnosed case one control at random from those in the cohort still free of asthma, matched on age, sex, community, and other factors, and counter-matched on maternal smoking while pregnant. For studying genetic factors, a case-parent-triad design (Chapter 5) was used, comparing the genotypes of asthma cases to the other genotypes the case could have inherited from their parents (Chapter 12). A variant of this approach is also available for continuous traits like lung function changes. This design is particularly useful for investigating both main effects of genes, as well as gene-environment interactions, in a way that overcomes confounding by unmeasured genetic or other personal risk factors.

Data from such epidemiologic studies—both acute and chronic—have had a major influence on regulatory policy for air pollution standards. Obviously, the single most relevant summary of the epidemiologic results for this purpose is the estimate of the excess risk of various endpoints per unit exposure, which can then be used by policy makers in combination with economic cost-benefit analysis and other considerations to evaluate the appropriateness of current standards or support new regulations. Some endpoints, like a reduction in lung function, are difficult to interpret in terms of clinically significant disease or prognosis for future adverse effects. The evidence for low-dose linearity may be equivocal and the possibility of a threshold impossible to exclude. Acute and chronic effects estimates could differ by an order of magnitude and their implications for life shortening be fundamentally impossible to determine without making untestable assumptions. Such issues will be addressed in Chapter 15.

Another important function of environmental epidemiology is evaluation of the effectiveness of interventions (Chapter 14). While epidemiology is basically an observational science, there are occasional opportunities to exploit "natural experiments" to evaluate the adequacy of current standards or the effect of changes in them—the latter has been called "accountability." The obvious comparison of health endpoints before and after the introduction of some new regulation is easily confounded by the many other changes that could have happened about the same time, although an abrupt and lasting improvement could provide compelling evidence of a benefit. More convincing yet would be when similar regulations are introduced in different places, since the timing of changes in confounders is unlikely to occur at the same times in each location. An example of a natural experiment occurred in the Utah Valley in 1967–68, when the workers at a copper smelter went on strike for about 9 months, shutting down the plant and removing the major source of air pollution. The resulting reduction in mortality (2.5% with 95% CI 1.1–4.0%)— and their subsequent return to previous levels—provide some of the most compelling evidence yet of the health benefits of cleaner air. Similar results were seen two decades later during a steel mill strike in the same general area. As this volume is going to press, a similar investigation is underway on the changes in respiratory and cardiovascular health indicators in panels of children, young adults, and elderly before, during, and after the Beijing 2008 Olympics, where considerable efforts are planned to reduce air pollution levels to those in other major Western cities.

Beyond the public policy implications, environmental epidemiology can also help elucidate biological mechanisms. Indeed, policy makers are often hesitant to act in the absence of at least some understanding of the mechanism underlying an epidemiologic association—particularly in the absence of an experimental animal model—lest the association not be truly casual. A potential mechanism for a broad range of air pollution health effects is thought to be oxidative stress. Ozone, particulate matter, and other gaseous pollutants are powerful oxidizing agents, capable of producing free radicals like hydroxyl radicals, peroxides, and super-oxides. These can cause damage to the respiratory epithelial lining fluid, inducing various host responses leading to chronic inflammation. The production of free radicals, the damage and repair process, the inflammation and its long-term sequellae can be modified by various genes, host factors (antioxidant intake and compromising disease conditions like asthma), and other exposures like metals in particulate matter that catalyze the Fenton reaction. Only by studying all these factors in combination can we hope to develop a comprehensive causal model for air pollution health effects, as discussed in Chapter 13.

Ultimately, the elucidation of causal mechanisms requires a multidisciplinary collaboration between epidemiologists, toxicologists, geneticists, physiologists, clinicians, and other disciplines. Such collaborations need to be a two-way street. Hypotheses suggested by epidemiologic data may be tested in various experimental ways—by acute human challenge studies, long term or high-dose animal toxicology, cell culture experiments—and conversely, confirmation can be sought for predictions of mechanistic studies in epidemiologic data. Ultimately, it is hoped that studies can be designed and statistical analysis methods developed that will allow the two to be more intimately connected. For example, a substudy within the CHS is currently underway using toxicological methods to characterize various measures of biological activity of particles sampled from each community on cells with different genes inactivated; these biological activity measures will then be incorporated into the analysis of the epidemiologic data.

With this overview of the scope of this book, illustrated through the radiation and air pollution stories, we now turn to a more systematic treatment of the principles of study design and analysis for environmental epidemiology. The emphasis in the ensuing chapters is on the methods, using our two tales (and others as needed) merely to illustrate principles, rather than to provide a comprehensive review of these two monumental literatures.

2 Basic epidemiologic study designs

Much of this book is concerned with dichotomous disease outcomes with variable age at onset. Continuous outcomes are usually studied with simple random samples, using standard linear models for normally distributed random variables. But disease traits—particularly, rare diseases—require the use of specialized sampling designs and methods of survival analysis. Here we provide a brief introduction to these principles, as they apply to the study of independent individuals. [See the companion volume (Thomas, 2004) for a discussion of dependent data arising in the context of family studies.] This introduction will be somewhat terse, and the reader who wishes a more in depth treatment of these topics might be referred to standard epidemiology (Kleinbaum et al. 1982; Rothman and Greenland 1998) and biostatistics (Breslow and Day 1980; 1987) textbooks.

Experimental

The gold standard for inference about cause–effect relationships is an experimental study. In clinical research, this would typically be a double-blind randomized controlled trial, in which subjects are randomly allocated to the treatments being compared and followed in the same manner to determine their outcomes, with neither the study subjects nor the investigators being aware of the specific treatment assignments for individuals so as to avoid bias. In prevention or program evaluation research, such studies might instead be conducted in a group-randomized fashion. Of course, in environmental epidemiology, one is generally concerned with the effects of hazardous exposures, so ethical concerns preclude deliberately exposing individuals to potential harms and an experimental design is seldom feasible, except perhaps for studying mild short-term reversible effects, as in challenge studies of asthma exacerbations or other chamber studies. Most of environmental epidemiology therefore involves observational studies, in which one passively observes the effects of exposures that have happened without any intervention by the investigator. The rest of this book is primarily concerned with these kinds of studies, except for a brief discussion of the kinds of causal inferences that are possible from experimental and observational studies in Chapters 14 and 16.

Descriptive

Epidemiologists distinguish two basic types of observational studies—
"descriptive" and "analytical." Descriptive studies are based on examination of routinely collected disease rates in relation to basic demographic
characteristics (age, gender, race/ethnicity, place of residence, etc.) in the
hopes of getting clues to possible risk factors. A commonly used approach
is the so-called ecologic correlation study, in which groups rather than
individuals are the unit of analysis and one studies the correlation between
disease rates and the prevalence of some characteristic of interest. An
example discussed in Chapter 1 was the correlation between lung cancer
rates and average household radon concentrations, which we will revisit
in greater detail in Chapter 10. Such studies are often useful for generating hypotheses, but are subject to numerous sources of bias and cannot
provide rigorous tests of hypotheses.

Nevertheless, various routinely collected sources of information can be
useful for generating hypotheses. Beyond such obvious demographic characteristics as age, gender, and race/ethnicity, clues can be obtained from
occupation, place of residence, or date of diagnosis, if combined with
available information about how various potential environmental agents
might be distributed across such factors. Thus the category of descriptive
studies used in environmental epidemiology can be broadly subdivided
into two main types of comparison, geographical and temporal.

Geographic comparisons

International comparisons of age-adjusted breast cancer rates in relation to average levels of fat intake have been used to support a causal
connection between the two (Prentice and Sheppard 1990; 1991), as
discussed further in Chapter 10. The Atlas of Cancer Mortality (Pickle
et al. 1987) at the county level has been widely examined in search
of environmental hypotheses to account for variation in rates. Freemen
(1987) reviews a number of examples of the use of cancer maps to generate etiologic hypotheses, including the possible association of high rates
of non-Hodgkin's lymphoma among males in the central United States
with herbicide exposure, of lung cancer among males in coastal areas
with asbestos exposure in shipyard workers, and of oral cancer among
females in the rural south with smokeless tobacco use. At an even finer
level, an elevated incidence of nasopharyngeal cancer in census tracts bordering the 710 freeway in Los Angeles suggests a connection with the
heavy diesel truck traffic from the Ports of Los Angeles and Long Beach
(Mack 2004).

Temporal comparisons

Time trends are frequently invoked to support hypotheses relating to novel or increasingly common environmental exposures. Asthma rates, for example, have been increasing, potentially due to factors related to urbanization—air pollution, crowding, allergens, etc. Of course, some of this increase could simply be due to changes in diagnostic practices or data collection methods, since asthma is not a notifiable disease. Since so many factors are changing simultaneously, it is impossible to pinpoint any one of them as causal solely on the basis of long-term trends. Short-term correlations between daily fluctuations in mortality, hospitalization, or absenteeism and daily fluctuations in air pollution levels are much more specific and hundreds of such studies have been reported (see Schwartz 1994a; Bell et al. 2004b for reviews). Of course, even such short-term correlations are subject to potential confounding, such as by changes in weather, requiring careful control in the analysis. See Chapter 8 for further discussion of such designs, including those like the National Morbidity and Mortality Air Pollution Study (Samet et al. 2000a; Bell et al. 2004a) that combines geographical and temporal comparisons. Nevertheless, it must be appreciated that short-term temporal comparisons are more useful for demonstrating potential acute effects than long-term temporal comparisons are for demonstrating chronic effects.

Prevalence surveys

A cross-sectional study aims to describe the prevalence of disease at a particular point in time in some population. Typically, that population is geographically defined, such as the entire United States or some city. A random sample is drawn and an inquiry is made into disease status of the individuals in the sample and various demographic and risk factors of interest. Because exposure and disease are measured at the same time, it is not possible to establish their temporal sequence, so any associations may not be causal (i.e., exposure could have followed disease development, not caused it). Furthermore, disease rates are measured in terms of *prevalence* (the proportion of the population currently living with the disease) rather than *incidence* (the rate of development of new cases). Since prevalence is approximately the incidence rate multiplied by the average duration of the disease, a positive association of a factor with prevalence could indicate that that factor increases the rate of new disease or it increases the length of time individuals are afflicted by it. For example, a factor that causes more rapid case fatality (shorter duration) could appear to be protective against the disease when assessed by a prevalence survey, when in fact it has no effect on the incidence rate or is even positively associated with incidence.

Despite these difficulties of interpretation, prevalence surveys like the National Health and Examination Survey (NHANES) can be an important source of descriptive information to help in the generation of hypotheses to be tested using analytical studies. In particular, unlike the geographical or temporal correlation studies described above, the unit of analysis in a prevalence survey is the individual, allowing control of more factors than is possible when only aggregate data are compared. NHANES uses a sophisticated multi-stage sampling strategy involving the selection of municipalities, followed by subsampling of blocks within tracts to identify individual homes to be enrolled in the survey (Ezzati et al. 1992). In the first stage, the 13 largest primary sampling units (mainly counties were selected with probability one, and an additional 68 were selected at random from 2,812 in the United States; some of the largest were further subdivided, to make a total of 89 primary sampling units. Within each of these, subsamples of city blocks or other geographically defined areas were selected, a total of 2,138 areas. All addresses within these areas were then enumerated and a random sample of about 100,000 homes was surveyed to determine the presence of individuals meeting the eligibility criteria for the survey. Sampling probabilities proportional to size were used at the various stages so that the final sample would be "self-weighting," that is, so that unweighted statistics could be used to estimate national parameters. The final sample comprised about 40,000 individuals, of whom about 35,000 were interviewed and 30,000 were examined. Similar techniques are discussed under the heading of control selection later in this chapter and in Chapter 5. They are also widely used in post-disaster epidemiologic surveys (Chapter 17).

In addition to their role in generating hypotheses and providing useful descriptive and administrative data, cross-sectional surveys can be useful in their own right for testing hypotheses. Prevalence is a natural measure of effect for some chronic, nonfatal diseases like asthma, and the relevant exposure could be lifetime or usual concentrations, so associations with environmental factors for such conditions can be usefully investigated with prevalence studies. As will be discussed in Chapter 14, the effect of an intervention to change the exposure of a population (e.g., regulations aimed at curbing air pollution) might be studied by repeated cross-sectional surveys before and after the intervention.

Analytical

To avoid the pitfalls of drawing causal inferences from descriptive studies, environmental epidemiology typically relies on analytic studies involving collection of original data on individuals in order to test specific hypotheses in a controlled manner.

The two principal analytical study designs used in traditional risk-factor epidemiology are *cohort* and *case-control* designs. These are distinguished primarily by the direction of inference: cohort studies reason forward in time from an exposure to disease, while case-control studies reason backward in time from disease back to possible causes. It is important to understand that it is this direction of inference—not the temporal sequence of data collection—that is conceptually important. Either cohort or case-control studies can be conducted "retrospectively" (using records from the past) or "prospectively" (collecting new observations as they occur in the future). Some authors have used the terms prospective and retrospective to refer to cohort and case-control designs, leading to a confusion we shall try to avoid by restricting these terms to the direction of data collection, not inference. (Other authors have used the terms "historical" and "concurrent" to refer to the direction of data collection.)

Cohort study design

Conceptually, the fundamental design in epidemiology (if not the most commonly used one) is the cohort study. In this approach, a cohort of at-risk individuals (currently free of the disease under study) is identified, characterized in terms of their baseline risk factors, and followed over time to identify which subjects develop disease. The risk of disease is then estimated in relation to these baseline characteristics. During the follow-up period, changes in risk factors might also be recorded, but this is not an essential element; the important element is that exposure is recorded before disease occurs. This design is generally felt by epidemiologists to be less subject to the selection and information biases than case-control studies are prone to, as discussed below. Nevertheless, for studies of anything but the most common diseases, a cohort study is an ambitious undertaking, generally requiring enrollment of a large cohort (sometimes hundreds of thousands of individuals) and follow-up for many years, with the consequent difficulties of tracing subjects over time and completely ascertaining disease outcomes. For these reasons, the preferred design for most rare diseases is the case-control design discussed in the following section. Use of historical records (retrospective data collection) avoids one of the fundamental challenges of cohort studies, namely the long period of observation needed. The feasibility of this option depends on the availability of a suitable sampling frame for defining the cohort members in the past, as well as mechanisms for tracking the current status of individuals (including those who have died or moved away in the interim).

The atomic bomb survivor study, the Colorado Plateau uranium miners study, and the Children's Health Study described in Chapter 1 provide examples of different kinds of cohort studies.

A central design issue in any cohort study is whether the primary comparison will be between the cohort as a whole and some external reference population or internal to the cohort. While the former can be useful for putting the overall experience of the cohort in a broader perspective, such comparisons can be biased by various factors influencing selection into or out of the cohort. In the context of occupational studies, these are generally referred to as the "healthy worker effect" (Bell and Coleman 1987), in which healthier individuals are preferentially selected for employment from the general population, and the "healthy worker survivor effect" (Robins 1987; Arrighi and Hertz-Picciotto 1994), in which individuals who are more resistant to the noxious effects of exposure are more likely to remain employed in the industry. These problems can be partially overcome by making comparisons between exposed and unexposed subcohorts or across a gradient of exposure within the cohort.

Still, identifying a comparable unexposed control group can be difficult, as illustrated in analyses of the atomic bomb survivors using the subgroup of individuals living in the two cities at the time of the bombing, but far enough away to have been assigned zero dose (Cologne and Preston 2001). Including various geographically defined subgroups in the analysis led to estimates of excess relative risk ranging from 6% higher to 8% lower than those excluding the zero dose group. This bias could be removed by including an indicator variable for zero dose in the model, thereby allowing better estimation of the effects of modifying factors like age, sex, and city on baseline rates because of the larger sample size.

Case-control study design

The case-control design begins with ascertainment of a representative series of cases of the disease and a comparable group of individuals from the same population who are free of the disease, and inquires into aspects of their past history that might account for their different outcomes. Frequently, controls are selected by individually matching to each case on established risk factors that are not of particular interest, such as age, gender, and race. The inquiry into possible risk factors might be done by questionnaire, structured personal interview, or retrieval of records. These should be designed in such a way to avoid any lack of comparability between the quality of information for the two groups, for example; by blinding interviewers to whether subjects are cases or controls.

The great advantage of the case-control design is that it does not require enormous sample sizes or a long period of follow-up—only enough to

accrue a sufficient number of cases and a comparable number of controls—
so is ideal for studying rare diseases. Hence more resources can be devoted
to data quality, for example, verifying the diagnosis of cases and unaffected status of controls, and obtaining much more detailed information
on exposure than is typically possible in a cohort study.

Probably, the biggest challenge in designing a case-control study is the
choice of controls. Before this can be done, a rigorous definition of the
case series is needed. Ideally this is population-based, for example all cases
identified by a registry covering some population defined by space and
time, or for mortality, all deaths in a similarly defined population. Many
diseases, however, are not routinely recorded in any population-based
registry, so one must resort to hospital- or clinic-based series or a special
survey of the population. In such cases, it may be more difficult to identify
the source population from which the identified cases arose, as hospitals
typically do not have well-defined catchment areas from which all cases
would come and different cases from the same area may go to different
hospitals. Nevertheless, the basic principle is that controls should represent
this "base population" that gave rise to the case series.

Having defined this population, there may be multiple ways of sampling from it. The ideal would be a random sample, possibly stratified
by such factors as age and gender to match the corresponding frequency
distribution of cases, but this would require a sampling frame listing all
the people eligible to be selected. While some countries (Australia and the
Scandinavian countries, for example) maintain such population registers,
many others do not, making this impractical, or have confidentiality policies precluding access for research purposes. Some alternatives that have
been widely used include

- *Neighborhood controls*: a census of the neighborhood surrounding each
  case is made by the investigator and an eligible control is selected at random from that set. For example, a field worker may start at the some
  pre-determined location near the case's residence (say the corresponding location one block away, to preserve the anonymity of the case) and
  walk the neighborhood in a spiral pattern to obtain a list of potential
  control residences. At each door, the walker either asks for the identity
  of individuals who might be eligible as controls or leaves a request to
  call the investigator with this information if no one answers; the walk
  continues until some predetermined number of residences has been surveyed. Once complete, the first eligible person who agrees to participate
  is used as the control.
- *Random digit dial controls*: Starting with the first few digits of a case's
  phone number, the remaining digits are selected at random and dialed.
  If a suitable control is available and willing to participate, he or she

is included in the list of available controls to be sampled, or the first available control is selected;
- *Friend controls*: Each case is asked to list the names and contact information for a number of friends and one of these is selected at random,
- *Spouse controls*: For diseases of adulthood, the spouse of the case is selected;
- *Sibling controls*: A sibling who has attained the age of the case, still free of the disease under study, is selected as the control;
- *Birth registry controls*: For diseases of childhood, the immediately preceding or following birth on the registry in which the case appears may be used.
- *Hospital controls*: For hospital-based case series, individuals attending the same hospital or practice for some condition or conditions thought not to be related to the risk factors under study are selected.

None of these is ideal (Wacholder et al. 1992b). Wacholder et al. (1992a) begin their series of three papers on the subject by discussing three general principles for choosing between potential control sources: representativeness of the study base population; freedom from confounding; and comparability of data quality. They also discuss the relative efficiency of different control sources and whether it is advisable for cases and controls to have equal opportunity for exposure (Poole 1986).

Neighborhood controls are likely to best represent the source population of cases, but are labor intensive and it may be impractical to survey some dangerous neighborhoods. Furthermore, for geographically determined exposures like air pollution, cases and controls would tend to have similar exposures, leading to what epidemiologists call "overmatching," with consequences that will be explored in the following chapter. Random digit dial controls are also labor intensive, typically requiring on average about 40 calls to identify each control and may be subject to various selection biases related to phone availability. The viability of random digit dialing is likely to become more and more difficult with the increasing prevalence of answering machines, caller id, and cell phones (Link and Kresnow 2006; Kempf and Remington 2007). Spouse controls are, of course, of the opposite gender, so unsuitable for studying exposures that are sex related. Many cases may also not have an eligible spouse or sibling control. Hospital controls are afflicted by some other condition that led them to the hospital, and it can be difficult to choose control conditions that are truly unrelated to the factors under study. Friend controls can be subject to various biases relating to differences between cases in their number of friends, the representativeness of their exposures, willingness of cases to name friends, and the risk of overmatching (Flanders and Harland 1986; Siemiatycki 1989; Robins and Pike 1990; Thompson 1990; Wacholder et al. 1992b; Kaplan et al. 1998; Ma et al. 2004). These are more likely to pose a problem for environmental risk factors, particularly those related to social behaviors like smoking, than for genetic factors, however (Shaw et al. 1991).

Inherent in many of the control selection strategies described above is the idea of individual matching. Beyond providing a convenient way of selecting controls, it also serves the important function of ensuring the comparability of case and control series on factors that are not of particular interest but may be important risk factors for the disease. Matching can be done either individually or by strata. For the former, the control who most closely matches the case on the set of factors under study is selected (typically this may entail prioritizing the various factors, e.g., gender, followed by age within 5 years, followed by race, followed by education, etc.). For the latter, cases are divided into mutually exclusive strata and an equal number of controls are selected at random from those eligible in each stratum. This approach is commonly called *frequency matching* or *stratum matching*.

Individual matching is generally done by defining a sequence of criteria to be matched upon, beginning by requiring an exact match (for discrete variables, or for continuous variables a match within either strata or some caliper (Austin et al. 1989)) on the most important criteria, and then seeking the closest available match on less critical factors, relaxing the closeness of matching as needed to find an acceptable control. For prospective case-control studies, this might be done one at a time as cases accrue; for a nested case-control study, the order in which cases are considered is randomized. In either situation, once selected as a control for one case, that individual is usually not eligible to be paired with another more closely matching case, so that the resulting set of pairs could be less than optimal overall. In any event, such rules can be difficult to implement in practice when several risk factors are to be matched for simultaneously. An attractive alternative is the use of optimal matching designs (Rosenbaum 1989), typically involving minimization over all possible case-control pairs of some measure of multivariate distance on the set of matching factors. This minimization can be accomplished without having to enumerate all possible pairings using an efficient network algorithm. Cologne and Shibata (1995) empirically compared two such approaches in the design of a nested case-control study of liver cancer within the atomic bomb survivor cohort, where hepatitis-B infection was considered as a potentially strong confounder or modifier. One was based on the propensity-score (the probability of being a case given the matching factors, as estimated from the entire pool of cases and potential controls (Rosenbaum and Rubin 1985)), the other on a variance-weighted Euclidean distance between case-control pairs (Smith et al. 1977). They concluded that the weighted-distance method produced better overall closeness across matched sets because the propensity score tended to be poorly estimated.

A particular advantage of individual matching in environmental epidemiology is that for evaluation of time-dependent exposure variables, cases and controls can be assigned comparable "reference dates," such as one year before diagnosis of the case or the corresponding interval before interview, age, or calendar date for the matched control. Unless cases' and controls' references dates are similar, the comparison of variables like cumulative exposure can be biased by lack of comparability of their times over which exposure is accumulated or their "opportunity for exposure." This is more difficult to accomplish for unmatched or frequency-matched case-control studies, for which it would be necessary to assign reference dates for controls corresponding to the distribution of reference dates for all cases in the same stratum; this would be difficult to accomplish until after all the cases have been enrolled.

The second major challenge in case-control study design is exposure assessment, particularly ensuring comparability of the quality of information from cases and controls. "Recall bias" is a particularly important challenge if exposure information is to be obtained by interview or questionnaire: cases may be more inclined to over-report exposures they think could have caused their disease or to deny exposure about which there is some stigma. Cases with advanced disease may be too ill to respond accurately to questions or it may be necessary to obtain exposure information from a proxy (e.g., next of kin) if the case is dead or too sick to respond (Nelson et al. 1990; Wacholder et al. 1992b). For this reason, Gordis (1982) suggested selecting dead controls for dead cases, so that exposure information would be obtained by proxy for both, but this violates the principle of controls' representativeness of the base population (McLaughlin et al. 1985), even if it does tend to promote comparability of data quality.

Nested case-control and case-cohort designs

One of the major expenses of a cohort study is assembling the exposure information on the entire cohort, when perhaps only a very small portion of the cohort will develop the disease. For example, in an occupational study, obtaining exposure information on a large cohort can be very expensive indeed. To minimize these costs, an efficient compromise can be to obtain this information only on the cases and a random sample of the rest of the cohort. There are two principal variants of this idea.

In the *nested case-control design*, controls are individually matched to each case by random sampling from the set of subjects who were at risk at the time that case occurred; the data are then analyzed as a matched case-control study. In this scheme, it is possible for a subject to be sampled as a control for more than one case, and for a case to serve as a control for an earlier case. Lubin and Gail (1984) show that this scheme leads to unbiased estimation of the relative risk parameter, while the alternative of excluding cases from eligibility to serve as controls for other cases (Hogue et al. 1983) would lead to a bias away from the null (Greenland et al. 1986).

In *case-base sampling* (Mantel 1973), the controls are a random sample of the entire cohort (the "subcohort") at the time of enrollment, irrespective of whether or not they later became cases; the analysis then compares the cases as a group to the controls. Thus, some cases will appear in the subcohort, some outside it, but all cases are used. The original *case-base analysis* (Kupper et al. 1975; Miettinen 1982b; 1985; Flanders et al. 1990; Langholz and Goldstein 2001) uses standard methods for estimating risk ratios (rather than rate ratios), with an adjustment to the variance to allow for the overlap between subjects appearing both as cases and as controls. The *case-cohort analysis* is aimed instead at estimating rate ratios for the same reasons they are generally preferred in cohort studies to deal with censoring, using a variant of the standard Cox regression model described in Chapter 4 (Prentice 1986). Advantages of the case-base sampling (with either analysis) are that the same control group can be used for comparison with multiple case groups and that obtaining the baseline data on the subcohort can be done early in the study while the cases are accumulating (for example, blood specimens could be obtained and genotyping or assays of serum biomarkers of exposure started for controls without waiting to see who became cases). The main disadvantages are that a more complex analysis is required and it can be less efficient than a nested case-control study for long duration studies with many small strata (Langholz and Thomas 1990).

One way to think about either nested case-control or case-base designs is as an analytic strategy for sampling from within an established cohort, as opposed to the standard case-control design entailing drawing separate samples of cases and controls from their respective populations. Either can be conducted in matched or unmatched fashion, the nested case-control design being the matched version (Langholz and Goldstein 1996), the case-base design being the unmatched version (Langholz and Goldstein 2001), paralleling the analogous analyses for population-based designs, matched (Breslow 1981) and unmatched (Prentice and Breslow 1978; Prentice and Pyke 1979).

Interpretation of epidemiologic associations

As an observational science, epidemiologists do not generally have the opportunity to test hypotheses by conducting controlled experiments relying on randomization to ensure comparability of the groups compared

(Greenland 1990). Hence, the associations between risk factors and disease found in epidemiologic studies are subject to a wide range of potential biases and do not necessarily indicate causation. Epidemiologists have developed a series of criteria for judging when an inference of causality is warranted from an observed association, of which the most famous are those outlined by Sir Austin Bradford Hill (1965), as described in most epidemiology textbooks: dose-response, temporal sequence, strength, lack of other explanations, consistency across multiple studies, coherence across types of evidence, and so on. Bates (1992) has elaborated upon the last of these in the context of air pollution, comparing 11 indices of acute and chronic effects in terms of their logical interrelationships and the extent to which 44 epidemiologic studies produce the expected pattern of coherence.

Particularly relevant to environmental epidemiology is freedom from biases, which are generally classified into three types:

1. *Selection bias*: any of several study design aspects that would tend to make the groups sampled unrepresentative of their respective source populations (e.g., using hospital controls to represent the population of unaffected individuals);
2. *Information bias*: various study design aspects that would tend to make the quality of the information obtained on subjects noncomparable between the groups compared (e.g., recall bias in a case-control study, where cases might tend to recall past exposures differently from unaffected individuals); and
3. *Confounding*: distortion of a true relationship by the action of another variable that is associated with exposure in the source population and, conditional on exposure, is also an independent risk factor for disease.

In addition to such potential "study biases" in the design and analysis of particular studies, one should also be aware of what we might call "meta-biases" in the interpretation, publication, and synthesis of evidence and the conceptualization of new hypotheses. For example, meta-analysis of the epidemiologic literature on a particular association depends for its validity on access to the entirety of evidence on the question. But individual investigators are more likely to submit findings for publication—and journals to accept reports—if they are deemed "significant," a phenomenon known as "publication bias" (Begg and Berlin 1988). Various methods have been suggested to overcome this problem in quantitative summaries of scientific evidence (Dickersin and Berlin 1992; Berlin et al. 1993; Greenland 1994b; Blair et al. 1995; Stroup et al. 2000; Greenland 2005). For example, it is generally agreed that the problem is more severe for initial reports of a novel association than for subsequent attempts at

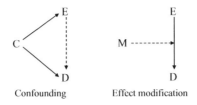

Figure 2.1. Schematic representation of confounding and effect modification.

replication, so exclusion of the first report may somewhat attenuate the problem. Of course, scientists are only human and their prior beliefs will inevitably color their interpretation the literature and their selection of hypotheses and approaches to testing them. Rigorous attention to the principles of good epidemiologic practice is thus essential to avoid having such meta-biases spill over into the actual conduct of epidemiologic studies.

Sackett (1979) has provided an exhaustive catalog of potential study biases and meta-biases in observational epidemiology and a discussion how sensitive different study designs are to each. Generally, cohort studies are felt to be less susceptible to selection and information bias than case-control studies, because the cohort is enrolled as a single group and everyone followed in the same way, and because the exposure information is obtained before the onset of disease. But they are not immune to bias; for example, usually some individuals are lost to follow-up, and the probability of being lost may depend upon exposure or disease status.

Epidemiologists attempt to control confounding by matching, stratified analysis (e.g., the SIR discussed in the following chapter), covariate adjustment (e.g., the logistic model discussed in Chapter 4), or restriction to homogeneous subpopulations. It is also possible that a factor could be related to exposure but not be an independent risk factor for disease (no C-D arrow in the left-hand side of Figure 2.1), that is, associated with disease only indirectly through exposure. Controlling for such a factor is not advisable and would be called "overmatching" or "overadjustment" (Day et al. 1980), not because it would yield a biased estimate of the E-D association but because it would inflate its variance and reduce power by unnecessarily restricting the variability in E conditional on C. This concept is related to the question of whether cases and controls should be required to be comparable in terms of "opportunity for exposure" (Poole 1986; Wacholder et al. 1992a).

Confounding must not be confused with *effect modification* or *interaction*, as discussed in the following chapter. As illustrated in Figure 2.1, an effect modifier M is a variable that alters the magnitude of the association between exposure E and disease D, for example, if disease rates differ between males and females. Such a variable need not be associated with either E or D directly. In contrast, a confounder C is associated with both E and D and thereby distorts the direct relationship between E and D that would be observed for any particular value of C. The concept of effect modification will be addressed in greater detail in Chapters 3, 4, and 12.

3   Basic statistical methods

Basic probability theory

Before describing the basic approaches to the analysis of epidemiological data, we need to establish some fundamental concepts from probability theory. Let us denote a *random variable* (a quantity that varies between observational units—individuals, times, and so on—in some manner that is not completely predictable) by the symbol Y and use the corresponding lower case letter y to denote the specific value a particular observation may have. Three kinds of random variables that will be particularly important in epidemiology are binary (dichotomous), count, and continuous variables. A binary variable is one with only two possible values, for example, present ($Y = 1$) or absent ($Y = 0$). A count variable is one which can take any nonnegative integer value, such as the number of cases of disease in a group. A continuous variable can take any real value (although possibly over some restricted range, such as only positive numbers).

We refer to the set of probabilities for the values a random variable can take as its *probability density function*. For example, for a binary variable, we might write $p = \Pr(Y = 1)$ and $q = 1 - p = \Pr(Y = 0)$. We say two random variables, say X and Y, are *independent* if the value of one does not affect the probability distribution of the other. In this case, we can write $\Pr(X = x$ and $Y = y) = \Pr(X = x) \times \Pr(Y = y)$. (From here on, we will write the former simply as $\Pr(X = x, Y = y)$, with the comma implying "and".) For example, if p is the probability that any random individual in a particular group will have a disease, then the probability that two independent individuals will both have the disease is $p \times p = p^2$. If the disease were infectious, however, and two individuals were in close contact with each other, then their outcomes would probably not be independent, as exposure to one could have caused the other; likewise, if two individuals are living far apart, then their risks of disease are probably independent, but if they are living in the same or neighboring households, then they could be dependent because of shared exposures to unmeasured risk factors. (Dependent data also arise in family studies due to shared genotypes.) These examples illustrate two different ways dependent data can arise: as a direct causal connection in the case of transmission of an infectious disease; or as a result of a shared *latent* (unobserved) variable in the case of shared exposures.

Because we are particularly interested in describing the *relationships* between variables that are not independent (or testing hypotheses about whether they are independent), we are frequently interested in their *conditional* probabilities, denoted $\Pr(Y|X)$, that is, the probability of Y for a specific value of X ("the probability of Y given X"). For example, we might be interested in the risk of disease Y in an individual exposed to X, that is, $\Pr(Y = 1|X = 1)$. The fundamental law of conditional probability states that

$$\Pr(Y|X) = \frac{\Pr(Y, X)}{\Pr(X)}$$

If the two variables were independent, the numerator would become $\Pr(Y)\Pr(X)$, so $\Pr(Y|X) = \Pr(Y)$, that is, the probability of Y would not depend upon X. By cross-multiplication, the law of conditional probability can also be written as $\Pr(Y, X) = \Pr(Y|X)\Pr(X)$. Of course, we could just as well write $\Pr(Y, X) = \Pr(X|Y)\Pr(Y)$. Equating these two expressions and solving for $\Pr(Y|X)$ leads to the famous *Bayes theorem*,

$$\Pr(Y|X) = \frac{\Pr(X|Y)\Pr(Y)}{\Pr(X)}$$

which defines the relationship between the two conditional probabilities.

Statistics is the study of the distribution of random variables, using random samples of observations from some larger unobserved population to which we wish to generalize. In particular, statisticians are concerned with two types of activities:

1. *Estimation*: finding simple summaries of a distribution, like its mean and variance, together with some measure of the uncertainty of these estimates; and
2. *Hypotheses testing*: determining the probability that two or more distributions differ.

For example, we might wish to estimate the risk of disease p in some population by obtaining a sample of N individuals from that population and observing the number Y who are affected. The sample proportion Y/N, denoted $\hat{p}$, is an *estimator* of the unknown true population risk p. Likewise, for a continuous variable, the sample average $\bar{Y} = \sum_{i=1}^{N} Y_i/N$ is an estimator of the population mean μ. *Confidence intervals* describe the likely range of estimates that could have been generated by hypothetical replication of the study using the same procedures (e.g., with the same design, sample size, method of sampling, measurement, etc.). It is not, however, the same as the likely range of the true population parameter given the data, as we shall see below.

In general, we are particularly interested in testing a hypothesis about some particular *null hypothesis* value of a population parameter, again based on a finite sample of observations. For example, we might want to know whether the risk of disease p_1 in a particular subgroup of the population (say, exposed individuals) differs from that in the general population p_0. Then we would write the null hypothesis as H_0: $p_1 = p_0$. We would then compute some summary statistic (such as the chi square statistic $(Y - E)^2/E$ where $E = Np_0$, described further below) for the observed data and compare that value to the distribution of possible values of the statistic that would be expected in hypothetical replications of the study if the null hypothesis were true. If the observed value is quite unlikely—say, fewer than 5% of these hypothetical replications would be expected to lead to more extreme values—then we would call the result "statistically significant" and quote as the *p*-value the probability of obtaining by chance a result as or more extreme than that actually observed. Note, that the *p*-value should not be interpreted as the probability that the null hypothesis is true. The latter would require the *prior* probability that the null hypothesis was true (the probability we would have assigned before seeing the data), which would be difficult to get a consensus about. Therein lies the difference between classical ("frequentist") and Bayesian statistics, which we will discuss further below. These basic concepts of probability and statistics are explained in greater detail in virtually any elementary statistics textbook, but this cursory treatment of the subject should be sufficient for the particular applications in environmental epidemiology discussed below.

Two summary statistics for a distribution that we will encounter repeatedly are its mean and variance. The mean, denoted $\mu = E(Y)$ for "expected value of", is the theoretical average of values from the entire population or of an infinitely large sample from it. Thus, if the range of possible values for Y is finite, we would define the mean as

$$\mu = \sum_y y \Pr(Y = y)$$

or if it is continuous, as

$$\mu = \int y \Pr(Y = y) \, dy$$

The variance, denoted $\sigma^2 = \text{var}(Y)$, is a measure of the spread of a distribution and computed as

$$\sigma^2 = \int (y - \mu)^2 \Pr(Y = y) \, dy$$

or the analogous summation for a discrete distribution. Thus, the variance can also be thought of as $E[(Y - \mu)^2]$.

The mean and variance of a sum S of independent random variables Y_i is simply the sum of their respective means and variances, $E(S) = \Sigma_i \mu_i$ and $\text{var}(S) = \Sigma_i \sigma_i^2$. In particular, if S is the sum of ν *independent identically distributed* (*iid*) random variables, then $E(S) = \nu\mu$ and $\text{var}(S) = \nu\sigma^2$.

Probability distributions used in epidemiology

Binomial distribution

For count data, the two most important distributions are the binomial and Poisson distributions, depending upon whether the number of cases is expressed relative to the number of people at risk (the resulting proportion affected being a *probability*) or the person-time at risk (the resulting fraction being a *rate*).

Suppose we have a group of $i = 1, \ldots, N$ individuals, each of whom is characterized by a binary random variable Y_i and let $Y = \Sigma_i Y_i$ be the number of "cases" in the group (the number with $Y_i = 1$). Then if each individual's value is independent with the same probability p, the probability distribution function for Y is given by the Binomial distribution

$$\Pr(Y = y) = \binom{N}{y} p^y (1 - p)^{N-y} = \frac{N!}{y!(N - y)!} p^y (1 - p)^{N-y}$$

where $N! = 1 \times 2 \times \cdots \times (N-1) \times N$. The mean of this distribution is readily shown to be pN and its variance $Np(1 - p)$. The binomial distribution puts probability mass at a finite set of integer values from 0 to N. When N is large (as is typical of most epidemiologic applications), the distribution resembles a continuous Normal distribution with mean pN. However, as the distribution is bounded at 0 and N, this is only an approximation and when p is small (as is also typical of many epidemiologic applications), the distribution will be markedly skewed (see Figure 3.1). For convenience, we write a binomially distributed random variable as $Y \sim \text{Binom}(N, p)$. In epidemiology, the binomial distribution most commonly arises in the analysis of disease prevalence data or case-control studies.

Poisson distribution

Since epidemiologic studies seldom follow individuals over their entire lives, or even over a uniform interval of time, the element of time at risk and under study ("in view") is central. Suppose we have some large population in which events (say, disease) occur only rarely at rate λ per unit time

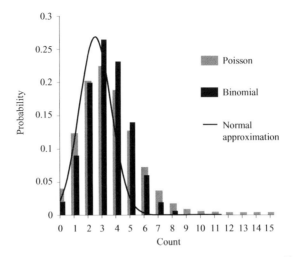

Figure 3.1. Binomial (black bars) and Poisson (gray bars) distributions and the Normal approximation (smooth line) for $N = 10$ and $p = 1/3$ or $\lambda = 3.33$ respectively, both having the same expectation.

and let Y denote the number of cases over some total amount of person-time T (the sum of all their individual times at risk). Now the probability distribution function for Y is given by the Poisson distribution

$$\Pr(Y = y) = e^{-\lambda T}(\lambda T)^y / y!$$

which we write in shorthand as $Y \sim \text{Poisson}(\lambda T)$. The Poisson distribution (see Figure 3.1) can be thought of as the limit of the Binomial distribution as N becomes very large and p becomes very small, so that their product converges to $pN = \lambda T$. Like the binomial distribution, as λT becomes large, the distribution takes on the appearance of a Normal distribution, although somewhat skewed. Characteristic of the Poisson distribution is that its mean and variance are the same, λT.

The Poisson distribution puts nonzero probability on the infinite set of nonnegative integers. Because the distribution is unbounded, it could include values greater than N, the total number of people at risk! Of course, if λ is small, then the probability associated with values greater than N will be trivial. Furthermore, if the outcome does not lead to termination of the period at risk (like death does), then individuals could have multiple events (e.g., asthma attacks) and so the total number of events could indeed be larger than the total number of individuals at risk. For events that do lead to termination of time at risk, the resolution of this paradox relies on the realization the total person-time T is also random, so it is the pair (Y, T) that needs to be modeled, using appropriate techniques for censored survival analysis, as discussed in the following chapter.

The remarkable observation is that by conditioning on T and treating Y as having a Poisson distribution, the resulting likelihood has the same dependence on λ [or the parameters of some more complex model for $\lambda(t, Z)$] as the real survival-time likelihood, and thus leads to identical tests, estimates, and confidence intervals, without requiring any assumption about disease rarity. The Poisson distribution commonly arises in the analysis of grouped cohort study data, such as from the atomic bomb survivors study, as well as in time-series studies or panel studies of acute effects, such as for air pollution.

Normal distribution

For continuous distributions, the most important probability distribution function is the Normal distribution, with density

$$\varphi(y)dy = \Pr(y \leq Y \leq y + dy) = \frac{1}{\sqrt{2\pi}\sigma} \exp\left(-\frac{(y-\mu)^2}{2\sigma^2}\right) dy$$

where μ is the mean and σ^2 is the variance of Y. In general, we use the notation $N(\mu, \sigma^2)$ to denote this distribution. The cumulative distribution function cannot be expressed in closed form, but is written as

$$\Phi(y) = \Pr(Y < y) = \int_{-\infty}^{y} \varphi(u)\, du$$

Chi square distribution

Unlike the three distributions described above, the chi square distribution generally arises not as the distribution of the data itself but rather of some summary statistic for a distribution or some hypothesis test about the data. For example, if $Y \sim N(0, 1)$, then $Z = Y^2$ has a *central* chi square distribution on 1 degree of freedom, with probability given by

$$\Pr(Z = z) = \frac{z^{-\frac{1}{2}} \exp(-z/2)}{\sqrt{2\pi}}$$

This distribution has mean 1 and variance 2. The sum of v independently distributed chi squares has a chi square distribution with degrees of freedom (df) v, so it follows from general result above about sums of random variables that its mean is v and its variance is $2v$. This general chi square distribution on v df is given by

$$\Pr(Z = z|v) = \frac{z^{v/2-1} \exp(-z/2)}{2^{v/2}\Gamma(v/2)}$$

Although this connection with Normal distribution defines the chi square distribution, it arises in practice more commonly in connection with the large sample approximation of the binomial or Poisson distributions. Thus, for example, if $Y \sim \text{Poisson}(E) \approx N(E, E)$, where $E = \lambda T$ is the expected number of cases, then $Z = (Y - E)^2 / E$ has approximately a chi square distribution on one degree of freedom. The analogous expression for a binomial random variable would be $Z = (Y - E)^2 / \text{var}(Y) = (Y - E)^2 [N/E(N - E)]$.

Frequently, we will wish to compare observed and expected events across several groups $j = 1, \ldots, v$, so following the general principles above, the total chi square statistic, $Z = \Sigma_{j=1}^{v} (Y_j - E_j)^2 / E_j$ has a chi square distribution on v df, where v is the number of groups compared, assuming the number of expected events are computed using a known external standard, e.g, $E_j = \lambda T_j$. Frequently, however, the rates are estimated from the ensemble of all the data, for example, $\hat{\lambda} = \sum_{j=1}^{v} Y_j / \sum_{i=1}^{v} T_j$, in which case the df is reduced by one (i.e., to $v - 1$) to allow for this conditioning on the total number of events.

The central chi square distribution applies under the null hypothesis $H_0: \mu = 0$, $\sigma^2 = 1$ for Normally distributed random variables, or, for the chi square statistic given above for Poisson variables, when H_0: $E(Y) = E$, $\text{var}(Y) = E$. Under the alternative hypothesis, the chi square statistic will have a *noncentral chi square distribution* with noncentrality parameter λ being the sum of the expectations of the original variables, $\lambda = \sum_{j=1}^{v} (E(Y_j)/\sqrt{\text{var}(Y_j)})$. This distribution has a complex expression, but is readily computed and has important applications in statistical power calculations. (Power is the probability that a real effect of some hypothesized size will be detected at a specified level of significance by a particular study design and analysis method. An investigator typically aims to choose the sample size for a particular design in a particular population so as to attain some minimum power, say, 90% for detecting a relative risk of 2 at a 5% level of significance. The details of the calculation of power thus depend upon the specific hypotheses, designs, analysis methods, and population characteristics under consideration, so are not treated in detail here. Such methods are covered in most of the standard epidemiology and biostatistics textbooks.)

Measures of disease frequency and risk

We now use the binomial and Poisson distributions defined in mathematical generality above to develop the concepts of risks and rates that epidemiologists use to describe disease outcomes. We begin by defining measures of absolute risk as applied to a single population or subgroup, and then discuss measures of relative rates used to compare disease outcomes between groups. As described earlier, we distinguish between the parameter value for a population and the estimator of that parameter obtained from the data on a finite random sample of individuals from the population. In this chapter, we assume that all the members of a particular group are "exchangeable," in the sense that their disease risks or rates are the same (at least within distinguishable subgroups, such as males and females), and hence the population parameter applies equally to any representative member of the group. In the following chapter, we will develop regression models for estimating the rate for a particular individual whose characteristics may be unique, by analysis of all the data on outcomes and predictors. This estimate could then be used as a risk predictor for new individuals whose unique combination of predictors may not even have existed in the sample used to estimate it.

Absolute risk

Risk

Risk or *incidence*, is defined as the probability of new disease occurring during some defined time period among a group of individuals free of the disease at the start of the period. Throughout this book, we will not distinguish between disease incidence and mortality, as both can occur only once—one cannot have a first event more than once! Of course, the study of recurrent events and prevalence (without the restriction to new cases) are of interest in epidemiology as well, but are of less use in studying disease etiology, although they do arise in, say, panel studies of asthma exacerbations due to air pollution.

Denoting the time period $[0, t]$, we might write the risk as $p(t) = \Pr[Y(t) = 1 | Y(0) = 0]$. Any of a number of time scales might be relevant, for example, age, time since start of observation, calendar year. In etiologic research, age is usually taken as the primary time scale, with year of birth and year of start of observation being treated as covariates. Being a probability, risk is a dimensionless quantity ranging between 0 and 1. For a given time interval, the population parameter is a fixed quantity we shall denote simply as p. If in a random sample of N observations from some population at risk, we observe Y cases developing during the at-risk period, then we would treat the random variable Y as having a Binomial distribution, $Y \sim \text{Binom}(N, p)$, and estimate the risk by the sample proportion $\hat{p} = Y/N$ with variance $\text{var}(\hat{p}) = Y(N - Y)/N^3$. The chi square distribution can be used to test the significance of the departure of the observed frequency from some null value H_0: $p = p_0$ (say, the population risk, for comparison with that observed in the cohort)

$$\frac{(\hat{p} - p_0)^2}{\text{var}(\hat{p})} = \frac{(Y - p_0 N)^2}{N p_0 (1 - p_0)} \sim \chi_1^2$$

Rates

In contrast to risk, a *rate* is a probability density, expressed in units of person-time at risk (typically, as a rate per 100,000 person-years). Rates are the preferred measure of disease frequency when observation is extended over time, so that not all subjects are at risk for the same duration due to "censoring" by competing risks or loss to follow up. They are also useful for describing disease frequency that varies over age or other temporal variables, as discussed further in Chapter 6. The probability density of disease occurrence at any instant of time within the period of observation amongst individuals free of the disease the instant before is defined as

$$\lambda(t) = \lim_{dt \to 0} \frac{\Pr(Y(t + dt) = 1 | Y(t) = 0)}{dt}$$

Viewed as a function of time, $\lambda(t)$ is known as a *hazard function* (equivalent terms include *incidence density*, *failure rate*, and (for death) *force of mortality*). Incidence and mortality rates used in epidemiology are a form of hazard rate, often expressed as step functions, treated as constant over some arbitrary grid of age and time intervals. For the moment, we consider a single such interval and assume the hazard rate λ is constant across time and across individuals, dropping the qualifier (t). Suppose the $i = 1, \ldots, N$ individuals have each been observed from time 0 to t_i (where t_i denotes to time of disease occurrence, loss to follow up, or termination of the study, whichever comes first), and let $T = \Sigma_i t_i$ be the total person-time of observation. Then the number of events has a Poisson distribution, $Y \sim \text{Poisson}(\lambda T)$, and we would estimate the incidence or hazard rate as $\hat{\lambda} = Y/T$ with variance $\text{var}(\hat{\lambda}) = Y/T^2$. The null hypothesis $H_0 : \lambda = \lambda_0$ can be tested by the chi square $(y - \lambda_0 T)^2 / \lambda_0 T \sim \chi_1^2$.

Survival analysis

Risks and rates are linked by the fundamental relationships:

$$S(t) = \exp\left(-\int_0^t \lambda(u)\, du\right)$$

$$\lambda(t) = -\frac{dS(t)/dt}{S(t)}$$

(3.1)

where $S(t) = 1 - p(t) = \Pr[Y(t) = 0 | Y(0) = 0]$ is known as the *survival function*. If the rate is assumed to be constant over some interval of time $(0, t)$, these expressions reduce to $p = 1 - \exp(-\lambda t)$ and $\lambda = -\ln(1 - p)/t$, respectively.

The survival function can be estimated in a number of ways. Using individual data, the *Kaplan–Meier* or *product-limit* estimator of the survival curve is given by

$$\hat{S}(t) = \prod_{(i | t_i < t)} \left(1 - \frac{Y_i}{R_i} \right),$$

where Y_i is an indicator for whether subject i developed the disease and R_i is the number of subjects who are still at risk (free of disease and under observation) at age t_i. This estimate is a step function, with discontinuous drops at the observed event times. Alternatively, one could start with a grouped time estimator of the incidence rate $\hat{\lambda}_k = Y_k / T_k$ for some categorization of the time axis (say, five-year intervals Δ_k), and compute the survival function as

$$\hat{S}(t_k) = \exp \left(-\sum_{j \le k} \hat{\lambda}_k \Delta_k \right).$$

Both these approaches estimate the probability of remaining free of disease absent *competing risks*, that is, assuming that one does not die of some unrelated cause first (sometimes called the "gross probability"). To estimate the lifetime risk of disease or the probability of developing disease at any specific age allowing for competing risks (the "net probability"), the *lifetable* method is used. Let μ denote the risk of dying of causes other than the disease of interest and assume that the two causes "compete independently." Then the probability of developing the disease at age t is

$$p(t) = \lambda(t) \exp \left(-\int_0^t [\lambda(u) + \mu(u)] \, du \right)$$

and the cumulative risk of disease to age T is $P(T) = \int_0^T p(t) \, dt$. Thus, the lifetime risk is $P(\infty)$. The quantity $\Lambda(t) = \int_0^t \lambda(u) \, du$ is known as the "cumulative hazard" and can be interpreted as the expected number of cases per person at risk (as opposed to the probability) up to time t in the absence of competing risks. The rationale for this expression is as follows: the probability of the disease of interest during the interval $(t, t + dt)$ is the probability of surviving to t, multiplied by the conditional probability of dying of that cause in the next instant; the former is the probability of surviving both the cause of interest and competing causes, given by the exponential factor; the latter is simply the hazard rate for the cause of interest at time t.

These calculations are illustrated for lung cancer incidence in Table 3.1. The column headed "Probability of surviving interval" is computed as $s_k = \exp[-5(\lambda_k + \mu_k)]$, where λ_k is the lung cancer incidence rate and μ_k the mortality rate for all other causes in age interval k. The column headed "Probability of surviving, cumulative" is the running product $S_k = \Pi_{j=1}^{k} S_j = S_{k-1} s_k$, representing the probability of surviving to the end of each interval. Thus, $P_k = S_k - S_{k-1} = S_{k-1}(1 - s_k)$ is the probability of either developing lung cancer or dying of other causes during the interval. "Risk of lung cancer" is computed as $P_k \lambda_k / (\lambda_k + \mu_k)$, the proportion of all these events in the interval that are incident lung cancer cases, the lifetime risk being given by the total at the bottom of the column. The column "Expected years of life remaining, survivors" column is $y_k = \int_{t_{k-1}}^{t_k} S(t)\,dt = [1 - \exp(-5\lambda_k)]/\lambda_k$ the average person-time during the interval among those surviving to the start of the interval (five years for those who survive to the next interval, plus slightly less than half that among those who die during the interval). The final column $Y_k = \sum_{k}^{\infty} y_k S_{k-1}$ is the remaining expectation of life among those alive at the start of the interval. The final row (age 90+) is calculated in a similar way, assuming the incidence and mortality rates remain constant thereafter (details not shown). Thus, by age 90, the cumulative risk of dying of lung cancer has reached about 7.8%; continuing the calculations beyond this age (assuming the death rates remain constant), the lifetime risk eventually converges to 8.0%. The expectation of life at birth is 72.4 years. We will revisit these calculations in Chapter 15, where we compare the effect of a particular exposure pattern on such quantities as excess lifetime risk of cancer or loss of life expectancy.

Relative risk

The term *relative risk* (RR) is used in a variety of ways, depending on the context, but generally refers to the ratio of risks or rates between groups differing with respect to some measurable risk factor, for example, between groups exposed and unexposed to some environmental agent. In its simplest form, we might define the relative risk as a ratio of risks, $p_1(t)/p_0(t)$, comparing individuals with or without some risk factor, or as a ratio of rates (*hazard ratio* or *incidence rate ratio*) $\lambda_1(t)/\lambda_0(t)$. A constant RR model assumes that these quantities do not vary over time, although there are numerous examples in which they do—for example, the marked variation in relative risks of cancer from ionizing radiation with age and latency, mentioned in Chapter 1.

Table 3.1. Calculation of age-specific and lifetime risk of dying of lung cancer using the lifetable method (U.S. white males, 1973–2004, from **http://www.seer.cancer.gov/canques/**)

Age interval	Incidence rate of lung cancer	Mortality rate for all other causes	Probability of surviving		Risk of lung cancer during interval	Years of life	
			Interval	Cumulative		Interval (survivors)	Cumulative
0–4	0.0	237.2	0.9882	0.9882	0.00000	4.970	72.427
5–9	0.0	25.7	0.9987	0.9869	0.00000	4.997	67.457
10–14	0.0	32.0	0.9984	0.9854	0.00000	4.996	62.519
15–19	0.2	116.2	0.9942	0.9796	0.00001	4.985	57.588
20–24	0.2	153.6	0.9923	0.9721	0.00001	4.981	52.676
25–59	0.4	147.6	0.9926	0.9650	0.00002	4.982	47.796
30–34	1.5	166.4	0.9916	0.9569	0.00007	4.979	42.954
35–39	4.9	211.0	0.9893	0.9466	0.00023	4.973	38.149
40–44	14.2	290.4	0.9849	0.9323	0.00067	4.962	33.390
45–49	37.6	429.6	0.9769	0.9108	0.00173	4.942	28.693
50–54	84.5	654.8	0.9637	0.8777	0.00378	4.909	24.085
55–59	162.1	1028.4	0.9422	0.8270	0.00691	4.854	19.614
60–64	276.7	1627.5	0.9092	0.7519	0.01091	4.769	15.354
65–69	404.3	2497.3	0.8650	0.6504	0.01415	4.654	11.409
70–74	514.0	3850.3	0.8040	0.5229	0.01502	4.492	7.910
75–79	572.1	5963.0	0.7213	0.3771	0.01276	4.265	4.988
80–84	553.4	9464.1	0.6060	0.2285	0.00821	3.933	2.758
85–89	446.5	17478.1	0.4081	0.0933	0.00337	3.302	1.275
90+	446.5	17478.1	0.0000	0.0000	0.00232	5.579	0.520
Totals					0.08017		

Case-control studies

In case-control studies, a more useful measure of association between some risk factor and disease is the *odds ratio*

$$\text{OR} = \frac{p_1(1 - p_0)}{p_0(1 - p_1)}$$

the ratio of the *odds* $p/(1 - p)$ of disease between those with and without the risk factor. It is readily seen by application of Bayes formula that this quantity is the same as the odds of exposure between cases and controls, since

$$\text{OR}(Y|Z) = \frac{\text{odds}(Y|Z = 1)}{\text{odds}(Y|Z = 0)} = \frac{\dfrac{\Pr(Y - 1|Z = 1)}{\Pr(Y = 0|Z = 1)}}{\dfrac{\Pr(Y = 1|Z = 0)}{\Pr(Y = 0|Z = 0)}} = \frac{\dfrac{\Pr(Z = 1|Y = 1)}{\Pr(Z = 0|Y = 1)}}{\dfrac{\Pr(Z = 1|Y = 0)}{\Pr(Z = 0|Y = 0)}}$$

$$= \frac{\text{odds}(Z|Y = 1)}{\text{odds}(Z|Y = 0)} = \text{OR}(Z|Y)$$

an attractive feature for its use in case-control studies. For a rare disease, the OR approximates the risk ratio, p_1/p_0. While in general one could combine estimates of the OR and the proportion of controls exposed from a case-control study with external information on the population average risk to derive an estimate of the risk ratio, a more useful observation is that, under appropriate circumstances, the OR provides a consistent estimator of the hazard ratio even *without the rare disease assumption* (Greenland and Thomas 1982). In particular, when matched controls are sampled from the individuals at risk at the time of each case ("incidence density sampling"), then no rare disease assumption is needed to justify the interpretation of the odds ratio as an estimator of the hazard ratio.

Epidemiologic data from cohort or case-control studies are frequently presented in the form of a 2×2 contingency table, as in Table 3.2. Using these data, a test of the null hypothesis H_0: OR = 1 is given by the chi

Table 3.2. Presentation of data from a cohort study or unmatched case-control study of a binary risk factor

Risk factor	Disease status	
	Unaffected ($Y = 0$)	Affected ($Y = 1$)
Absent ($Z = 0$)	A	B
Present ($Z = 1$)	C	D

square test

$$X^2 = \frac{[A - E(A)]^2}{E(A)} + \frac{[B - E(B)]^2}{E(B)} + \frac{[C - E(C)]^2}{E(C)} + \frac{[D - E(D)]^2}{E(D)}$$

$$= \frac{(AD - BC)^2 N}{(A + B)(C + D)(A + C)(B + D)}$$

where $E(A) = (A + B)(A + C)/N$ and so on. Under the null hypothesis, X^2 has asymptotically (i.e., in large samples) a chi square distribution on 1 degree of freedom (df) (henceforth, we write statements like this as $X^2 \sim \chi_1^2$). A continuity correction of $N/2$ is usually subtracted from $|AD - BC|$ in the numerator to allow for the discreteness of the possible values of the chi square test for a given sample size. In small samples, a more appropriate procedure for significance testing is Fisher's exact test, which takes the form

$$\Pr(X > A | A + B, A + C, N) = \sum_{X = \min(A+B, A+C, A-D)}^{N} \Pr(X | A + B, A + C, N)$$

where

$$\Pr(X | A + B, A + C, N) = \frac{A!B!C!D!}{X!(A + B - X)!(A + C - X)!(D - A + X)!N!}$$

The OR is estimated as AD/BC and its asymptotic variance as

$$\text{var}(\ln \widehat{OR}) = \frac{1}{A} + \frac{1}{B} + \frac{1}{C} + \frac{1}{D}$$

A confidence interval with $1 - \alpha$ percent coverage is then given by

$$\exp\left[\ln \widehat{OR} \pm Z_{\alpha/2} \sqrt{\text{var}(\ln \widehat{OR})} \right]$$

Of course, there could be more than two categories of "exposure," for example, none, light, moderate, or heavy exposure; the relevant degrees of freedom of the chi square test would then be one fewer than the number of categories.

Stratum-matched (frequency-matched) case-control studies are presented as a series of 2×2 tables, each in the same form as Table 3.2. The stratum-specific measures of association would then be summarized across subtables s in some fashion, the most-commonly-used method being the Mantel–Haenszel odds ratio and test described in Chapter 12.

Table 3.3. Presentation of data from a matched case-control study

Control risk factor	Case risk factor		
	Absent	Present	Total
Absent	a	c	C
Present	b	d	D
Total	A	B	N

For pair-matched case-control studies, the data needs to be presented in a form that keeps the matched pairs intact, as displayed in Table 3.3, rather than as independent individuals. Here the McNemar estimator of the OR is given by c/b with asymptotic variance

$$\text{var}(\ln \widehat{\text{OR}}) = \frac{1}{b} + \frac{1}{c}$$

and significance test under H_0

$$\frac{(b-c)^2}{b+c} \sim \chi_1^2$$

The appropriate continuity correction to $|b-c|$ in the numerator is -1. The corresponding exact test would be

$$\Pr(X > B|B+C) = \sum_{X=B+1}^{B+C} \text{Binom}\left(X|N, \frac{1}{2}\right)$$

$$= \sum_{X=B+1}^{B+C} \frac{(B+C)!}{X!(B+C-X)!} 2^{B+C}$$

Extensions to categorical exposure variables with more than two levels (Pike et al. 1975) or to binary variables with more than one control per case Miettinen (1969) are available, but are complex. These turn out to be special cases of conditional logistic regression, as described in the following chapter. With the widespread availability of logistic regression programs in standard statistical packages, there is no longer any need for explicit formulae for dealing with these special settings.

Only variables that are related both to exposure and to disease (conditional on exposure) need to be controlled in case-control studies by stratification or matching. There is no benefit from controlling for a risk factor that is independent of exposure, but no penalty for doing so if one is uncertain whether or not the variable is really related to exposure. On the other hand, adjusting for a variable that is related to exposure but not to disease will lead to unnecessary inflation of the variance of the OR and a loss of power. This situation is commonly known as "overmatching" or "over-adjustment" (Day et al. 1980). Provided the correct matched or stratified analysis is done, however, the estimate will still be unbiased and the nominal size of the significance test will be preserved.

Cohort studies

For a cohort study, in which individuals might be followed over a long and variable length of time and a range of ages, the standard method of data analysis is the *standardized incidence ratio* (SIR, or for mortality, the standardized mortality ratio, SMR). One begins by tabulating each individual's time at risk over a two-dimensional array of ages and calendar years (e.g., five-year intervals), known as a Lexis diagram (Figure 3.2) to obtain the total person-time T_{zs} in each age–year stratum s and exposure category z. Next, one tabulates the number of observed cases Y_z in each exposure category. These are then compared with the corresponding numbers expected E_z based on a set of standard age-year specific incidence rates λ_s, obtained by multiplying the rates by the total person-time

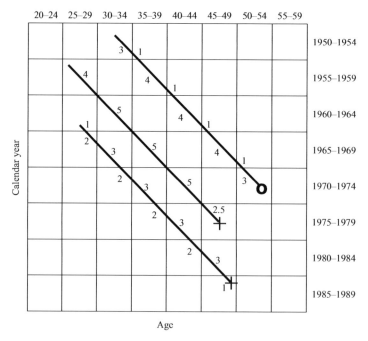

Figure 3.2. Lexis diagram, illustrating the calculation of events and person-time at risk in a cohort study.

Table 3.4. Presentation of data from a cohort study using the standardized incidence ratio (SIR) method

Exposure status	Cases		Standardized incidence rate	Standardized rate ratio
	Observed	Expected		
Unexposed	Y_0	$E_0 = \Sigma_s \lambda_s T_{0s}$	$SIR_0 = Y_0/E_0$	1
Exposed	Y_1	$E_1 = \Sigma_s \lambda_s T_{1s}$	$SIR_1 = Y_1/E_1$	SIR_1/SIR_0

in each stratum and summing over strata, $E_z = \Sigma_s \lambda_s T_{zs}$. Standard rates might come from some external source, like national death rates or incidence rates from a disease registry, or internally from the cohort as a whole, ignoring the exposure classification.

The data are typically displayed as shown in Table 3.4. The indirectly standardized incidence ratio ($SIR^{indirect}$) for each exposure category is defined as the ratio of observed to expected events $SIR_z^{indirect} = Y_z/E_z$ and the Standardized Rate Ratio (SRR) as the ratio of SIRs between exposure categories.

Unfortunately, the ratio of indirect SIRs, which we have loosely called the SRR, is not an unbiased estimator of the ratio of hazard rates, even if these stratum-specific rate ratios are constant across strata, for reasons explained by Breslow and Day (1987). An alternative estimator, known as the directly standardized incidence ratio, is defined as the ratio of the number of cases that would have occurred in the standard population if the cohort rates applied versus the standard rates, $SIR_s^{direct} = D_0/\Sigma_s T_{0s} \lambda_s$. The ratio of these directly standardized incidence rates *is* a consistent estimator of a constant hazard ratio, but has larger variance than the ratio of indirect SIRs, and is much less commonly used. As we shall see in the next chapter, the ratio of indirect SIRs turns out to be a special case of multivariate methods of survival analysis like Cox regression.

Attributable risk and other measures of public health impact

Suppose we had a simple dichotomous classification of exposure as "exposed" or "unexposed" with a relative risk for exposure of RR and a proportion p_E of the population exposed. *Attributable risk* is a measure of the proportion of disease in the population that is caused by exposure, that is, that would be eliminated if the exposure were eliminated. This quantity is easily estimated for the exposed population by recognizing that the disease rate attributable to exposure AR_E is simply the difference between the rates in the exposed and unexposed populations, $AR_E = \lambda_1 - \lambda_0 = \lambda_0(RR - 1)$. Expressed as a proportion of the total risk in the exposed population $\lambda_1 = \lambda_0 RR$, the baseline rate λ_0 cancels out and this becomes the quantity known as the "attributable or etiologic *fraction* among the exposed" $AF_E = (RR - 1)/RR$ (Miettinen 1974). (Note that the term "attributable risk" is used somewhat ambiguously in the epidemiologic literature to refer either to the excess *rate* or the excess *fraction* of the total rate; to avoid this confusion, we will used the terms "excess rate" to refer to the former and "attributable fraction" for the latter.) Of course, among the unexposed, none of the cases can be attributed to exposure, so the population attributable fraction is obtained by dividing the excess $p_E AR_E$ by the total population rate $p_E \lambda_1 + (1 - p_E)\lambda_0$ to obtain

$$ PAF = \frac{p_E(RR - 1)}{p_E RR + (1 - p_E)RR} $$

(Levin 1953). We will revisit the quantity AF_E in Chapter 16 when it is applied to individuals as the "probability of causation" and defer until then a treatment of the assumptions required for either the population or individual attributable fractions to be estimable.

For categorical exposure variables with more than two levels or for continuous exposure variables Z, the population attributable fraction is computed in a similar fashion, summing or integrating over the population distribution of exposure $p(z)$:

$$ PAF = \frac{\int p(z)[RR(z) - 1]\,dz}{\int p(z)RR(z)\,dz} $$

The PAF can also be decomposed into portions attributable to each of several risk factors and their interactions, as described in Chapter 16. These various components, along with the fraction unexplained by any of them, must add to 100%. The total attributable to any particular risk factor would include its interactions with other factors, since if that factor were eliminated, all of the cases attributed to both its main effect and its interaction with other factors would be eliminated. Thus, the sum of these total attributable fractions could easily add up to more than 100%, due to double counting of interaction effects.

This derivation assumes that the relative risk is a constant over time. In general, the calculation of lifetime risk and quantities derived from it, like the lifetime population attributable risk or the population loss of life expectancy requires the use of lifetable methods, as described above and in greater detail in Chapter 15.

For a chronic disease like asthma, one might wish to distinguish between acute exacerbations that are caused directly by exposure and those for which exposure is the cause of the underlying disease. Kunzli et al. (2008) decomposed the attributable risk of bronchitic symptoms among CHS participants in Long Beach and found that 33.5% were due to exposure to traffic-related pollutants (3.1% among those whose underlying asthma was also due to air pollution, 30.5% among those whose asthma was due to other factors). However, an additional 6.2% of bronchitic symptoms attributable to other factors occurred *among children whose chronic disease was due to air pollution*, for a total of 39.8%. It is this latter contribution that would be neglected in the usual calculation of attributable risk for acute effects.

Interaction

By *interaction* or *effect modification*, we mean a variation in some measure of the effect of an exposure on disease risks across the levels of some third variable M, known as a *modifier* (Figure 2.1). For example, among the atomic bomb survivors, the excess risk of leukemia declines by about 6.5% per year after 5 years following exposure (the start of observation) (Preston et al. 1994). This decline is highly significant ($p < 0.001$) and decreases more rapidly in those exposed at younger ages (Figure 3.3). Hence, we would call latency a modifier of the radiation risk.

The definition of interaction depends upon the measure of association used. For solid cancers, the *excess* risk (the difference in annual incidence rates between radiation exposed and unexposed subjects) increases with age, but the relative risk declines. Hence, age is a modifier of both measures of association, but in very different ways. In general, the absence of interaction *on a scale of relative risk* corresponds to a multiplicative model for the joint effect of the two factors, that is, $p(Z, M) = p_0 \times RR_Z \times RR_M$. The

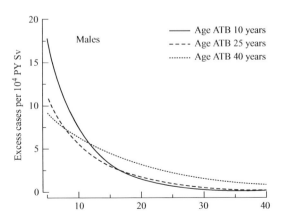

Figure 3.3. Excess absolute risks of leukemia in males by years following exposure to the atomic bomb. (Reproduced with permission from Preston et al. 1994.)

absence of interaction *on a scale of excess risk* corresponds to an additive model $p(Z, M) = p_0 + ER_Z + ER_M$. Here, $p(Z, M)$ represents some measure of risk (lifetime risk, hazard rate, etc.), p_0 the risk in the absence of exposure to both factors, RR the relative risk for the indicated factor, and ER the corresponding excess risk.

There is a philosophical debate in the epidemiologic literature about the meaning of the word *synergy*; some reserve this term to mean any joint effect of two factors that is greater than additive (Rothman and Greenland 1998) whereas others apply it to any departure from some biological model for independent effects (Siemiatycki and Thomas 1981). We distinguish between *qualitative* interaction, where the effect of one variable is completely absent in one stratum or the other or even goes in opposite direction, from *quantitative* interaction, where the effect is in the same direction in all strata but varies in magnitude.

Simple descriptive analyses of interaction effects can be accomplished by stratifying the analysis into categories of the modifier, estimating the association and its variance in each stratum, and comparing the magnitudes of the effect sizes between the two strata. For example, suppose in a case-control study the odds ratios in two strata of the modifier were OR_1 and OR_2 with variances of ln OR being V_1 and V_2, respectively. Then we might take the ratio of the two ORs, $\psi = OR_1/OR_2$ as a measure of interaction. The variance is thus $\text{var}(\ln \psi) = V_1 + V_2$, so we could test the null hypothesis of no interaction (constancy of the OR) $H_0: \psi = 1$ by testing $\ln \widehat{\psi}/\sqrt{\text{var}(\ln \widehat{\psi})}$ against a normal distribution. Confidence limits on $\widehat{\psi}$ would be obtained in the usual way as $\exp[\ln \widehat{\psi} + Z_{1-\alpha/2}\sqrt{\text{var}(\ln \widehat{\psi})}]$ for large enough sample sizes (i.e., *asymptotically*).

Note, however, that the finding of a significant effect in one stratum but not in another does not necessarily imply interaction: the sample size in one of the subgroups might simply be too small for that OR to be significant, even though the magnitude of its effect was not appreciably different from the significant one; on the other hand, the differences in the estimated ORs might appear large, but be quite unstable due to inadequate sample sizes and not be significant. Hence a claim of an interaction should be reserved for situations where the two estimates are actually judged to differ from each other by an appropriate significance test.

Basic principles of statistical inference

Let us denote a body of "data" by $\mathcal{D}$. In this book, this will generally mean the raw observations from a single epidemiologic study, but in some contexts might include a meta-analysis of many such studies, or more

broadly a compilation of all relevant information, observational or experimental, in humans, animals, cell cultures, or other systems. We wish
to compare various alternative "hypotheses" (H), of which we single out
one—the "null hypothesis" (H_0) that there is no association—for special
attention. The remaining possibilities we will denote generically as the
"alternative hypothesis" (H_1). This could be simply the converse of H_0
(there is some association), or a whole family of alternative hypotheses,
indexed by some parameter θ, say the true relative risk RR (in some cases,
θ might represent a whole vector of parameters, some of which might be
the parameter(s) of primary interest (e.g., the risks associated with several
pollutants), others "nuisance" parameters needed to describe other parts
of a model such as the dependence of risk on confounders like age). Classical "frequentist" statistical inference is based on the likelihood function,
denoted $L(\theta) = \Pr(\mathcal{D}|H_\theta)$, generally viewed as a function of the parameters
of some statistical model, as described in greater detail in the following
chapter. It is important to realize that the likelihood is *not* the probability
that a given hypothesis is true in light of the data (see below), but rather
the probability of observing the actual data if the various hypotheses were
true.

From the likelihood, one can compute the "likelihood ratio", LR =
$\Pr(\mathcal{D}|H_1)/\Pr(\mathcal{D}|H_0)$, where the numerator is the maximum likelihood over
all hypotheses in the family of models under consideration. From this, it is
evident that the LR is necessarily greater than or equal to one, that is, that
one can always find a specific hypothesis, such that $\Pr(\mathcal{D}|H_1) \geq \Pr(\mathcal{D}|H_0)$.
The real question is whether the LR is *compellingly* large, in the sense
that it is bigger than can be explained simply by chance. Another way
of expressing this observation is the maxim that "one can never prove
the null hypothesis;" instead, science proceeds by finding evidence that
would lead one to reject the null hypothesis in favor of some alternative
hypothesis.

This attempt at rejection is known as significance testing. In classical frequentist statistics, we judge whether the evidence is "significant" based on
whether the observed values of the LR (or some other test statistic) would
be highly unlikely if the null hypothesis were true. Thus, in likelihood-
based methods, the p-value is defined as the probability $\Pr(\text{LR} > T|H_0)$
for some critical value T, which can be computed from a model for the
distribution of the data in theoretical random samples from the population
or by randomly scrambling the data in some appropriate way. (Details of
this and subsequent calculations are deferred to the following chapters.)
By convention, scientists declare a result "significant" if $p < 0.05$ (i.e.,
there is only a 1 in 20 chance that such an extreme result could have
occurred by chance), but we recognize the arbitrariness of such a convention and, depending upon the context, other thresholds might be adopted.
The essential point is that a finding of statistical significance does not tell

one whether H_0 or H_1 is more likely to be true in light of the data, only that the data were highly unlikely to have occurred if H_0 were true. The calculation of these significance tests is described in the following chapter.

The kinds of judgment that one might really like to make are more in the spirit of Bayesian inference, which is based on the "posterior probability" $\Pr(H|\mathcal{D})$, namely the probability that hypothesis H is true given the observed data. One could then, in principle, answer the question "Is H_1 more likely than H_0 in light of the data?" We can express this question formally as "Is $\Pr(H_1|\mathcal{D}) > \Pr(H_0|\mathcal{D})$?" or equivalently, "Is the Posterior Odds, $\Pr(H_1|\mathcal{D})/\Pr(H_0|\mathcal{D}) > 1$?" The computation of these posterior probabilities is given by Bayes' formula, which can be expressed as posterior odds $=$ LR $\times$ prior odds, where prior odds $= \Pr(H_1)/\Pr(H_0)$. This clarifies the fundamental difficulty that to answer the question we are really interested in, we must consider the relative *prior* credibility of the competing hypotheses. By its nature, these prior odds are a matter of judgment, about which scientists are likely to disagree. Nevertheless, this is essentially what scientists are doing—albeit not in a mathematical way—when, individually or in expert committees, they evaluate the totality of the evidence regarding causality.

In addition to testing hypotheses, investigators are frequently interested in estimating the parameters of a model and putting some confidence bounds on them. Again deferring the details of the calculations to the following chapter, we would call the parameter value $\hat{\theta}$ that maximizes the likelihood function (i.e., the value for which the observed data would be the most likely) the "maximum likelihood estimate" (MLE) and a confidence limit or interval estimate the set of such values that would be most likely to be observed in hypothetical replications of the study. They should not be interpreted as the most likely range of the true parameter given the observed data, which would be obtained in Bayesian inference, however. That posterior mode estimate $\tilde{\theta}$ would be the value which maximizes the posterior probability $\Pr(\theta|\mathcal{D})$ and the posterior mean would be $\bar{\theta} = \int \theta \, \Pr(\theta|D) \, d\theta$. "Credibility intervals" are then the range of θ values that encompasses most of the distribution $\Pr(\theta|\mathcal{D})$. As with Bayesian tests, the calculation of these quantities requires the specification of prior distributions $\Pr(\theta)$.

Although in general the specification of prior distributions is a matter of judgment based on substance matter expertise, one important use of Bayesian inference is to combine the data at hand with knowledge from previous literature. Thus, if one had a set of previous studies that yielded estimates of $\hat{\theta}$ and confidence limits on each, one could transform each of these into data of the same form as the study at hand and simply supplement the observed dataset with these "pseudodata" and analyze the ensemble with conventional methods (Greenland 2007).

4 Multivariate models

The previous chapter introduced the elementary measures of disease risk and rates per unit time, and methods for estimating and comparing them in groups of subjects. Any group is necessarily a somewhat artificial assemblage of a set of heterogeneous individuals, even though they may share some characteristics of interest. Some of this heterogeneity may be quantifiable, some not. (Of course, here we need only be concerned with characteristics that relate in some way to their risk of disease.) To the extent that this heterogeneity can be characterized, the epidemiologist could, of course, define further subgroups, but soon one is confronted with a multitude of groups, each with more unstable estimates of risk as their sample sizes get smaller, and it can become quite a challenge to discern the patterns in the data. For this purpose, this chapter introduces a general regression framework for modeling disease risks in individual or grouped data. The specific methods are special cases of what is known as the general linear model (GLM) framework, which is applicable to more than dichotomous disease data, but we will defer treatment of continuous and longitudinal data to Chapter 7, where we will introduce the further generalization of general linear mixed models (GLMMs).

To set the stage, before presenting the GLM in its full generality, we begin with a brief review of ordinary linear regression models for continuous outcome data. Then we describe three specific methods for binary outcome data: logistic regression for risks in individual binary outcome data; Cox regression for rates in individual censored survival data, and Poisson regression for rates in grouped data. All these models are fitted using the techniques of maximum likelihood, which will be described later in this chapter. Since not only disease rates, but also exposure, can vary over time as a cohort is followed, special consideration is needed for time-dependent data. This will be introduced in the context of Cox regression, but developed in greater depth in Chapter 6. Here we restrict attention to particular forms of the exposure–response relationship that are mathematically convenient, the so-called canonical links appropriate to each data structure. Chapter 13 will extend to more general forms that may be motivated by biological or physical considerations.

Linear regression for continuous data

Suppose we have a continuous outcome variable Y and a vector $Z = (Z_1, \ldots, Z_P)$ of predictors (covariates). We might propose a linear model of the form

$$Y_i = \beta_0 + \sum_{p=1}^{P} Z_{ip}\beta_i + e_i = Z'_i\beta + e_i, \quad \text{where } e_i \sim N(0, \sigma^2)$$

or equivalently

$$E(Y_i|Z_i) = Z'_i\beta \quad \text{and} \quad \text{var}(Y_i|Z_i) = \sigma^2$$

or

$$Y_i \sim N(Z'_i\beta, \sigma^2)$$

(From here on, for notational simplicity, we omit the intercepts β_0 by adopting the convention $Z_0 \equiv 1$ and letting the summation in $Z'\beta$ run from 0 to P.)

There are two approaches we could take to fitting this model: ordinary least squares (OLS) or maximum likelihood (ML). OLS seeks to minimize the total squared deviations between the observed Y values and their predicted values, that is, to find β that minimizes

$$SS(\beta) = \sum_{i=0}^{N}(Y_i - Z'_i\beta)^2$$

whereas ML seeks to maximize the probability of the observed Y data, that is, to find β that maximizes

$$L(\beta) = \prod_{i=1}^{N} \varphi\left(\frac{Y_i - Z'_i\beta}{\sigma}\right)$$

Both approaches solve this problem by taking derivatives of their respective criteria with respect to β, setting them equal to zero, and solving for β. Remarkably, the two yield identical solutions. This is not a coincidence, but merely the consequence of $SS(\beta)$ and $\ln L(\beta)$ having the same dependence on $Y_i - Z'_i\beta$. The resulting solution can be found in closed form as

$$\hat{\beta} = \left(\sum_{i=1}^{N} Z'_i Z_i\right)^{-1} \left(\sum_{i=1}^{N} Y_i Z_i\right) = (Z'Z)^{-1}(YZ)$$

As we shall see below, the models for binary data will generally not lead to closed-form solutions, but will require iterative methods. Also, OLS and ML may not yield the same answers, but by incorporating variance-based weights, weighted least squares (WLS) will turn out to be equivalent to ML, a result that lies at the heart of the GLM framework.

Multivariate models for binary data

In the previous chapter, we introduced risk p as the basic measure of disease risk for individuals followed for a fixed period of time and the hazard rate $\lambda(t)$ as the basic measure of disease rates per unit time. For case-control data, the odds $p/(1 - p)$ is the natural measure of risk. We now extend the estimation of these basic measures for groups of people and tests of their similarity across groups to regression models involving person- or group-specific covariates. We begin with methods for simple binary outcomes at the individual and group levels (logistic and binomial regression respectively), then introduce regression models for individual censored survival-time data (parametric as well as the semi-parametric Cox regression model), and conclude this section with a discussion of Poisson regression for grouped survival-time data.

Logistic regression

Suppose that during a fixed period of follow-up time T, each person i either experiences the event under study ($Y_i = 1$) or does not ($Y_i = 0$). When follow-up times vary between individuals, the techniques of censored survival analysis described in the next section would be needed. Also suppose each individual is characterized by a vector of personal covariates $Z_i = (Z_{i1}, \ldots, Z_{iP})$. Now we have all the ingredients needed to model risk at the individual level.

The quantity we wish to model is now the risk, $p_i = \Pr(Y_i = 1 | Z_i)$. The *observed* outcome is either 0 or 1—not particularly useful for discerning patterns with respect to Z—but it is the model for the *true* probabilities that we are interested in. Since a probability cannot be negative or exceed unity, some transformation of the regression model to the unit interval is needed to model them. There are many functions that could be used for this purpose, including

Logistic: $\quad \Pr(Y = 1 | Z) = e^{Z'\beta}/(1 + e^{Z'\beta})$

Probit: $\quad \Pr(Y = 1 | Z) = \Phi(Z'\beta) = \int_{-\infty}^{Z'\beta} \varphi(u)\, du$

Complementary log–log: $\quad \Pr(Y = 1 | Z) = \exp[-\exp(-Z'\beta)]$

Arctangent: $\quad \Pr(Y = 1 | Z) = \arctan(Z'\beta)/\pi + 1/2$

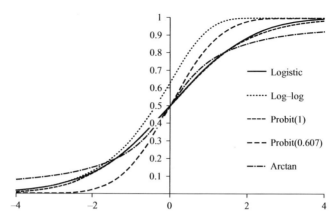

Figure 4.1. Alternative transformations from linear predictors on the real line (X-axis) to the probability interval [0, 1] (Y-axis).

These four functions are shown in Figure 4.1. All four are asymptotic to 0 and 1 as $Z'\beta$ goes to minus and plus infinity respectively, and all but the complementary log–log are symmetric about $Z'\beta = 0$. The latter two have not been widely used in modeling binary outcome data. The first two differ only in slope and, with the scaling factor of 0.607, they become almost identical to each other. The logistic function is mathematically easier to work with, as it does not involve the integral $\Phi(\cdot)$ that cannot be expressed in closed form. Nevertheless, the probit form has some advantages when dealing with latent variables, as we shall see in Chapter 11. For the remainder of this chapter, we will not be concerned with latent variables, so we will confine attention to the logistic model. In some cases it will be convenient to express this model in terms of the logistic transformation

$$\text{logit}[\Pr(Y = 1|Z)] = Z'\beta$$

where $\text{logit}(u) = \ln[u/(1 - u)]$. The inverse logistic transformation we will denote by $\text{expit}(u) = e^u/(1 + e^u)$, so the model could be written equivalently as $\Pr(Y = 1|Z) = \text{expit}(Z'\beta)$.

To fit the model, we use ML, forming a likelihood function by multiplying each individual's probability of their observed disease status given their covariates, assuming they are independent. Thus,

$$L(\beta) = \prod_{Y_i=1} \frac{e^{Z'_i\beta}}{1 + e^{Z'_i\beta}} \times \prod_{Y_i=0} \frac{1}{1 + e^{Z'_i\beta}} = \prod_{i=1}^{N} \frac{e^{Y_i Z'_i\beta}}{1 + e^{Z'_i\beta}}$$

For cohort studies with extended and possibly variable lengths of follow-up (say, because of loss to follow-up, deaths due to competing causes, or staggered entry to the cohort), one could subdivide the total period of observation into fixed-length intervals and build up a likelihood by multiplying each subject's conditional probabilities for each interval, given that he or she survived the previous interval disease-free and was still under observation. Thus, letting R_{ik} be an indicator for whether individual i was at risk in time-interval k, Y_{ik} an indicator for disease status in that interval, and Z_{ik} the covariate values at the start of that interval, the likelihood would become

$$L(\boldsymbol{\beta}, \alpha) = \prod_i \prod_{k|R_{ik}=1} \Pr(Y_{ik}|Z_{ik}) = \prod_i \prod_{k|R_{ik}=1} \frac{e^{Y_{ik}(\alpha_k + Z'_{ik}\boldsymbol{\beta})}}{1 + e^{\alpha_k + Z'_{ik}\boldsymbol{\beta}}}$$

where α_k are intercept terms for the baseline risk in each time interval. This provides a natural way to accommodate time-dependent covariates. However, since in practice, individuals may be at risk for only part of an interval, and the interval boundaries are somewhat arbitrary, this approach should be viewed as only an approximation to the survival analysis techniques discussed in the next section. Indeed, in the limit as the time intervals become infinitesimal, this reduces to Cox regression.

Although originally introduced in 1967 by Truett et al. (1967) as a means of analyzing multiple risk factors for coronary heart disease incidence in the Framingham cohort study, it is less often used for cohort studies because individuals' times at risk generally vary, requiring the use of censored survival analysis techniques, as discussed in the next section. Instead, the major application of logistic regression is to case-control data.

In principle, case-control analysis would appear to require a retrospective likelihood of the form $\Pr(Z|Y)$, since the disease status (case or control) is fixed by design and the observed random variables are the Zs. The retrospective likelihood, however, would require complex modeling of the joint population distribution of the covariates. Breslow and Powers (1978) showed that with increasing degree of stratification, the prospective and retrospective models became similar, and recommended the use of the prospective model in the case where there were continuous or many covariates. Using Bayes formula, $\Pr(Z|Y) = \Pr(Y|Z)\,\Pr(Z)/\Pr(Y)$, Prentice and Pyke (1979) showed that using the prospective probability $\Pr(Y|Z)$ as if it were the likelihood for the data yielded consistent estimators of the regression coefficients, that is, the two likelihoods estimate the same population log relative risk parameters. The only exception is the intercept term, which no longer estimates the log baseline odds of disease in the population (the odds for a subject with $Z = 0$), but rather that in the sample, which is inflated by the different sampling fractions π_1 and π_0 for cases and controls respectively. Indeed, it is easily shown that the intercept term β_0 estimates $\beta_0 + \ln(\pi_1/\pi_0)$. Prentice and Pyke's elegant proof requires no parametric assumptions about the joint distribution $\Pr(Z)$, relying instead on nonparametric density estimation techniques where that distribution is assumed to be concentrated at the observed covariate locations, and using the constraint that the denominator $\Pr(Y = 1) = \int \Pr(Y = 1|Z) \Pr(Z) \, dZ$ is the population rate to estimate the probability masses in $\Pr(Z)$ at each observed Z value. Ultimately, their proof relies on the same invariance of the odds ratio (OR) mentioned in the previous chapter, namely that the odds ratio comparing Z to Z_0 can be written in either of the following two forms:

$$\frac{\Pr(Y = 1|Z)/\Pr(Y = 0|Z)}{\Pr(Y = 1|Z_0)/\Pr(Y = 0|Z_0)} = \frac{\Pr(Z|Y = 1)/\Pr(Z|Y = 0)}{\Pr(Z_0|Y = 1)/\Pr(Z_0|Y = 0)}$$

See Langholz and Goldstein (2001); Arratia et al. (2005) for formal treatments of the asymptotic distribution theory justifying the use of the prospective likelihood in terms of sampling from the study base population.

The preceding discussion applies to unmatched case-control studies, where only the numbers of cases and controls are fixed by design (possibly within broad strata defined by confounders like age, sex, or race). Most case-control studies, however, are conducted in matched fashion, each case being individually matched to one or more controls on a number of potential confounders. Recall how, for this design, the data are represented as matched sets, as in Table 3.3, rather than as individuals. The likelihood is formed by treating each matched set as the observational unit and computing the prospective probability of the observed outcomes of all the members of the matched set conditional on it containing exactly the numbers of cases and controls specified by the design. Letting i subscript the matched sets and j the members of the set, the likelihood would be written as

$$L(\beta) = \prod_{i=1}^{N} \Pr\left(\mathbf{Y}_i|\mathbf{Z}_i, \sum_{j=1}^{m_0+m_1} Y_{ij} = m_1\right)$$

where m_0 and m_1 are the numbers of controls and cases in each set. For example, for case-control pairs, this becomes

$$L(\beta) = \prod_{i=1}^{N} \Pr(Y_{i1} = 1, Y_{i0} = 0|Z_{i0}, Z_{i1}, Y_{i0} + Y_{i1} = 1)$$

$$= \prod_{i=1}^{N} \frac{\Pr(Y_{i1} = 1|Z_{i1}) \Pr(Y_{i0} = 0|Z_{i0})}{\Pr(Y_{i1} = 1|Z_{i1}) \Pr(Y_{i0} = 0|Z_{i0}) + \Pr(Y_{i1} = 0|Z_{i1}) \Pr(Y_{i0} = 1|Z_{i0})}$$

$$
= \prod_{i=1}^{N} \frac{\left(\dfrac{e^{\alpha_i+Z'_{i1}\beta}}{1+e^{\alpha_i+Z'_{i1}\beta}}\right)\left(\dfrac{1}{1+e^{\alpha_i+Z'_{i0}\beta}}\right)}{\left(\dfrac{e^{\alpha_i+Z'_{i1}\beta}}{1+e^{\alpha_i+Z'_{i1}\beta}}\right)\left(\dfrac{1}{1+e^{\alpha_i+Z'_{i0}\beta}}\right)+\left(\dfrac{1}{1+e^{\alpha_i+Z'_{i1}\beta}}\right)\left(\dfrac{e^{\alpha_i+Z'_{i0}\beta}}{1+e^{\alpha_i+Z'_{i0}\beta}}\right)}
$$

$$
= \prod_{i=1}^{N} \frac{e^{Z'_{i1}\beta}}{e^{Z'_{i1}\beta}+e^{Z'_{i0}\beta}}
$$

(Note that in the third line of this derivation, we are making the same independence assumptions across individuals within a matched set as that discussed below for Poisson regression.) The logic is similar for larger case-control sets, except that there are more terms to be included in the denominator, specifically the number of possible ways to select m_1 cases out of a set of m_0+m_1 cases and controls. Perhaps the best way to think about this likelihood is as the conditional probability of *which* members of each matched set are the cases, given their covariate values and how many cases there are in each set. For reasons that will become clear shortly, this is essentially the same form of likelihood as that used in the analysis of censored survived data, thus clarifying the relation between nested case-control and cohort studies, so we defer further discussion to the following section.

As an example, we turn to the analysis of asthma incidence in relation to air pollution in the Children's Health Study (CHS). This study has both cross-sectional and longitudinal elements, and is complicated by the fact that the effects of air pollution are mainly assessed at the community and temporal levels, rather than the individual level. We will return to these complications in Chapters 7 and 9–11. To illustrate the use of logistic regression, we focus here on an analysis of the asthma prevalence data in the initial cross-sectional survey in relation to an individual assessment of NO_2 exposures based on measurements conducted on a subsample of 208 of the participants' homes (Gauderman et al. 2005), which will be described in greater detail in Chapter 9. Figure 4.2 shows the measured NO_2 concentrations at each of the participants' home, showing the general tendency for those with asthma to be at the upper end of the distribution for their respective community. The risk of prevalent asthma was fitted to a logistic regression model for various measures of traffic exposure as well as measured NO_2 concentrations in winter and summer seasons, with adjustments for sex, race/ethnicity, and year of enrollment. Indicator variables for each of the study communities were also included as covariates, so that the relative risks for exposure variables can be interpreted as effects *within* community.

Significant associations were found with measured NO_2 (OR $=1.83$, 95% CI 1.04–3.21 per inter-quartile range (IQR) of 5.7 ppb), as well as for distance to the nearest freeway (OR/IQR $= 1.89(1.19 - 3.02))$ and

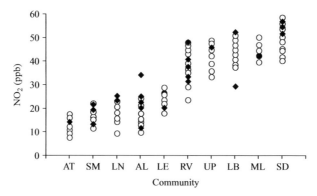

Figure 4.2. Four-week average concentration of NO_2 measured at homes of CHS participants in 10 communities, with asthma cases shown in black. (Reproduced with permission from Gauderman et al. 2005).

model-based prediction of pollution from freeways (OR/IQR = 2.22 (1.36 − 3.63)), but not from model-based prediction of pollution from surface streets (OR/IQR = 1.00(0.75 − 1.33)). Associations with traffic exposures in the entire cohort were found to be stronger in children without a family history of asthma (McConnell et al. 2006).

McConnell et al. (1999) analyzed the prevalence of bronchitic symptoms (chronic cough, wheeze, phlegm, etc.) in children with and without asthma in the entire CHS cohort in relation to individual risk factors such as the presence of mildew, cockroaches, gas stoves, or smokers in the home using logistic regression, adjusted for community as above. However, associations with ambient air pollution required a second level of analysis in which the coefficients for community (the community log-odds, adjusted for personal risk factors) were regressed on ambient exposure. This second-stage regression demonstrated significant associations of bronchitis with ambient PM_{10} and of phlegm with ambient NO_2 only in asthmatics.

Cohort analyses of the incidence of newly diagnosed asthma were performed using Cox regression, as described later in this chapter.

Binomial regression for grouped binary data

Now suppose we have only data on the *proportions* $P_g = N_g/Y_g$ of individuals i in group g with $Y_i = 1$. In some applications, such as individual chromosome abnormality data, one might have data on the proportion of cells $p_i = n_i/y_i$ from an individual i that are affected. In other cases, the observations might be of the number of affected offspring i in a litter g. Such data are structurally similar to grouped individual data, so we will use the same notation, with the subscript g now representing subjects or litters, with i representing single cells or offspring that are either affected or not.

The obvious analysis of grouped binary data would be the same as for individual data, with $\text{logit}(P_g) = Z'_g\beta$, simply treating the groups as independent and multiplying their likelihood contributions

$$L(\beta) = \prod_g \left(\frac{\exp\left(Z'_g\beta\right)}{1 + \exp\left(Z'_g\beta\right)} \right)^{Y_g} \left(\frac{1}{1 + \exp\left(Z'_g\beta\right)} \right)^{N_g - Y_g}$$

The problem with this naïve analysis, however, is that it ignores the overdispersion that can result from failure to fully account for all sources of variability *within* groups in the regression model, so that the Y_g are not really binomially distributed. A parametric analysis of grouped binary data might assume some model for the distribution of residual risks within group. Since the group risks must still be bounded by 0 and 1, a natural choice would be the Beta distribution, $\Pr(P) = P^{a-1}(1 - P)^{b-1}/B(a,b)$ with parameters a and b, where $B(a,b) = \Gamma(a)\Gamma(b)/\Gamma(a + b)$ is the Beta function that normalizes the distribution to integrate to 1. The beta distribution is the conjugate prior for the binomial, leading to a posterior distribution for P given N and Y that is another beta distribution. This is convenient, as it allows a closed-form expression for the likelihood of the parameters (a, b) for a single group, or for a set of parameters for comparing two or more discrete groups. Specifically, for a single group, the likelihood is

$$L(a,b) = \left(\frac{\Gamma(a + b)}{\Gamma(a)\Gamma(b)} \right)^N \prod_{i=1}^{N} \left(\frac{\Gamma(a + n_i)\Gamma(b + y_i - n_i)}{\Gamma(a + b + y_i)} \right)$$

The beta/binomial model cannot readily incorporate continuous covariates, however. A simpler approach in this case is to use a marginal model for the mean and variance of the distribution, where the mean is still given by $P_g = \text{expit}(Z'_g\beta)$ and the usual binomial variance $V_g = P_g(1 - P_g)/N_g$ is inflated by a multiplicative factor $[1 + (N_g - 1)\sigma^2]$, where σ^2 is the "extra-Binomial variance", the variance of the residual within-group risks. Fitting this model can be accomplished using generalized estimating equations, as described later in this chapter.

Censored survival analysis

Now suppose we have a cohort study with individual data involving extended periods of follow-up—perhaps decades long—with individuals entering and leaving observation at different times and their exposures and other covariates varying over time. For example, in the Colorado Plateau uranium miner cohort that we shall use to illustrate this section, individuals join the cohort on their date of first examination by the U.S. Public

Health Service (sometime after first employment in any of the mines) and they leave observation either on their date of death (from lung cancer, the cause of interest coded $Y = 1$, or any other cause), date last known to be alive, or closing date of follow-up for analysis purposes (here Dec. 31, 1990). Throughout this period of observation, as long as they were working in a uranium mine, they were continuing to accumulate exposure. In what follows, we will define $Z(t)$ as cumulative dose of radiation up to 5 years before time t to allow for latency. Chapter 6 will explore models for exposure–time–response relationships incorporating temporal modifiers like age at exposure and latency in more detail.

We take the quantity to be modeled to be the hazard rate $\lambda(t, Z)$. Before proceeding further, we need to clarify the meaning of "time." In clinical trials, where the techniques we are about to discuss were first developed, t is generally defined as time since start of treatment. This is a natural choice in that context, since the investigator is generally interested in making statements about prognosis under alternative treatments. In the context of cohort studies, however, time since start of observation—for example in the Framingham Heart Study, some arbitrary calendar time when the study began—may not be of particular interest. Furthermore, we are generally more interested in the effects of covariates than time, considering the latter as a confounder to be adjusted for. Of the various time scales that might be considered—age, calendar year, time since start of exposure, or observation—age is generally by far the most powerful confounder, so it is natural to adopt this as the time scale for modeling purposes. This is not to say the other time scales are unimportant, but they can generally be incorporated into the model for covariate effects or controlled by stratification.

Thus, the raw data for each subject i can be summarized as $\{Y_i, T_i, E_i, S_i, Z_i(t)\}$, where Y_i denotes the final disease status at end of observation at age T_i ($Y_i = 1$ for affected, $Y_i = 0$ for censored, that is, not yet known to be affected), E_i denotes age at start of observation, S_i denotes stratum membership (if any stratification is to be done, say by gender or year of birth interval), and $Z_i(t)$ the covariate history (possibly time-dependent).

We wish to describe the random variability in the observed outcomes in terms of a model for the true underlying rate $\lambda(t, Z)$ in relation to age t and covariates Z. We do this in two stages, first specifying the manner in which these two factors combine, then the specific form for each factor separately. In terms of their joint effects, three models are in common use:

$$
\begin{array}{ll}
\text{Multiplicative:} & \lambda(t, Z) = \lambda_0(t)r(Z) \\
\text{Additive:} & \lambda(t, Z) = \lambda_0(t) + e(Z) \\
\text{Accelerated:} & S(t, Z) = S_0[r(Z)t]
\end{array}
$$

where $S(t, \mathbf{Z}) = \Pr(T > t | \mathbf{Z}) = \exp[-\int_0^t \lambda(u, \mathbf{Z})\, du]$ is the survival function. The multiplicative model, also known as the proportional hazards model, assumes that at any age, the effect of covariates is to multiply the baseline hazard rate (the rate for subject with $\mathbf{Z} = 0$) by some relative risk $r(\mathbf{Z})$ that depends only on covariates (but recall that these covariates can be time-dependent, either intrinsically as in cumulative exposure or by virtue of including interactions with temporal modifiers). Likewise, the additive hazard model assumes that the effect of covariates is to add some excess hazard rate $e(\mathbf{Z})$ depending only on the covariates to the baseline age-specific rate. The accelerated failure time model is quite different, assuming that the effect of covariates is to accelerate the rate of time flow by some factor $r(\mathbf{Z})$ to produce a form of premature aging.

Breslow and Day (1987, chapter 2) provide a thorough discussion of arguments, both theoretical and empirical, in support of the general utility of the multiplicative model. Broadly speaking, across a range of exposure, time scales, and confounding factors, relative risks tend to be more nearly constant than are excess risks, allowing more parsimonious modeling. Furthermore, as we shall see shortly, the multiplicative model allows a form of analysis of covariate effects on relative risks $r(\mathbf{Z})$ that does not require any assumptions about the form of $\lambda_0(t)$, a great advantage since the latter can be very strong and not of particular interest. For these reasons, we will restrict attention in this chapter to the multiplicative model. Chapter 6 will explore some of the alternatives in greater depth.

Next we must specify the forms of $r(\mathbf{Z})$ and $\lambda_0(t)$. Since hazard rates, and hence relative risks, must be nonnegative, it is convenient to adopt the loglinear form $r(\mathbf{Z}) = \exp(\mathbf{Z}'\boldsymbol{\beta})$. Here, the parameters β_p represent the logarithms of the change in the hazard rate (rather than the odds of disease as in the previous section) per unit change of Z_p, or for a binary covariate, the log rate ratio, ln RR. Note that the model assumes that hazard rates depend multiplicatively on multiple covariates and exponentially on each continuous covariate. These assumptions can be tested and relaxed if needed by adding additional covariates, say a quadratic term to test for log-linearity of a single continuous covariate or a product term to test for multiplicatively of two or more covariates, as discussed further in Chapter 12, or by adopting some alternative form entirely, as discussed in Chapters 6 and 13.

The baseline hazard $\lambda_0(t)$ can be specified parametrically or eliminated entirely using partial likelihood. We defer the latter possibility for the moment and suppose that we have adopted some specific mathematical form, such as a constant $\lambda_0(t) \equiv \alpha$, Weibull $\lambda_0(t) = \alpha_0 t^{\alpha_1}$, Gompertz $\lambda_0(t) = \exp(\alpha_0 + \alpha_1 t)$, or step-functions $\lambda_0(t) = \alpha_k$ for $\tau_k \leq t < \tau_{k+1}$. The full likelihood is then obtained by multiplying the conditional probabilities for each individual's outcome Y_i, given their periods of observation and covariate histories. (This assumes that censoring is uninformative, in the sense that the probability of censoring events like deaths from competing causes do not depend upon the parameters of the hazard rate for the cause of interest, although they *can* depend upon some of the same covariates.) These probabilities contain two parts: first, the probability that the individual remained alive and disease free up until their event or censoring time t_i; second, if they experienced the event, the probability that it occurred at that instant, given that he or she was still at risk. Mathematically we write this as

$$L_i = \Pr(T_i = t | \mathbf{Z}_i, \ E_i)$$

$$= \Pr(T_i \geq t | \mathbf{Z}_i, \ E_i) \Pr(T_i = t | T_i \geq t, \ \mathbf{Z}_i, \ E_i)$$

$$= S(t | \mathbf{Z}_i, \ E_i) \lambda(t | \mathbf{Z}_i)$$

where $S(t | \mathbf{Z}, \ E)$, as before, is the probability of surviving disease-free from time E to time t (we assume that all cohort members were disease-free at entry). Under the assumption of independent competing risks—that those who died of other causes or were lost to follow-up were at the same risk of the cause of interest as those who did not—then $S(t | \mathbf{Z}, \ E)$ can be decomposed into the product of the probability that they escaped the cause of interest and the probability that they escaped the competing risks (including loss to follow-up). This is where the assumption of uninformative censoring is required, as the real survival probability is $S(t) = \exp[-\Lambda(t) - M(t)] = S_\lambda(t) S_\mu(t)$, where $\Lambda(t) = \int_0^t \lambda(u) \, du$ is the cumulative hazard for the cause of interest and $M(t) = \int_0^t \mu(u) \, du$ is the cumulative hazard for censoring As we assume the latter does not depend on any of the parameters we wish to estimate (the relative risk parameters β or baseline hazard parameters α for the cause of interest), this part is essentially a constant and can be ignored. Now substituting the relative risk model, we obtain the full likelihood

$$L(\alpha, \beta) \propto \prod_{i=1}^{N} \lambda(T_i | \mathbf{Z}_i) S_\lambda(T_i | \mathbf{Z}_i, E_i)$$

$$= \prod_{i=1}^{N} \left(\lambda_0(T_i) e^{\mathbf{Z}_i(T_i)'\boldsymbol{\beta}} \right)^{Y_i} \exp \left(- \int_{E_i}^{T_i} \lambda_0(t) e^{\mathbf{Z}_i(t)'\boldsymbol{\beta}} \right) \qquad (4.1)$$

Maximization of this likelihood jointly with respect to α and β is conceptually straightforward using the techniques of ML described below, although it can be computationally burdensome. In most cases, no closed-form solution is possible, so numerical methods are needed. One special case is worth mention, however: if $\lambda_0(t)$ is specified as a step-function with fixed intervals, then it is possible to express the MLE of λ as a function of β leading to $\hat{\alpha}_k(\beta) = Y_k / \Sigma_i T_{ik} \exp(Z'_i \beta)$ where T_{ik} is the time individual i was at risk during time interval k. Substituting this into the full likelihood yields a profile likelihood $L_P(\beta) = L[\hat{\alpha}(\beta), \beta]$ that depends only upon β.

In order to avoid making any assumptions about the form of the baseline hazard, Cox's (1972) seminal paper proposed a "semi-parametric" model for $\lambda(r, Z)$ and a partial likelihood that does not involve $\lambda_0(t)$ at all. The semiparametric model is of the multiplicative form $\lambda_0(t) \exp(Z'\beta)$, with $\lambda_0(t)$ assumed to be zero everywhere except at the times t_k when events (cases or deaths, but not censoring times) occur. The masses at the observed event times, λ_k, can be thought of as infinitely dense Dirac δ-functions (zero everywhere except at the origin, where the masses are infinite, integrating to one), so the baseline hazard rate can be written as $\lambda_0(t) = \Sigma_k \lambda_k \delta(t - t_k)$, leading to a baseline survival function $S_0(t) = \exp(-\Sigma_{t_k \leq t} \lambda_k)$, a step-function with drops at each event time.

To fit this model, Cox proposed what he later (Cox 1975) called a partial likelihood by treating each event time as independent and multiplying the conditional probabilities of who was the case, given the set $R(t_k)$ of subjects at risk at that time:

$$
L(\beta) = \prod_{i|Y_i=1} \Pr\left(Y_i = 1 \,\middle|\, \{Z_j(t_i)\}_{j \in R'_i}, \sum_{j \in R_i} Y_j = 1\right)
$$

$$
= \prod_{i|Y_i=1} \frac{\lambda(t_i|Z_i(t_i))}{\Sigma_{j \in R_i} \lambda(t_i|Z_j(t_i))} = \prod_{i|Y_i=1} \frac{\exp(Z_i(t_i))'\beta}{\Sigma_{j \in R_i} \exp(Z_j(t_i)'\beta)} \qquad (4.2)
$$

This partial likelihood is exactly the limit of the profile likelihood for the step-function model described earlier as the interval widths became infinitesimal (Holford 1976). Formal justification for multiplying these conditional probabilities as if they were independent, when in fact the set of subject of risk at any time depends on who were the cases at previous times, was given by Andersen and Gill (1982). The classical theory is described in Kalbfleisch and Prentice (1980) and other survival analysis texts. For a more formal treatment using modern "counting process" theory, see Fleming and Harrington (1991) or Andersen et al. (1993).

This likelihood accommodates time-dependent covariates by evaluating numerator and denominator using the covariate values at each successive event time. This is computationally more intensive than for time-constant covariates, however. For fixed covariates, the summation in the denominator need be computed only once for each iteration, accumulating terms across subjects in a single pass. For time-dependent covariates, these summations must be computed separately for each event time.

Data from the follow-up portion of the CHS was analyzed using Cox regression. McConnell et al. (2002a) found associations of newly diagnosed asthma with humidifier use and pets (especially dogs). As with the cross-sectional data discussed above, associations with ambient pollution required a multi-level analysis. Stratifying on air pollution, McConnell et al. (2002b) found an association of incident asthma with participation in team sports only in high-ozone communities; the association with sports was not modified by other pollutants, however. In a separate analysis, McConnell et al. (2003) exploited the year-to-year variation in pollution levels to demonstrate a stronger within-community association of bronchitic symptoms among the subcohort of asthmatics with annual variation in NO_2 and organic carbon levels than for the same associations of long-term average exposures between communities.

Nested case-control studies

The same partial likelihood can also be used for nested case-control studies (Liddell et al. 1977), simply replacing the summation over the risk set $R(t)$ in the denominator by the set $\tilde{R}(t)$ comprising the case and matched controls, assuming the latter are a random sample from $R(t)$. Goldstein and Langholz (1992) provide the asymptotic distribution theory underlying this approach, and generalizing it to other methods of case and/or control sampling, such as the counter-matched design described in Chapter 5.

A rough rule of thumb for the *asymptotic relative efficiency* (ARE)—the ratio of variances of different estimators of $\hat{\beta}$, or equivalently, the ratio of sample sizes required to attain the same precision—for nested case-control analysis relative to the full cohort is $ARE = M/(M + 1)$ where M is the number of controls per case (Ury 1975). Thus, 1:1 matching yields 50% efficiency (i.e., variances of the log RR double those from the full cohort), 1:2 matching improves this to 67%, 1:3 to 75%, and diminishing returns thereafter. This result is strictly true only for small relative risks (RRs), however. For estimating strong effects or fitting complex models, additional controls can indeed be more helpful than this relationship would imply. A general expression for the ARE can be derived for a binary covariate with control frequency p_0 from the Fisher information,

$$I_M(\hat{\beta}) = E(\text{OBS}) \sum_{m=1}^{M} \frac{\pi_m m e^\beta (M - m + 1)}{M - m + 1 + m e^\beta}$$

where

$$\pi_m = p_1 \binom{M}{m-1} p_0^{m-1}(1 - p_0)^{M-m+1} + (1 - p_1)\binom{M}{m} p_o^m (1 - p_0)^{M-m}$$

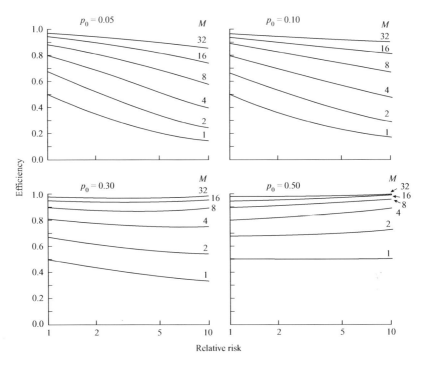

Figure 4.3. Efficiency of matched case-control designs with M controls per case relative to a full cohort analysis, as a function of the relative risk and prevalence p_0 of exposure. (Reproduced with permission from Breslow et al. 1983).

and

$$p_1 = \frac{p_0 e^\beta}{p_0 e^\beta + (1 - p_0)}$$

(Breslow et al. 1983). This expression is plotted in Figure 4.3. At a relative risk of 1, these efficiencies follow the simple $M/(M+1)$ relationship, irrespective of the exposure proportion, but as the relative risk increases, more and more controls are needed to attain the same efficiency, particularly for rare exposures.

The following chapter provides a comparison of full cohort, nested case-control, and a novel design known counter-matching applied to the U.S. uranium miner cohort, illustrating the relative efficiency of these designs with varying numbers of controls per case.

Within the CHS, a nested case-control study of incident asthma is currently underway. Each newly diagnosed case is matched with one control who is free of the disease at the time the case occurred, selected at random from the cohort and matched on community, age, and gender. Intensive measurements of air pollution at the cases' and controls' homes, along with lung function measurements and clinical examinations to better characterize the cases' and control's phenotypes, which would be infeasible for the entire cohort, are being performed. A nested case-control study of prevalent asthma (Langholz and Goldstein 2001; Salam et al. 2004) focused on early life exposures. This study used countermatching to assess the effect of maternal smoking in pregnancy and in early life, as described in the following chapter.

Poisson regression

Now suppose instead of individual data, we had only the numbers of events and person-time at risk from a cohort in various strata s defined by intervals of age, calendar time, and various discrete exposure, confounding, and modifying variables. As in the case of individual data, the target of interest is still the hazard rates as the basic measure of event rates per unit of person-time. For a particular group of individuals followed for a particular period of time (hereafter called a stratum, s), we seek to estimate their average event rate λ_s. The raw data for such estimation is the observed number of events (incident cases, deaths, recurrences, etc.) D_s and the total person-time at risk T_s in that stratum, computed as described in the previous chapter from each individuals' time at risk during each age/year slice of observation. The observed hazard rate estimates are thus $\hat{\lambda}_s = D_s/T_s$ with variance $\text{var}(\hat{\lambda}_s) = D_s/T_s^2$. For finely stratified data, however, these observed rates can become quite unstable and it is really the "true" rate we are interested in estimating, under the general hypothesis that similar groups should have similar rates. To describe what we mean by "similar," let us characterize each stratum by a vector of potentially relevant characteristics $\mathbf{Z}_s = (Z_{s1}, \ldots, Z_{sP})$, usually person-time weighted averages of individual Z_{ip} for continuous covariates. These are thus the ingredients needed to build a regression model for disease rates across all the strata.

Recall from the previous chapter that, conditional on the observed person-time at risk, the number of events follows the Poisson distribution $\Pr(D) = e^{-\lambda T}(\lambda T)^D/D!$. For recurrent events like asthma attacks, this assumes that the rates are homogeneous across individuals, so the number of events depends only the total person-time at risk, irrespective of how this time is distributed across individuals. For nonrecurrent events like incidence or mortality, the use of the Poisson distribution might seem questionable, since the only possible outcomes for any individual are zero or one, so the total number of events is bounded by the number of individuals at risk, whereas the Poisson distribution is unbounded. Of course, for a rare disease, this should be a good approximation. But more importantly, without requiring any rare disease assumption but only the piecewise constancy of the baseline hazard, the likelihood for the parameters of a Poisson regression model for $\lambda(t, Z)$ has exactly the same form (to within some constants) as the model for censored survival data described in the previous section. Specifically, assuming the baseline hazard is a constant α_s within each stratum, the likelihood is obtained by treating the strata as independent and multiplying the Poisson probabilities of the observed numbers of events, conditional on the person-times at risk. Thus, the group-data form of Eq. (4.1) becomes

$$L(\alpha, \beta) = \prod_{s=1}^{s} \left(\alpha_s e^{Z_s \beta} \right)^{D_s} \exp \left(-\alpha_s e^{Z_s \beta} T_s \right) \tag{4.3}$$

The model is fitted by maximizing the probability of the observed data (the "likelihood") over the values of the parameters α and β, as described below.

In particular, if Z_s were a categorical variable that distinguished each stratum—a fully saturated model—then using the maximization techniques described below, it is easy to see that maximum of the likelihood is obtained by $\beta_s = \ln(D_s/T_s)$, the log of the observed rate. In restricted models, where the number of parameters P is less than the number of strata s, $\hat{\lambda}_s = \exp(Z_s'\hat{\beta})$ provides an estimate of the rates in each stratum that provide the best fit to the observed data across all the strata.

The most recent comprehensive analysis of mortality from solid cancers in the atomic bomb survivor cohort (Preston et al. 2003) analyses a cohort of 86,572 individuals who were within 10 km of the hypocenters and for whom dose estimates were available. Over the period from 1950 to 1997, there were over 3 million person-years at risk and a total of 9355 solid cancer deaths in this cohort. Most of the analyses of these data have used an excess relative risk model of the form

$$\lambda(t, Z, \mathbf{W}) = \lambda_0(t, \mathbf{W})[1 + \beta Z e^{\alpha' \mathbf{W}}] \tag{4.4}$$

where t denotes age, Z dose, and $\mathbf{W}$ a vector of baseline risk factors and/or modifying factors, such as sex, city, calendar year, or age at exposure, and $\lambda_0(t, \mathbf{W})$ is some general function to be estimated along with the other parameters. We will consider such models in greater detail in Chapter 6, focusing on methods for studying the shape of the dose–response relationship and its modification by temporal factors, but here we restrict attention to the estimation of the coefficient β for excess relative risk per unit dose. To fit the model, the numbers of deaths D_s, person years Y_s, and person-year-weighted means of the doses $\overline{Z}_s$ and modifiers $\overline{\mathbf{W}}_s$ were finely cross-classified by the two cities, two genders, 23 dose categories, 17 5-year attained age categories, 14 5-year age at exposure, 11 5-year calendar-time categories, and two categories of distance from the hypocenter—a total of about 37,000 cells with nonzero person-time. Needless to say, with on average about 80 person years and

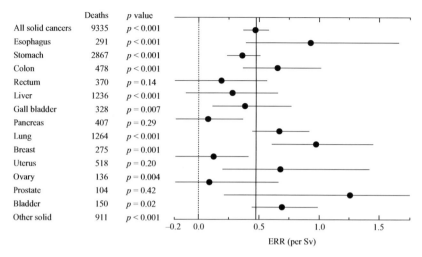

Figure 4.4. Site-specific estimates of excess relative risk (ERR) per Sv in the atomic bomb cohort. (Reproduced with permission from Preston et al. 2003).

only 0.25 deaths per cell, most of these cells are quite small, but the Poisson model can correctly handle such sparse data. For this model, the likelihood analogous to Eq. (4.3) can be written as

$$L(\beta, \alpha, \lambda_0) = \prod_{i=1}^{37,000} \lambda_s^{D_s} \exp(-\lambda_s Y_s)$$

substituting $\lambda_s = \lambda(\bar{t}_s, \overline{Z}_s, \overline{W}_s)$ from Eq. (4.4). The overall relative risk coefficient was estimated as 0.47 ± 0.06 per Sv (i.e., a RR at 1 Gy of 1.47) although this figure varies markedly with cancer site (Figure 4.4), sex, age at exposure, attained age, and other factors, as we will discuss in Chapter 6. Of the 9355 cancer deaths, this relative risk estimate translates into about 440 excess deaths in this cohort (or about 1100 extrapolated to all survivors). Thus, the vast majority of cancer deaths are probably not caused by radiation, emphasizing the importance of careful modeling of the background rates $\lambda_0(t, \mathbf{W})$ and correct specification of the form of the additional risk due to radiation.

A final word about the independence assumption: if the strata comprise disjoint sets of unrelated individuals, each observed over a single time period, then this would seem to be a reasonable assumption. It would be questionable, however, if the sample included genetically related individuals or if there was some form of person-to-person transmission, say by an infectious agent, or some shared unmeasured risk factors. If such related individuals appeared in the same stratum, then the observed rates

D_s would not be expected to follow the Poisson distribution and if they appeared in different strata, then their rates would be correlated and one couldn't simply multiply their Poisson probabilities together. Analysis of such dependent data is beyond the scope of this book (see the companion volume (Thomas 2004) for a treatment of genetic data, for example).

A subtler form of dependency might seem to arise if the strata are defined by time in such a way that the same individuals can contribute person-time to more than stratum. For survival data, this is not a problem, as the person-time at risk can be subdivided as much as one likes and the contrasts of numbers of cases against person-time at risk are orthogonal across intervals. For recurrent events data like asthma exacerbations, however, this would be a problem, as the number of events in one-time stratum would depend upon the individuals' past histories of events in the same or previous time strata or on their common un-modeled "proneness" to the outcome. Again, the analysis of recurrent event data is beyond the scope of this book and we shall assume that individuals can experience the outcome—incidence of new disease or death—only once.

A great advantage of Poisson regression is that it allows a very compact form of data presentation as $\{D_s, T_s, Z_s\}$ for sharing with other investigators, without the need to address the confidentiality issues that can arise in sharing individual data. For example, the aggregate data supporting each of the more recent publications from the atomic bomb survivor studies are freely available for download from http://www.rerf.jp/library/dl_e/index.html.

Extra-Poisson variation

In the same manner as for grouped binary data, the counts of events per group of person-time will also tend to be over-dispersed due to unmeasured risk factors that vary within groups (or under dispersed if the sample includes correlated observations). The solution to this problem is similar to that discussed earlier: either assume a parametric model for such residual variation—the natural choice being the gamma distribution, which is conjugate to the Poisson—or use a marginal model with an "extra-Poisson variance" term. The sampling variance of Y_s is $\mu_s = \lambda_s T_s$, to which one adds a term of the form $\sigma^2 \mu_s^2$. Again, fitting the model can be done using Generalized Estimating Equations methods. This thus provides a convenient approach to the problems of recurrent or dependent event data discussed above.

Principles of ML inference

Fitting of the Poisson, logistic, and Cox regression models, as well as the more complex variants described in subsequent chapters, is generally based on the principle of ML introduced at the end of Chapter 3. By a "model," we mean a probability density function $f(Y|\Theta)$ for a set of observations Y_i in relation to a vector of parameters Θ. For example, in ordinary linear regression, one might write $f(Y_i|Z_i;\Theta) = \varphi[(Y_i - Z_i'\beta)/\sigma]$, where Z_i is a vector of covariates for the ith subject, $\Theta = (\beta, \sigma^2)$ are the regression coefficients and residual variance one wishes to estimate, and φ denotes the normal density function. Then if a set of $i = 1, \ldots, N$ observations are independent, one would form the *likelihood function* as the product of these probabilities,

$$L(\Theta) = f(\mathbf{Y}|\mathbf{Z}; \Theta) = \prod_{i=1}^{N} \varphi\left(\frac{Y_i - Z_i'\beta}{\sigma}\right) = \frac{1}{\sqrt{2\pi}\sigma} \exp\left(-\sum_{i=1}^{N} \frac{(Y_i - \mu)^2}{2\sigma^2}\right)$$

which is now viewed as a function of the parameters Θ conditional on the observed data, rather than the other way around. We will see shortly that this yields an estimator of $\hat{\mu} = \Sigma_i Y_i/N$ and $\hat{\sigma}^2 = \Sigma_i(Y_i - \mu)^2/N$.

We concentrate now on their general uses for testing hypotheses about the parameters and for point and interval estimation. Thus, for the remainder of this section, we simply let $\mathbf{Y}$ denote the observed data and Θ a generic parameter (or vector of parameters) to be estimated.

The principle of ML states that for any model, we prefer the value of Θ for which the observed data is most likely (i.e., which maximizes the likelihood function). This value is known as the ML estimate (MLE) and denoted $\hat{\Theta}$. This is not the most likely value of Θ given the data, *unless all parameter values were equally likely before seeing the data*. The value which would maximize the posterior probability given the data, $\Pr(\Theta|\mathbf{Y})$, is called the *posterior mode*; its calculation would require specification of a prior distribution for parameters $\Pr(\Theta)$. Then by Bayes formula the posterior distribution is proportional to

$$\Pr(\Theta|\mathbf{Y}) = \frac{\Pr(\mathbf{Y}|\Theta)\Pr(\Theta)}{\Pr(\mathbf{Y})} \propto L(\Theta)\Pr(\Theta)$$

We shall return to this approach to estimation when we discuss Bayesian methods later, but for now will confine ourselves to ML methods.

In most cases, it turns out to be more convenient to work with the logarithm of the likelihood rather than the likelihood itself. Because the log transformation is monotonic, it follows that the value $\hat{\Theta}$ that maximizes the loglikelihood function $\ell(\Theta) = \ln[L(\Theta)]$ will also maximize the likelihood $L(\Theta)$ itself. Since most likelihood functions are products of contributions from a set of independent observations, the loglikelihood becomes a sum of independent contributions, whose derivatives are more easily found than the derivative of a product. (Since sums of independent random variables are asymptotically normally distributed, this also provides the basis for first-order asymptotic distribution theory.) Further-more, since often these contributions are members of what is known as the "exponential family" of distributions (defined more formally below), their loglikelihood contributions are of a simple form. The maximum of any function is found by setting its slope (derivative) equal to zero. The derivative of the loglikelihood is known as the score function, denoted $U(\Theta)$ and the expression

$$U(\Theta) = \frac{\partial \ell(\Theta)}{\partial \Theta} = 0$$

is called an *estimating equation*. In the example of the normally distributed outcome, to find the MLE of μ, we would solve this equation as follows:

$$\ell(\mu, \sigma^2) = -N \ln\left(\sqrt{2\pi}\sigma\right) - \left(\frac{1}{2\sigma^2}\right) \sum_{i=1}^{N} (Y_i - \mu)^2$$

$$U_\mu(\mu, \sigma^2) = \frac{\partial \ell(\mu, \sigma^2)}{\partial \mu} = \left(\frac{1}{\sigma^2}\right) \sum_{i=1}^{N} (Y_i - \mu) = 0$$

$$\sum_{i=1}^{N} Y_i - N\mu = 0$$

$$\hat{\mu} = \sum_{i=1}^{N} Y_i / N$$

the familiar sample mean estimator of the population mean. Similarly, setting $U_\sigma(\mu, \sigma^2)$ to zero yields the MLE for $\sigma^2 = \Sigma_i (Y_i - \mu)^2 / N$, with the sum of squares divided by N rather than by $N - 1$ as in the unbiased estimator. (MLEs are theoretically "Fisher consistent," meaning that they converge to the true value as the sample size goes to infinity, but are not guaranteed to be unbiased in small samples.)

When the estimating equations cannot be solved in closed form, finding the MLE requires iterative methods. The most commonly used technique is the *Newton–Raphson method*, which basically constructs a better estimate at each cycle by moving along a straight line in the direction of increasing likelihood, as viewed from the perspective of the current estimate. The Newton–Raphson method uses the *Fisher information*, the negative of the expectation of the matrix of second derivatives,

$$i(\Theta) = -E\left[\frac{\partial^2 \ell(\Theta)}{\partial \Theta^2}\right]$$

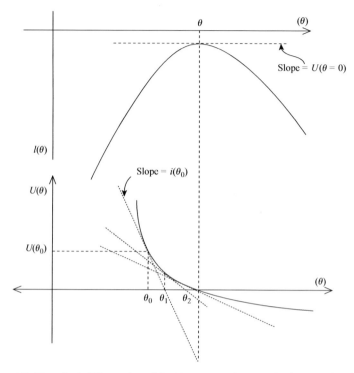

Figure 4.5. Hypothetical illustration of the Newton–Raphson method. (Reproduced with permission from Thomas 2004.)

replacing the current estimate of Θ by a new estimate

$$\Theta' = \Theta + U(\Theta)i^{-1}(\Theta)$$

and continues by computing new values of $U(\Theta')$ and $i(\Theta')$ and repeating this updating procedure until no further changes result. This process is illustrated in Figure 4.5. Starting with an initial guess θ_0, we fit a tangent line through the curve $U(\theta)$ at that point, which has a height $U(\theta_0)$ and slope $i(\theta_0)$. This tangent line crosses the θ-axis at the point θ_1. Drawing a new tangent line at θ_1 and extending it back to the axis yields a new estimate θ_2 and so on.

All three models discussed earlier in this chapter require such iterative methods for fitting. In each case, it can be shown that the score equation can be written in a form like

$$U(\beta) = \sum_{i=1}^{N} Z_i \left[Y_i - p_i(\beta) \right] = 0$$

where $p_i(\beta) = \Pr(Y_i = 1|Z_i)$ is given by the particular model, for example, $\mathrm{expit}(Z_i'\beta)$ for the logistic model. For that model,

$$\ln p_i(\beta) = Y_i(Z_i'\beta) - \ln[1 + \exp(Z_i'\beta)],$$

so

$$U_i(\beta) = \frac{\partial \ln p_i}{\partial \beta} = Y_i Z_i - \frac{Z_i e^{Z_i'\beta}}{1 + e^{Z_i'\beta}} = Z_i [Y_i - p_i(\beta)]$$

For the Poisson model, Y_i would be the observed number of cases in stratum i and $p_i(\beta)$ the corresponding predicted number. For the full survival likelihood, $p_i(\boldsymbol{\beta}) = \Lambda_0(t_i) \exp(Z_i'\boldsymbol{\beta})$, where $\Lambda_0(t_i) = \int_0^{t_i} \lambda_0(t)\, \mathrm{d}t$. The score equation from the Cox conditional likelihood and the conditional logistic likelihood takes a somewhat different form, interchanging the roles of Y and Z,

$$U(\beta) = \sum_{i=1}^{N} Y_i \left[Z_i - E\left(Z_i | \{Z_j\}_{j \in R_i} \right) \right] = 0$$

where

$$E\left(Z_i | \{Z_j\}_{j \in R_i} \right) = \frac{Z_i e^{Z_i'\boldsymbol{\beta}}}{\sum_{j \in R_i} Z_j e^{Z_j'\boldsymbol{\beta}}}$$

Having found the MLE, we wish to put confidence bounds around it and to test the statistical significance of various hypotheses. Confidence limits are usually determined in one of two ways. The first assumes that the model is correctly specified, so that asymptotically the likelihood function is approximately normal, and uses the asymptotic variance derived from Fisher information, specifically,

$$\mathrm{var}(\hat{\Theta}) = i^{-1}(\hat{\Theta})$$

(We will revisit the case of misspecified models in the section on generalized estimating equations below.) For a multiparameter model, the standard error of the pth component is $SE(\hat{\Theta}_p) = \sqrt{i^{pp}(\hat{\Theta})}$, the pth diagonal element of the inverse of the i matrix. For a one-parameter model, the standard error is simply $1/\sqrt{i(\hat{\Theta})}$. Thus, the asymptotic $(1-\alpha)$ confidence limits on $\hat{\Theta}_p$ are simply

$$\hat{\Theta}_p \pm Z_{\alpha/2} SE(\hat{\Theta}_p) = \hat{\Theta}_p \pm Z_{\alpha/2} \left[i^{-1}(\hat{\theta}) \right]_{pp} \tag{4.5}$$

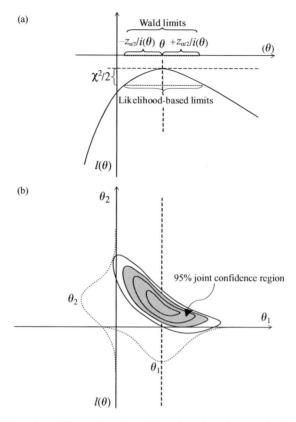

Figure 4.6. Hypothetical illustration of confidence limits based on the likelihood function. (a) A one-parameter model, showing Wald limits (dashed interval) and LR limits (dotted interval). (b) A two-parameter model, showing the joint confidence region. (Reproduced with permission from Thomas 2004.)

[see Figure 4.6(a)]. These are commonly known as Wald confidence limits because of their connection with the Wald test described below; by construction they are always symmetric around the MLE. *Likelihood-based limits* are defined as those values for which the likelihood ratio test (given below) is exactly significant. They have better coverage than Wald limits, but require further iterative search and are not widely used in epidemiology because of their computational difficulty. In multi-parameter models, it may be important to show the full confidence region if the components are highly correlated (Figure 4.6(b)), but this is difficult to do in more than two dimensions.

Hypothesis testing is done to compare alternative models where one is a special case of the other. Thus, for a given model form, we view the specific models generated by different values of the parameters as a family of models and wish to determine whether we can reject a specific null hypothesis $H_0 : \Theta = \Theta_0$. There are three tests available to do this based on likelihood theory, the *likelihood ratio* (LR) test, the *score test*, and the *Wald test*:

$$G^2 = 2\ln(L(\hat{\Theta})/L(\Theta_0)) \qquad \text{LR test}$$
$$X^2 = U'(\Theta_0)\, i^{-1}(\Theta_0)\, U(\Theta_0) \qquad \text{Score test}$$
$$Z^2 = \hat{\Theta}'\, i(\hat{\Theta})\, \hat{\Theta} \qquad \text{Wald test}$$

All three are asymptotically equivalent—not in the sense that they give identical answers on any particular data set, but in the sense that in large enough samples (*asymptotically*) they will have the same test size and power. The LR test is generally the best behaved in small samples. The score test generally converges to its asymptotic distribution faster than the Wald test, which can be seriously misleading when the likelihood is highly skewed. However, the Wald test has the appeal that there is a one-to-one correspondence between the significance test and confidence limits based on Eq. (4.5). The score test has the attraction that no iterative calculations are required to obtain $\hat{\Theta}$. Furthermore, most simple chi square tests based on a comparison of observed and expected events can be derived as score tests from the appropriate likelihood. All three tests are distributed as a chi square with df equal to the difference in dimensions between Θ and Θ_0.

In some circumstances, the null hypothesis value of a parameter is located on the boundary of the permissible parameter space. An example of this arises in variance components models, where the null value for a variance is zero and negative values are not permissible. In this case, the usual asymptotic distribution theory does not apply and typically the null distribution of the LR and other tests is a mixture of chi square distributions with differing df (Self and Liang 1987).

The chi square distribution is only appropriate for comparing *nested* models, meaning one of these models is a subset of a more general model. For example, in linear regression, model $M_1: Y = \alpha + \beta_1 X_1 + \varepsilon$ is more general than model $M_0: Y = \alpha + \varepsilon$ and $M_2: Y = \alpha + \beta_1 X_1 + \beta_2 X_2 + \varepsilon$ is even more general; hence, we can compare the likelihood ratio statistics for M_1 with M_0 or M_2 with M_1 as chi squares on 1 df or M_2 with M_0 directly as a chi square on 2 df. However, we cannot directly compare M_2 with $M_3: Y = \alpha + \beta_1 X_2 + \beta_3 X_3$ because they are not nested—neither is a submodel of the other. In order to make such a comparison, we can nest both models in an even more general one that includes both as submodels, for example, $M_4: Y = \alpha + \beta_1 X_1 + \beta_2 X_2 + \beta_3 X_3 + \varepsilon$. Now all the models $M_0 - M_3$ are submodels of M_4, so in particular we could compare M_2 and M_3 separately to M_4 as 1-df tests. If, for argument sake, we were to reject M_2 and fail to reject M_3, we could then conclude that M_3 fits significantly better.

Nevertheless, we frequently wish to compare nonnested models to each other directly, without having to compare them each separately against some more general alternative. This is often done in a descriptive way using the *Akaike Information Criterion* (AIC) $= 2\ln(L) - 2p$ (Akaike 1973), where p is the number of free parameters in the model. The subtraction of p is intended to reward more parsimonious models, since obviously with each additional parameter the likelihood can only increase. We therefore prefer the model with the highest value of the AIC. A number of alternative criteria have been suggested, including Mallows (1973) C_p, the Bayes Information Criterion (BIC) $= 2\ln(L) - p\ln(N)$ (Schwartz 1978), and the Risk Inflation Criterion (RIC) $= 2\ln(L) - 2\ln(p)$ (Foster and George 1994), each of which have certain theoretical advantages (see George and Foster (2000) for a review), but the AIC is the most widely used.

Bayesian methods

Rather than relying on the likelihood as the basis for inference, Bayesian methods are based on posterior probabilities, $\Pr(\Theta|Y)$. As shown in Chapter 3, these are computed by multiplying the likelihood, $\Pr(Y|\Theta)$ by the prior probability of the parameters $\Pr(\Theta)$, and normalizing by dividing by $c(Y) = \Pr(Y) = \int \Pr(Y|\Theta)\Pr(\Theta)d\Theta$. The analogue of the MLE is the posterior mode estimate $\tilde{\Theta}$, the value of Θ that maximizes $\Pr(\Theta|Y)$. Maximization of this quantity does not require the normalizing constant $c(Y)$, but it does involve the prior distribution $\Pr(\Theta)$, so different choices of priors would yield different posterior estimates. For hypothesis testing, the analogue of the likelihood ratio is a quantity known as the Bayes factor (BF)

$$\left(\quad \mathrm{BF}\left(\tilde{\Theta} : \Theta_0\right) = \frac{\Pr(\tilde{\Theta}|Y)}{\Pr(\Theta_0|Y)} \div \frac{\Pr(\tilde{\Theta})}{\Pr(\Theta_0)} \right)$$

which can be thought of as the relative increase provided by the data in the odds of one model over another beyond their prior odds. For the full model, this of course simply reduces to the likelihood ratio, but interest generally is focused on a particular parameter θ in a complex model that may involve other nuisance parameters or latent variables φ. For example, θ may index a series of alternative model forms (e.g., choices of variables to include in a regression model) and φ the regression coefficients in the corresponding models. In this case, the posterior mode is found by integrating over these other variables, and the BF becomes

$$\mathrm{BF}\left(\tilde{\theta} : \theta_0\right) = \frac{\Pr(\tilde{\theta}|Y)}{\Pr(\theta_0|Y)} \div \frac{\Pr(\tilde{\theta})}{\Pr(\theta_0)} = \frac{\int \Pr(Y|\tilde{\theta}, \varphi)\Pr(\varphi|\tilde{\theta})\,d\varphi}{\int \Pr(Y|\theta_0, \varphi)\Pr(\varphi|\theta_0)\,d\varphi}$$

This will not generally be the same as the likelihood ratio, as it involves integration rather than maximization over φ. It is, however, generally less sensitive to the choice of priors than is the ratio of posterior probabilities. The Bayes factor does not have the repeated sampling distribution theory that likelihood ratios have, so its interpretation is more qualitative. Kass and Raftery (1995) suggest the following subjective criteria: values less than 1 would be evidence against model Θ relative to Θ_0; values between 1 and 3 would be "barely worth a mention"; $3 - 20$ "positive evidence in favor of model m"; $20 - 150$ "strong evidence"; greater than 150, "very strong evidence."

Fitting of Bayesian models has historically been plagued by the computational difficulty of evaluating these high-dimensional integrals over φ. The last few decades however saw a revolution due to the introduction of Markov chain Monte Carlo (MCMC) methods. These proceed by sampling each unknown (parameter or latent variable) in turn, conditional on the observed data and the current values of the other variables:

$$[\theta_1|\mathbf{Y}, \theta_2, \ldots, \theta_P],$$

$$[\theta_2|\mathbf{Y}, \theta_1, \theta_3, \ldots, \theta_P],$$

$$\ldots,$$

$$[\theta_p|\mathbf{Y}, \theta_1, \ldots, \theta_{p-1}, \theta_{p+1}, \ldots, \theta_P],$$

$$\ldots,$$

$$[\theta_P|\mathbf{Y}, \theta_1, \ldots, \theta_{P-1}]$$

and continuing in this manner for many iterations. After a sufficient number of "burn-in" iterations to allow the distributions to stabilize, subsequent sampled values are tabulated as the joint posterior distribution $\Pr(\Theta|\mathbf{Y})$. Each of the component "full conditional" distributions can be computed as $[\theta_p|\mathbf{Y}, \theta_1, \ldots, \theta_{p-1}, \theta_{p+1}, \ldots, \theta_P] \propto [\mathbf{Y}|\theta_1, \ldots, \theta_p, \ldots, \theta_P] \times [\theta_p|\theta_1, \ldots, \theta_{p-1}, \theta_{p+1}, \ldots, \theta_P]$, which may be relatively easy to sample from. For example, if θ_p is a mean parameter (e.g., a coefficient in a linear regression model) and the priors are independent normal distributions, then the conditional distribution is just another univariate normal distribution. Likewise, if θ is a residual variance with an inverse gamma prior distribution, then its full conditional distribution is also inverse gamma with shape parameter being the prior shape plus the degrees of freedom of the data and scale being the prior scale plus the residual sum of squares. In situations where the full conditional distribution is not easily sampled, one can use the Metropolis–Hastings algorithm (Chib and Greenberg 1995), proposing a move from θ to new value θ' from some distribution $Q(\theta \rightarrow \theta')$ that is easy to sample from and then accepting the new value with probability given by the Hastings ratio, $R = \min[1, \pi(\theta')Q(\theta' \rightarrow \theta)/\pi(\theta)Q(\theta \rightarrow \theta')]$, where $\pi(\theta)$ represents the true conditional distribution, otherwise retaining the previous value and sampling again. The appeal of this method is that one need only be able to evaluate the true conditional density at θ and θ', not the full distribution. Other approaches, such as adaptive rejection sampling (Gilks and Wilde 1992), are available and discussed in the book by Gilks et al. (1996). These techniques are implemented in the widely used software package WinBUGS (Spiegelhalter et al. 2003), freely available from http://www.mrc-bsu.cam.ac.uk/bugs/welcome.shtml.

Although Bayesian methods have not been widely used in environmental epidemiology, they are finding application to complex models such as arise in spatial statistics, exposure measurement error, and mechanistic models, as discussed in later chapters. They also provide a natural way to synthesize evidence from multiple studies or multiple disciplines where prior knowledge is available to be combined with the data at hand, as, for example, in the field of risk assessment. Most importantly, they provide a means for allowing for uncertainties about model form through the technique of Bayesian model averaging discussed in Chapter 12, and for uncertainties in computing risk estimates and probabilities of causation, as discussed in Chapters 15 and 16.

The general linear model

The specific methods discussed earlier in this chapter are all special cases of the GLM (Nelder and Wedderburn 1972) for the exponential family of distributions. The exponential family comprises all probability distributions that can be represented as

$$\Pr(Y) = \exp[(Y\theta - g(\theta))/\varphi]/c(Y, \varphi)$$

With appropriate specification of the functions $g(\theta)$ and $c(Y)$, most of the commonly used distributions, including the normal, Poisson, and binomial can be written in this form. For example, the normal density with mean parameter $\theta = \mu$ and scale parameter $\varphi = \sigma^2$ is obtained by setting

$$g(\theta) = \frac{\theta^2}{2} \quad \text{and} \quad c(Y) = \sqrt{2\pi\varphi} \exp\left(\frac{Y^2}{2\varphi}\right)$$

With these substitutions, we see that

$$\Pr(Y) = \frac{\exp\left(Y\theta - \dfrac{g(\theta)}{\varphi}\right)}{c(Y)} = \frac{\exp\left(Y_\mu - \dfrac{\mu^2/2}{\sigma^2}\right)}{\sqrt{2\pi}\sigma \exp\left(\dfrac{Y^2}{2\sigma^2}\right)} = \frac{\exp\left(-\dfrac{(Y-\mu)^2}{2\sigma^2}\right)}{\sqrt{2\pi}\sigma}$$

Likewise, for the Poisson distribution,

$$\theta = \ln(\lambda), \quad g(\theta) = \theta, \quad \varphi = 1, \quad \text{and} \quad c(Y) = 1/Y!$$

and for the binomial,

$$\theta = \text{logit}(p), \quad \varphi = 1, \quad g(\theta) = \ln(1 + e^{\theta}), \quad \text{and} \quad c(Y) = 1 \Big/ \binom{N}{Y}.$$

For purposes of regression modeling, we now set the "canonical parameter" θ in any of these models equal to the linear predictor $\mathbf{Z}'\beta$. Thus, the GLM can be written as $\theta = f[E(Y|\mathbf{Z})]$, where $f(.)$ is the link function relating θ for the appropriate distribution to its mean. For example, for logistic regression the canonical link for the binomial distribution is $g(x) = \text{logit}(x)$, while for Poisson regression, the canonical link is the log transform $g(x) = \ln(x)$. For both these distributions, the scale factor φ is fixed at unity, but allowing it to be a free parameter allows for the possibility of extra-binomial or extra-Poisson variation, that is, additional variation in $Y|\mathbf{Z}$ that is not accounted for by the model for $E(Y|\mathbf{Z})$ or the random variation in Y around the value expected under the corresponding distribution.

Generalized estimating equations

To fit the GLM, one can use the technique of generalized estimating equations (GEE), which exploits only the part of the model involving the canonical parameters, not the full distribution. Thus, the estimating equation is given by

$$U(\beta) = \sum_{i=1}^{N} (Y_i - \mu_i(\beta)) \, \mathbf{W}_i \frac{\partial \mu_i}{\partial \beta} = 0$$

where $\mu_i(\beta)$ denotes the expectation of Y and W_i is a weight that depends upon the distribution of Y. For example, for ordinary linear regression, $\mu_i(\beta) = \mathbf{Z}_i'\beta$, $W_i = \sigma^2$ (a constant with respect to β), and the derivatives $D_i = \partial \mu_i / \partial \beta$ are simply $\mathbf{Z}_i$, so this reduces to the standard estimating equation

$$U(\beta) = \sigma^{-2} \Sigma_i (Y_i - \mathbf{Z}_i'\beta) \mathbf{Z}_i = 0$$

from either OLS or ML fitting. For correlated outcome data, such as a vector of longitudinal observations for each subject (as discussed in Chapter 7), time series for different cities (Chapter 8), or spatially correlated observations (Chapter 9), $\mathbf{Y}_i$, μ_i, and $\mathbf{D}_i$ become vectors of correlated variables for each independent sampling unit (e.g., individuals in the former case, cities in the latter cases), and $\mathbf{W}_i$ becomes a weight matrix that describes the correlation structure of the deviations. Thus, the estimating equation is a sum over the independent sampling units of matrix-vector products of the form $(\mathbf{Y}_i - \mu_i)'\mathbf{W}_i\mathbf{D}_i$.

The variance of $\hat{\beta}$ is given by a formula known as the "sandwich estimator" (Huber 1967; Liang and Zeger 1986):

$$
\operatorname{var}\left(\hat{\beta}\right) = \frac{1}{N} \left(\sum_{i=1}^{N} \mathbf{Z}_i'\mathbf{W}_i\mathbf{Z}_i\right)^{-1} \left(\sum_{i=1}^{N} \mathbf{Z}_i'\mathbf{W}_i(\mathbf{Y}_i - \mu_i)\right.
$$

$$
\left.(\mathbf{Y}_i - \mu_i)' \mathbf{W}_i\mathbf{Z}_i\right) \left(\sum_{i=1}^{N} \mathbf{Z}_i'\mathbf{W}_i\mathbf{Z}_i\right)^{-1} \tag{4.6}
$$

This is known as the "robust" variance because it is a consistent estimator of the sampling variance of β even if the model is misspecified. In particular, although the most efficient estimator of β is obtained when one chooses $\mathbf{W}_i$ to be the inverse of var($\mathbf{Y}_i$), confidence intervals obtained using an incorrect weight matrix will still have the nominal coverage rate. This is a particularly attractive feature when one isn't sure of what the true correlation structure is. For example, one might suspect that a time series has some autocorrelation structure, but is unsure of whether AR_1 or AR_2 or something else is correct. One could simply assume an AR_1 structure, or an exchangeable correlation (all correlations the same), or even an independence model (all correlations zero), and still be assured of consistent estimates of both β and var(β).

Freedman (2006) notes that when the model is misspecified, one should be more concerned about bias than variance. The sandwich estimator may yield the right variance, but around an incorrect estimate of the parameter of interest or even of a meaningless parameter; but if the model is not seriously misspecified, there may be little difference between robust and naïve variances. While there is some wisdom in this advice, most of the applications considered here involve misspecification of parts of the model that are not of primary interest, but could affect the variance of the parameters of interest. For example, a regression model for the mean of a distribution may be correctly specified and the regression coefficients may be the quantities of interest to an investigator, but assumptions about the shape of the entire distribution (e.g., normality) or the covariance between related observations (e.g., repeated observations on the same subjects in a longitudinal or time-series study) could be misspecified. In this case, Freedman's critique of the sandwich estimator are less cogent.

This also applies when one has independent observations from some over-dispersed distribution, say a binomial or Poisson distribution with additional variation due to unmeasured variables. In standard logistic regression, one would set $W_i = 1/\mu_i(1 - \mu_i)$ and in standard Poisson regression, one would set $W_i = 1/\mu_i$. To allow for the possibility of overdispersion, one might simply multiply these weights by an unknown scale parameter φ to be estimated along with β, or use some other function motivated by theoretical considerations. For example, if one assumed that $Y_s \sim \text{Poisson}(\lambda T_s X_s)$, where X_s was an unobserved latent variable having gamma distribution with mean 1 and variance φ, then $\text{var}(Y_s) = \mu_s(1 + \varphi\mu_s)$, so one might take the inverse of this quantity as W_s. The overdispersion parameter can then be estimated from $\text{var}(Y_s - \mu_s)$. If the weight matrix involves parameters in the correlations, these too can be estimated by regressing the cross-products $C_{ij} = (Y_i - \mu_i)(Y_j - \mu_i)$ on the appropriate variables describing the correlation structure (Liang and Zeger 1986).

Missing data

Missing data is a fact of life in any study, even the best designed in the most cooperative population. In case-control or cohort studies of binary disease endpoints or cross-sectional studies with a single outcome variable, some exposure variables may be missing for one subject, other variables for different subjects. Thus, few if any subjects may have complete data on all variables and few variables may be complete for all subjects. In this case, restricting the analysis to subjects with complete data (known as "complete case analysis") or only to variables with no missing values is unlikely to be practical; worse, it might not even be immune from bias, depending on how the missingness arose. In studies with longitudinal measurements of outcomes, many subjects may have missed particular observations or dropped out before the end of the study. Here, we focus on the former situation, deferring treatment of missing values in longitudinal studies to Chapter 7.

Various alternatives to complete case analysis involving the use of missing indicators, fitting different parts of the model using all cases with the data needed for that part ("available case" analysis), or some form of imputation of the missing data have been developed. Some of these methods, like replacing missing values by the overall mean or predicted values based on the available data, are rather *ad hoc* and cannot be guaranteed to give valid answers. In this section, we begin by considering the conditions under which complete data analysis is valid and then describe some of the alternative approaches. There is a vast literature on this subject, which we review only briefly. For a recent review, including a discussion of available software see (Harel and Zhou 2007). Several textbooks (Rubin 1987; Little and Rubin 1989a; Schafer 1997) and other reviews (Little and Rubin 1989b; Little 1992; Greenland and Finkle 1995) provide more comprehensive treatments.

Missingness models

As before, let Y denote the observed data on a single outcome variable and let Z be a vector of exposure variables, some components of which may be missing for different subjects. (We assume the data on Y are complete, otherwise all the data on Z would be useless for those subjects!) For the purpose of the present discussion, we need not distinguish whether the data arise from a cross-sectional, case-control, or cohort study, but for simplicity let us assume that the covariates are not time-dependent. We may also have one or more additional variables W (completely observed) that could be related to the probability that a subject has missing data but are not thought to be relevant to the risk of disease; to simplify the notation, we sometimes omit the dependence on W below.

It is convenient to define a set of indicator variables $M = (M_1, \ldots, M_P)$ taking the value 0 if Z_p is observed for a given subject and 1 if it missing, and write $Z = (Z_{obs}, Z_{mis})$ for the observed and missing components of Z for any given individual. Thus, we can write the observed data for a given subject as (Y, Z_{obs}, M, W). Consider the probability of the complete data and model parameters α in the missingness model and β in the covariate distribution (for the time being, we ignore possible dependencies on Y and W):

$$\Pr(Z, M, \alpha, \beta) = \Pr(M|Z, \alpha) \Pr(Z, \beta) \Pr(\alpha, \beta) \qquad (4.7)$$

We say a variable Z_p is *missing at random* (MAR) if $\Pr(M_p = 1)$ does not depend on the value of Z_p given the other observed variables $Z_{(-p)}$, that is, $\Pr(M_p|Z) = \Pr(M_p|Z_{(-p)})$ where $Z_{(-p)}$ denotes the vector of all variables except p (Table 4.1). In a case-control study, for example, the probability of missingness *could* depend on case-control status, on the other components of Z_{obs}, or on additional completely measured variables W, but cannot depend upon the true value of the missing data Z_p. This is a weaker condition than *missing completely at random* (MCAR), which also requires that $\Pr(M_p = 1)$ be independent of *any* components of Z (although it could still depend upon Y or W). Data that are not MAR (and hence also not MCAR) are said to be *nonignorably missing*.

Table 4.1. Missing data models for covariate Z_p

Missing data model	Assumptions
Missing completely at random	$\Pr(M_p|Y, Z, W) = \Pr(M_p|Y, W)$
Missing at random	$\Pr(M_p|Y, Z, W) = \Pr(M_p|Y, Z_{(-p)}, W)$
Ignorable	MAR and prior independence of the parameters in $p_\beta(Y|Z)$ and $p_\alpha(M|Z, W)$
Nonignorably missing	None of the above (e.g., missingness can depend upon unobserved covariate values

Although the term *ignorably missing* is similar to MAR, it involves the additional requirement that none of the parameters β in the disease model $\Pr(Y|Z)$ are functionally related to any parameters α that may be needed to describe the probability of missingness $\Pr(M|Y, Z, W)$, that is, that $\Pr(\alpha, \beta) = \Pr(\alpha)\Pr(\beta)$ in the notation of Eq. (4.7).

Now suppose we are interested in estimating the parameters β for the regression of Y on Z, say the relative risk coefficients in a logistic regression model. If we restrict the analysis to those with complete data ($M = 0$), under what conditions will estimates of β be unbiased? If missingness does not depend upon Y, there will be no problem (Little 1992). To see this, consider the dependence of Y on Z and W among subjects with $M = 0$. By Bayes formula

$$\Pr(Y|Z, W, M = 0) = \Pr(M = 0|Y, Z, W)\Pr(Y|Z, W)/\Pr(M = 0|Z, W)$$

If M is independent of Y, the first term in numerator is the same as the denominator, so $\Pr(Y|Z, W, M = 0) = \Pr(Y|Z, W)$, that is, the complete case analysis is unbiased in this situation. It may, however, be highly inefficient if this restriction means throwing away a high proportion of the data. If this is unsatisfactory, or if M depends on Y, we need to consider the missing data mechanism. One way to do this is through full ML analysis, incorporating a model for the missing data mechanism; such analysis can yield valid estimates, provided the missingness model is correctly specified.

Key to the validity of the various imputation strategies discussed below is whether missingness is ignorable. If the data are nonignorably missing, then it is possible that subjects with missing observations have systematically larger or smaller values of Z than those actually observed. Unfortunately, it may not be possible to determine whether this is the case by simply looking to see whether the sets of missing and complete subjects differed in terms Y or W, or even the other components of Z unless Z_p was highly correlated with them. Nevertheless, by including in W enough variables that might be related to missingness, the MAR assumption can become more plausible.

Missing data imputation

As an alternative to complete case analysis, we could try filling in the missing Z values in some way and use them to analyze the entire data set. The "single imputation" method entails replacing missing values by $E(Z_{\text{mis}}|Z_{\text{obs}}, W)$. This gives valid estimates of β, but the usual variance estimator fails to adequately take account of the uncertainty in these imputations.

The "multiple imputation" method instead generates R data sets with random samples from $\Pr(Z_{\text{mis}}|Z_{\text{obs}}, W, Y)$, estimates $\beta^{(r)}$ from each imputed dataset, and then uses an average of the different estimates. (This, of course, generally requires a model for the distribution of the missing data, although this can be avoided using some of the semipara-metric methods described below.) The variance of the estimator is obtained by combining the "within sample" variance $\bar{V}_\mathrm{w}$ (the mean of the variance estimates $\mathrm{var}(\hat{\beta}^{(r)})$ from each imputed dataset) with the "between sample" variance V_b (the variance among the dataset-specific estimates of $\hat{\beta}^{(r)}$):

$$\mathrm{var}(\hat{\beta}) = \bar{V}_\mathrm{w} + \left(\frac{R+1}{R}\right) V_\mathrm{b}$$

In either case, $\mathrm{E}(Z|W)$ or $\mathrm{Pr}(Z|W, Y)$ entails fitting some model to the data from completely observed subjects. (See Box 4.1 for a theoretical justification of this procedure).

In the case of nonignorable missingness, the missing data model needs to be formally included in the likelihood. One might, for example, fit a logistic model for $p_\gamma(Y, Z_\mathrm{obs}, W)$, possibly including various interaction effects, and then add $\pi(Z_\mathrm{obs}, W) = \ln[p_\gamma(1, Z_\mathrm{obs}, W)/p_\gamma(0, Z_\mathrm{obs}, W)]$ as an offset term in the disease model

$$\mathrm{logit}\, p_\beta(Y|Z) = \beta'(Z_\mathrm{obs}, Z_\mathrm{mis}^*) + \pi(Z_\mathrm{obs}, W)$$

where Z_mis^* denotes an imputed value for the missing data, given (Z_obs, W). Paik (2004) describes the theoretical basis for such adjustments in the context of matched case-control studies, assuming the distribution of $Z_\mathrm{mis}|Y, Z_\mathrm{obs}, W$ is a member of the exponential family, and provides a simple way to impute missing values using standard software. For example, if Z is normally distributed, he shows that one can impute missing case and control values by their predicted means (given Z_obs and W), simply replacing Y by $1/2$ irrespective of their case/control status. See also Lipsitz et al. (1998); Satten and Carroll (2000); Rathouz et al. (2002); Sinha et al. (2005) for other parametric and semi-parametric approaches to missing data in matched case-control studies.

Regression methods are an obvious way to implement multiple impu-tation (Rubin 1987). Using subjects with complete data, one first builds a regression model for the variable(s) with missing values based on the variables with no missing values, for example, for a case-control study, one might use a linear regression model $Z_i \sim N(\alpha_0 + \alpha_1 W_i + \alpha_2 Y_i, \sigma^2)$. One then uses this model to sample random values for those with miss-ing data, using their predicted means and residual standard deviation. A variant of this approach uses single imputation with the predicted mean, adjusting the variance appropriately to allow for the uncertainty of these imputations (Schafer and Schenker 2000). Obviously a drawback of either approach is that they require a parametric model for the distribution of the missing data that would not otherwise be needed for analysis of the complete data.

4.1 *Theoretical justification for multiple imputation procedures*

Consider the likelihood for the observed data (Y, Z_{obs}, M), for simplicity, ignoring W:

$$
\begin{aligned}
\Pr(Y, M|Z_{obs}) &= \Pr(M|Y, Z_{obs}) \Pr(Y|Z_{obs}) \\
&= \Pr(M|Y, Z_{obs}) \int \Pr(Y, Z_{mis}|Z_{obs}) dZ_{mis} \\
&= p_\gamma(M|Y, Z_{obs}) \int p_\beta(Y|Z) p_\alpha(Z_{mis}|Z_{obs}) dZ_{mis} \quad (4.8)
\end{aligned}
$$

Now the integral can be approximated by taking $r = 1, \ldots, R$ random samples of Z_{mis} from its predictive distribution given Z_{obs},

$$
\int p_\beta(Y|Z) p_\alpha(Z_{mis}|Z_{obs}) dZ_{mis} \simeq R^{-1} \sum_{Z_{mis}^{(r)}|Z_{obs}, \hat{\alpha}^{(r)}} p_\beta\left(Y|Z_{obs}, Z_{mis}^{(r)}\right)
$$

$$(4.9)$$

where $\hat{\alpha}^{(r)}$ denotes a random sample of α estimates from their posterior distribution (i.e., centered around their MLE with the estimated sampling variance). To clarify the dependence on α further, we use the factorization given in Eq. (4.7) to compute the posterior distribution of the missing data as

$$
\Pr(Z_{mis}|Z_{obs}) = \int \Pr(Z_{mis}|Z_{obs}, \alpha) \Pr(\alpha|Z_{obs}) \, d\alpha \quad (4.10)
$$

where

$$
\Pr(\alpha|Z_{obs}) \propto P(\alpha) \int \Pr(Z|\alpha) dZ_{mis}
$$

(Harel and Zhou 2007). Thus, to impute the missing data, a fully Bayesian approach would use the following two-step approach: first sample random values of the parameters from $\Pr(\alpha|Z_{obs})$ using Eq. (4.11); then impute the missing data by sampling from $\Pr(Z_{mis}|Z_{obs}, \alpha)$ using Eq. (4.11). The first step, however, requires evaluation of the integral in Eq. (4.11), and might be more easily accomplish by MCMC methods, iteratively sampling α from $\Pr(\alpha|Z)$ using current assignment of Z_{mis}, then sampling new values of Z_{mis} using the current α value, and so on. Multiple imputation essentially approximates this process with small samples of parameter values and of missing data given these parameters.

The first term in Eq. (4.11) may affect the estimation of β, depending upon the missing data model. Suppose M depends upon $\mathbf{W}$ and or $\mathbf{Z}$, but not Y. Then the ignorability assumption that $p_\alpha(M|Y, \mathbf{Z}_{obs}, \mathbf{W})$ does not involve β means that the multiple imputation approximation to the likelihood [Eq. (4.11)] can be used without further consideration of the missing data model. Likewise, if M depends upon Y but not on $\mathbf{Z}$ or $\mathbf{W}$, then again this term can be ignored.

To avoid such ancillary assumptions, one could use a semiparametric approach known as the "propensity score" method (Rosenbaum and Rubin 1983). Here, the basic idea is to model the probability of missingness based on the observed data (their propensity score), then stratify the subjects into strata based on these scores, and sample values to impute to those with *missing* values from the distribution of *observed* values for other subjects in the same propensity score stratum. This neatly gets around the problem in nonignorable missingness where subjects with missing data could have different distributions of their true values than those with observed data.

Either of these methods has an important limitation in multivariate applications, namely that they require what is known as a "monotone missingness pattern," that is, that it is possible to order the variables in such a way that the subjects with missing values on one variable are a proper subset of those with missing values on all subsequent variables. In this situation, it is easy to build up models for $\Pr(\mathbf{Z}_{mis}|\mathbf{Z}_{obs}, \mathbf{W}, Y)$ (in the regression approach) or $\Pr(M|\mathbf{Z}_{obs}, \mathbf{W}, Y)$ (in the propensity score method) one variable at a time, using the subjects with complete data on all previous variables, along with imputed values for variables previously considered. If the data have an arbitrary (nonmonotone) missingness pattern, than more complicated techniques are required, such as using MCMC methods to sample each missing value conditional on the observed data and the current assignment of all missing values. We will return to the problem of nonmonotone missingness patterns in the context of longitudinal data in Chapter 7.

Missing indicator methods

Missing data can be particularly annoying in matched case-control studies, because a matched analysis would require that if a case has a missing value, his or her matched control also be excluded (even if the data were complete for that subject) and vice versa if the control has a missing value. The alternative of ignoring the matching and using all complete-data subjects is likely to be biased, as it would even if there were no missing data.

A simple, but much maligned method (see Greenland and Finkle 1995 and references therein) to overcome this problem is to include the missing value indicators **M** as covariates in a multiple regression along with the **Z** variables (with missing values set to any convenient constant value, usually the overall mean). The missing indicators then soak up any differences between subjects with and without a value for each variable; the missing indicator coefficients are of no particular interest themselves, but under appropriate conditions, the coefficient of the **Z** variables are unbiased estimators of their true associations. Of course, the method could be applied to unmatched data as well, but is particularly attractive for matched studies. Huberman and Langholz (1999) describe a version of this approach that entails combining the usual conditional likelihood for complete-data case-control pairs with unconditional likelihood contributions from pairs with exposure information missing on either the case or the control (but not, of course, both). The latter contributions can easily be implemented in standard software simply by including a missingness indicator in the model. They show that this is a valid method of analysis provided that the complete-pairs analysis is valid and that any confounding in the incomplete pairs has been adequately adjusted for. Li et al. (2004) compared this method to a complete-pairs analysis using simulation and found that the complete-pairs performed slightly better in terms of bias and confidence interval coverage, at the expense of some loss of power and efficiency, and recommended caution in the use of the missing indicator method, depending on the extent of confounding and the missing data model.

5 Some special-purpose designs

The cohort and case-control designs and their hybrids—nested case-control and case-cohort—are the work-horses of environmental epidemiology. There are circumstances, however, where special purpose designs are appropriate, either to control biases or improve statistical efficiency. Here, we review six such designs: two-phase case-control designs; counter-matched case-control designs; case-crossover designs; case-specular designs; case-parent triad and case-only designs for gene–environment interactions. One additional design, a hybrid of ecologic and individual-level designs, we defer to Chapter 10, after discussing the fundamental capabilities and limitations of ecologic inference.

Two-phase case-control designs

Multistage sampling designs have been used by survey statisticians for ages (Neyman 1938), but their introduction to the epidemiology community is generally ascribed to a seminal paper by Emily White (1982) concerning studies of the relationship between a rare disease and a rare exposure. The basic design idea was also contained in a paper by Walker (1982), but without the analytic sophistication of White's paper. In econometrics and other literature, the design goes by the name "double sampling." In the epidemiologic literature, the words "stage" and "phase" have often been used interchangeably, but recent usage tends to favor "phase," a convention we will adopt here. As is well known, the case-control design is the method of choice for studying associations of a rare disease with common exposures, and a cohort design preferred in the converse situation. If both factors are rare, neither is very efficient, as the precision of the relative risk estimate (and the power of the test) is limited by the number of exposed cases, which will be very rare in this circumstance.

Consider Table 5.1a, describing the joint distribution of exposure and disease in the source population. In a cohort study of a rare exposure (Table 5.1b), one might try to enroll most of the exposed population and sample a small fraction of the unexposed population. Both samples are then followed in the same way to determine their outcomes (i.e., assuming one is not simply relying on external population rates for comparison). Conversely, in a case-control design (Table 5.1c), one might try to ascertain most of the cases in some defined population and time period

Table 5.1. Schematic representation of different sampling schemes used in epidemiology

Exposure	(a) Population			(b) Cohort study			(c) Case-control study		(d) Two-phase study	
	$Y=0$	$Y=1$	Total	$Y=0$	$Y=1$	Total	$Y=0$	$Y=1$	$Y=0$	$Y=1$
$Z=0$	N_{00}	N_{10}	N_{+0}	n_{00}	n_{10}	$n_{+0}=s_0 N_{+0}$	n_{00}	n_{10}	$n_{00}=s_{00}N_{00}$	$n_{10}=s_{10}N_{10}$
$Z=1$	N_{01}	N_{11}	N_{+1}	n_{01}	n_{11}	$n_{+1}=s_1 N_{+1}$	n_{01}	n_{11}	$n_{01}=s_{01}N_{01}$	$n_{11}=s_{11}N_{11}$
Total	N_{0+}	N_{1+}	N_{++}	—	—	n_{++}	$n_{0+}=s_0 N_{0+}$	$n_{1+}=s_1 N_{1+}$	—	—

and a small fraction of unaffected individuals in that population to serve as controls; again, determination of exposure histories is done in a similar way for cases and controls. In both designs, the sample odds ratio estimates the population odds ratio, *provided the sampling of subjects is unbiased with respect to the variable being measured*—disease in a cohort study, exposure in a case-control study. A two-phase case-control design appears to violate this basic principle by sampling subjects *jointly* with respect to both variables (Table 5.1d). Whereas in case-control or cohort studies, the sampling fractions s_0 and s_1 cancel out in the calculation of the odds ratio, the key to the validity of two-phase designs is that the sampling fractions be explicitly taken into account in the analysis, either by estimating them from the first phase data or by analyzing the two phases together.

Multi-phase designs can be particularly useful in two important situations. Let Z represent the variable used for stratified sampling and X the additional data collected in the second phase Z may represent a categorical exposure variable of interest, but additional covariates X are to be collected as confounders or modifiers of the Y-Z relationship that is of primary interest. Alternatively, Z may represent an easily-determined categorical surrogate for the true exposure X, which is feasible to assess only on a small sample, and it is the Y-X relationship that is of primary interest. Other applications arise where one wishes to incorporate external information on population relationships among a subset of variables or where some variables have missing values (Cain and Breslow 1988). In either case, the analysis is essentially the same. One uses the data from the observed subsample, together with knowledge of the sampling fractions, to estimate the joint distributions in the source population; these are then used to estimate the parameters of interest. Variance estimation then requires that one take account of not just the sampling variation in the second-stage sample data, but also the variability in the sampling fractions themselves.

There are various ways this can be done. Consider first the expanded Table 5.2, illustrating the case of binary X, Y, and Z data. We could easily estimate the expected values of the unobserved quantities in the first-phase sample by dividing the corresponding entries in the second-phase sample by their sampling fractions: $E(N_{yzx}) = n_{yzx}/s_{yz}$. Thus, the stratified ORs would be given by

$$\mathrm{OR}(Y, X|Z = z) = \frac{E(N_{1z1})E(N_{0z0})}{E(N_{1z0})E(N_{1z0})} = \left(\frac{n_{1z1}n_{0z0}}{n_{1z0}n_{0z1}}\right) \div \left(\frac{s_{1z}s_{0z}}{s_{1z}s_{0z}}\right)$$

$$= \mathrm{or}(Y, X|Z = z)$$

where $\mathrm{or}(Y, X|Z = z)$ denotes the odds ratio estimated from the second-phase sample. Thus, the second-phase estimates of the conditional ORs

Table 5.2. Joint distribution of disease Y, sampling surrogate Z, and other risk factors X in a two-phase sampling design

Z	X	First-phase sample		Second-phase sample	
		$Y = 0$	$Y = 1$	$Y = 0$	$Y = 1$
0	0			n_{000}	n_{100}
	1			n_{001}	n_{101}
	Total	N_{00}	N_{10}	$n_{00} = s_{00} N_{00}$	$n_{10} = s_{10} N_{10}$
1	0			n_{010}	n_{110}
	1			n_{011}	n_{111}
	Total	N_{01}	N_{11}	$n_{10} = s_{10} N_{10}$	$n_{11} = s_{11} N_{11}$

are unbiased because, within strata defined by Z, the two sampling fractions cancel out. However, for the marginal odds ratio, $OR(Y, X)$, this cancellation does not occur and we have instead

$$OR(Y, X) = \frac{E(N_{1+1})E(N_{0+0})}{E(N_{1+0})E(N_{0+1})} = \frac{\left(\dfrac{n_{101}}{s_{10}} + \dfrac{n_{111}}{s_{11}}\right)\left(\dfrac{n_{000}}{s_{00}} + \dfrac{n_{010}}{s_{01}}\right)}{\left(\dfrac{n_{100}}{s_{10}} + \dfrac{n_{110}}{s_{11}}\right)\left(\dfrac{n_{001}}{s_{00}} + \dfrac{n_{011}}{s_{01}}\right)}$$

The variance of the log OR can be computed in the obvious way by the delta method, assuming the sampling fractions are fixed by design, leading to a somewhat messy expression. (Note, however, that this expression makes no use of the data from the first-phase sample, other than to compute the sampling fractions, so we would not expect it to be fully efficient.) From this expression, one can then compute the optimal sampling fractions that would minimize the var(ln(OR)) for a specified ratio of cost per subject in phases one and two, subject to a constraint on the total subsample size n_{+++}. Table 5.3 illustrates the results for a particular combination of disease and surrogate exposure frequencies, sensitivity and specificity of Z as a measurement of X, and the true Y-X odds ratio, computed from the full likelihood described below (Thomas 2007a).

One variant of this design that frequently proves to be nearly optimal is the so called "balanced design", in which the target cell sizes in the table of n_{yz+} are arranged to be nearly equal, that is, $n_{yz+} = n_{+++}/4$ (Cain and Breslow 1988; Reilly 1996; Holcroft and Spiegelman 1999). More generally, one might aim to keep the marginal frequencies in the second phase, n_{y++} and n_{+z+}, equal and adjust the marginal association between Y and Z in the sampling fractions so as to achieve an even more efficient design.

Table 5.3. Relative efficiencies of different two-phase sampling schemes

Design	Sampling fractions (%)	Subsample sizes (%)	ARCE (%)
No sampling	(100, 100, 100, 100)	(74, 6, 15, 5)	100
Random sampling	(22, 22, 22, 22)	(74, 6,15, 5)	146.8
Outcome-based	(23, 23, 19, 19)	(77, 6, 13, 4)	147.5
Determinant-based	(8, 64, 8, 64)	(44, 25, 9, 22)	285
Equal allocation	(6, 89, 33, 100)	(25, 25, 25, 25)	328
Optimal allocation	(3, 86, 19, 19)	(21, 44, 26, 8)	399

Cost ratio = 16, sensitivity = specificity = 80%; predictive value of surrogate = 90%, surrogate prevalence = 20%, baseline log odds = −3, log OR = 2. (Adapted from Thomas 2007a).

We now wish to generalize this approach in various ways: to continuous and multivariate X variables; to exploit the information on Y and Z in the first-phase sample; and to develop more efficient analysis procedures. The full likelihood of the data is

$$L(\alpha, \beta) = \prod_{i,j} \Pr(Z = j|Y = i)^{N_{ij}} \prod_{k=1}^{n_{ij}} \Pr(X = x_{ijk}|Y = i, Z = j) \quad (5.1)$$

These terms, of course, involve both the marginal distribution α of covariates X and stratifying variables Z, as well as the disease parameters β of real interest. We discuss below how the latter can be estimated without requiring assumptions about the former, but first consider the situation where X, Y, and Z are all binary, so the data can be completely represented by the counts given in Table 5.2. The full likelihood is then obtained by summing over all possible combinations of the unobserved cell counts N_{yzx}

$$L(\beta, \alpha) = \sum_{N_{yzx}|N_{yz+}} \Pr(N_{yzx}|N) \times \Pr(n_{yzx}|n_{yz+}, N_{yzx})$$

$$= \prod_{y,z} \Pr(N_{yz+}|N) \sum_{N_{yz}|N_{yz+}} \Pr(N_{yz1}|N_{yz+}) \times \Pr(n_{yz1}|n_{yz+}, N_{yz1})$$

where the summation is over the range of values that are compatible with the observed margins (Breslow and Holubkov 1997a). The first factor within the summation is a binomial with probability $\Pr(X|Y, Z)$ and the second factor is a central hypergeometric probability. The difficulty with practical applications of this likelihood (beyond the purely computational one) is that it explicitly involves the joint distribution of X and Z in the population, which is of no particular interest and could require many parameters and assumptions if there are multiple or continuous covariates. There are three alternative approaches available for the analysis of two-phase case-control studies with general covariates $\mathbf{X}$: weighted likelihood (Flanders and Greenland 1991; Reilly and Pepe 1995; Whittemore 1997), pseudolikelihood (Breslow and Cain 1988; Scott and Wild 1991; Schill et al. 1993), and constrained maximum likelihood (Breslow and Holubkov 1997a, b).

The simplest of these is the "weighted likelihood approach" derived from sampling theory, as in the simple method for $2\times2\times2$ tables described above. This entails weighting the loglikelihood contributions from each stratum inversely by their sampling fractions. The derivatives of this weighted likelihood yield what is known as the Horvitz–Thompson (1952) estimating equations. This approach uses only the data from the second phase, but reweights the score contributions for each stratum s (here based on categories of Y and Z) inversely by their estimated sampling fractions $\hat{f}_s = n_s/N_s$:

$$\mathbf{U}(\boldsymbol{\beta}) = \sum_s \hat{f}_s \mathbf{U}_s(\boldsymbol{\beta}) = 0$$

where the total score for stratum s is just $\mathbf{U}_s(\boldsymbol{\beta}) = \sum_{i\in n_s} \mathbf{U}_i(\beta)$. In the case of a logistic model for binary data, for example, the individual score contributions are given by $\mathbf{U}_i(\boldsymbol{\beta}) = (Y_i - p_i)\mathbf{X}_i$ where $p_i = \text{expit}(\mathbf{X}'_i\boldsymbol{\beta})$. The variance of $\hat{\boldsymbol{\beta}}$ is given by the sandwich estimator, which takes the form

$$\text{var}(\hat{\boldsymbol{\beta}}) = \left(\mathbf{I}(\hat{\boldsymbol{\beta}})\right)^{-1} \left\{ \sum_s f_s^{-2} \left[\sum_{i\in n_s} \mathbf{U}_i(\hat{\boldsymbol{\beta}})'\mathbf{U}_i(\hat{\boldsymbol{\beta}}) \right. \right.$$
$$\left. \left. - \frac{1-f_s}{n_s}\mathbf{U}_s(\hat{\boldsymbol{\beta}})'\mathbf{U}_s(\hat{\boldsymbol{\beta}}) \right] \right\} \left(I(\hat{\boldsymbol{\beta}})\right)^{-1}$$

(Breslow and Chatterjee 1999), thereby taking advantage of the additional information about β contained in the sampling fractions themselves, yielding a smaller standard errors than those derived from the inverse information alone.

Applying Bayes formula, the real likelihood, Eq. (5.1), can be rewritten as

$$L(\alpha, \beta) = \prod_{i,j} \Pr(Y = i | Z = j)^{N_{ij}} \prod_{k=1}^{n_{ij}} \Pr(Y = i | X = x_{ijk}, Z = j)$$
$$\times \prod_j \Pr(Z = j | S_1)^{N_{+j}} \prod_{k=1}^{n_{+j}} \Pr(X = x_{ijk} | Z = j, S_2) \qquad (5.2)$$

where S_1 and S_2 denote the events that a subject was sampled in phases 1 and 2, respectively. The "pseudolikelihood" approach ignores the terms on the second line, simply multiplying the conditional probabilities logit $\Pr(Y = 1 | Z = j, S) = \alpha_0 + \delta_j$ from the first phase and logit $\Pr(Y = 1 | \mathbf{X}, Z = j, S) = \alpha_j + \gamma_j + \beta' \mathbf{X}$ from the second phase (Fears and Brown 1986; Breslow and Cain 1988; Schill et al. 1993), where the αs are functions of the respective sample sizes. The resulting estimating equations can be shown to yield consistent estimators of the parameters of interest, despite this not being the true conditional likelihood. Breslow and Cain fit the model by estimating the stratum parameters δ_j in $[Y|Z]$ as $\ln(N_{1j}N_0/N_{0j}N_1)$ and substituting these into $[Y|\mathbf{X}, Z]$ to obtain a profile likelihood in β. This is easily implemented by including the logarithm of ratios of phase 1 and 2 sample sizes as offset terms in standard logistic regression software. Schill et al. instead maximize both parts of the pseudolikelihood jointly. In practice, the two estimators are very similar. The variance estimate based solely on the inverse information (as used by Fears and Brown) fails to incorporate the additional information from the first phase, however, even if the offset terms are included. Breslow and Cain provide two versions of the more appropriate sandwich estimator; one version is simpler but may not be positive definite if there is extensive first phase data; the other is based on the within-stratum covariances of $\mathbf{X}$ and guaranteed to be positive definite. See Breslow and Zhao (1988) for a comparison of the different variance estimators, showing the considerable reduction in standard errors that can result from properly incorporating the additional information from the first stage data.

The constrained maximum likelihood approach estimates the marginal distributions $\Pr(Z|S)$ and $\Pr(X|Z, S)$ in the second line of Eq. (5.2) nonparametrically and maximizes the resulting likelihood subject to the constraints that the marginal distribution of cases and controls in phase I and their conditional distributions in phase II are fixed by design. Thus, no assumptions are required about the distribution of Z or $\mathbf{X}$. For discrete covariates, a solution for random sampling in phase I was given by Scott and Wild (1991) and extended to case-control sampling in phase I by Breslow and Holubkov (1997a, b); Scott (2007) later provided a generalization for continuous covariates and established its semiparametric efficiency. The variance of $\hat{\beta}$ is given by the appropriate submatrix of the inverse information, without requiring a sandwich estimator.

An important special case of multi-phase case-control sampling arises in hybrid ecologic/individual level studies, in which N_{YZ} is not observed, but only the margins N_{Y+} and N_{+Z} in multiple populations. These are then used for sampling individuals to observe n_{yzx}. This design will be discussed further in Chapter 10.

There is an extensive statistical and epidemiological literature on multi-phase case-control sampling designs. Useful reviews and comparisons of methods include Zhao and Lipsitz (1992); Breslow and Holubkov

(1997a); Whittemore (1997); Breslow and Chatterjee (1999); Borgan et al. (2000); Scott et al. (2007). Further results on optimal designs using one or more of these estimation approaches are provided in Tosteson and Ware (1990); Weinberg and Wacholder (1990); Wacholder et al. (1994); Reilly (1996); Schaubel et al. (1997); we will revisit some of these in Chapter 11 in the context of measurement error models using a second-phase sample based on Y and a flawed surrogate Z to obtain measurements of true exposure X. Fears and Gail (2000) consider the additional complications when cluster sampling is used to obtain controls.

Counter-matching

The two-phase case-control designs described above are aimed at *unmatched* case-control samples. In the pair-matched situation with a dichotomous surrogate variable Z, one could imagine a two-phase design in which one sampled from within the 2×2 table of Z values for the case-control *pairs* (rather than individuals) in the first-stage sample. Since the pairs that are concordant for Z are likely to be less informative about the risk associated with true exposure X than those that are discordant (assuming Z is a good surrogate for X), it is natural that we would want to over-represent the discordant pairs in the second stage sample. Indeed, if the second-stage sample were restricted to *only* Z-discordant pairs, one would have a "counter-matched" design, every surrogate-exposed case being matched with a surrogate-unexposed control and vice-versa. We now generalize these ideas to a broader range of designs and discuss the analysis of such studies.

Consider a nested case-control study in which one wishes to study the relationship of some risk factor X with disease Y and one has available for all members of the cohort a surrogate Z for X. In the standard nested case-control design, one would sample for each case in the cohort one or more controls at random from that case's risk set, the set of all cohort members who were still at risk at that time. The disadvantage of this approach is that in some sampled risk sets, the case and control(s) will be concordant for X and hence those matched sets will be uninformative. The counter-matched design overcomes this difficulty by systematically *mismatching* cases and controls on the exposure surrogate Z, so as to maximize the variability in X within matched sets, thereby improving the efficiency of the design, often quite substantially. In general, if Z has k categories, a $1:(k-1)$ counter-matched design would have one member of each matched set in each category of Z.

In order to account for the bias in the sampling design, it suffices to include the sampling fractions as a function of Z as an offset in the standard conditional logistic likelihood:

$$L(\beta) = \prod_i \frac{\exp(X'_{i0}\beta - \ln(n_{i0}/N_{i0}))}{\sum_{j \in \tilde{R}_i} \exp(X'_{ij}\beta - \ln(n_{ij}/N_{ij}))}$$

where $\tilde{R}_j$ denotes the set comprising the case and his or her counter-matched controls, n_{ij} and N_{ij} the numbers of subjects in the $\tilde{R}_i$ and R_i respectively with counter-matching variable value Z_j. No adjustment to the variance of $\hat{\beta}$ is needed in this approach.

Although we have described this as a design for case-control studies nested within cohorts, it could in principle also be applied to case-control studies within a general population, provided it were possible to estimate the numbers at risk in relation to Z. Langholz and Goldstein (2001) discuss extensions of the approach to more general stratified case-control designs.

The WECARE study (Bernstein et al. 2004) used a counter-matched design to investigate the effects of ionizing radiation exposure and DNA repair genes on the risk of cancer in the contralateral breast in women surviving a first breast cancer. (For the purposes of the present illustration, we focus on the main effect of radiation, deferring discussion of gene–environment interactions to Chapter 12.) The exposure of interest X is the dose to the contralateral breast from radiotherapy for the first cancer, as reconstructed from phantom studies. The surrogate Z used for counter-matching is a binary indicator for whether the woman received radiotherapy or not, as recorded in the cancer registries through which the cohort of breast cancer survivors was identified. In addition to conventional matching on age and calendar year of diagnosis of the first cancer, survival to the time of the case's second cancer, race, and center, counter-matching was implemented by having each matched set comprise two radiotherapy patients and one without radiotherapy. Thus, if the case was exposed, she was matched with one exposed control and one unexposed, whereas if she was unexposed, she was matched with two exposed controls. Table 5.4 compares the theoretical relative efficiency of this design to a standard 1:2 matched case-control design, a 1:2 counter-matched study with only one exposed case in each set, and a cohort analysis in which X would be obtained on the entire cohort. Although no design based on sampling from within the cohort could be more efficient that the full cohort analysis, we see that the 1:2 counter-matched design can be nearly as powerful as a full cohort analysis and substantially more powerful than a standard nested design.

The counter-matched design is also quite efficient for testing interactions (Table 5.5), although there can be some loss of efficiency for testing

Table 5.4. Statistical power for various nested case-control and counter-matched designs using 700 sets of 1 case and 2 controls each. (Data from Bernstein et al. 2004.)

RR at 2 Gy	Full cohort	Nested case-control	Counter-matched (N unexposed:exposed)	
			1:2	2:1
1.4	79	61	70	63
1.5	87	69	84	78
1.6	97	86	95	92
1.7	100	95	99	97

Table 5.5. Asymptotic relative efficiency for the 1:1 counter-matched and 1:1:1 hybrid design with an additional randomly selected control, relative to a standard 1:1 matched design. (Langholz and Goldstein, 1996.)

$\exp(\beta_1)$	1:1 counter-matched			1:1:1 hybrid		
	Z	X	$Z \times X$			
0	2.00	0.19	1.00	2.00	1.05	1.82
4	4.35	0.62	2.09	4.35	1.32	3.21

confounders. The relative efficiency of counter-matching depends in large part on how good a surrogate Z is for X.

Figure 5.1 illustrates this for dichotomous Z and X variables across a range of sensitivities and specificities (Langholz and Borgan 1995). It is apparent that for testing the main effect of exposure and its interaction with a covariate, there can be considerable improvements in efficiency, particularly under the alternative. Conversely, there is a substantial loss of efficiency for the covariate effect, but this can be recovered by including a separate random control group (the "hybrid" design in Table 5.5).

Table 5.6 illustrates Cox regression and nested case-control analysis of radiation exposure in the Colorado plateau uranium miner cohort. Both analyses use the same series of 324 lung cancer cases and the same exposure variable, cumulative radon daughter exposure in "working level months" (WLM), up to five years previously. For the full cohort analysis, each case is compared with all subjects still alive at the case's age, born in the same year, and enrolled in the cohort by that date. For the nested case-control analysis, 1 or 3 controls were sampled at random from the risk set for each

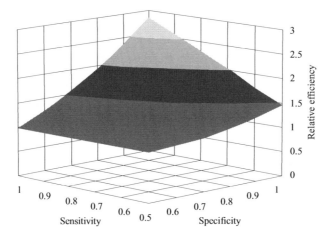

Figure 5.1. Effect of sensitivity and specificity of surrogate variable Z used for counter-matching on the efficiency for estimating the Y-X association, relative to a conventional nested case-control design. (Data from Langholz and Borgan 1995.)

case. The counter-matched analysis led to substantially smaller standard errors that the conventional case-control design for both the radon main effect and its interaction with smoking, but somewhat larger standard errors for the smoking main effect. A hybrid matched design using one random and one counter-matched control per subject was generally less efficient that the 1:3 counter-matched design, even for the smoking main effect (albeit with a smaller total sample size), although it was generally slightly more efficient at estimating all three effects than either of the 1:1 designs (albeit with a larger sample size).

Other applications of counter-matching have been discussed by Langholz and Goldstein (1996); Cologne (1997); Steenland and Deddens (1997); Cologne and Langholz (2003); Langholz (2003); Cologne et al. (2004), including studies of reproductive factors in breast cancer in the atomic bomb survivor cohort, lung cancer in the Colorado plateau uranium miners, and gold miners.

To investigate the effect of maternal smoking on asthma prevalence and its modification by genetic susceptibility, prevalent cases were counter-matched to controls from the Children's Health Study on a binary indicator for maternal smoking during pregnancy and additional information on smoking during and after pregnancy was obtained, along with DNA samples (Langholz and Goldstein 2001). This study also illustrates counter-matching with multiple cases per stratum. Maternal and grandmaternal smoking during pregnancy was positively associated with early-onset asthma, but postnatal exposure was not (after controlling for *in utero* exposure), nor was risk increased in those whose mothers quit smoking before pregnancy (Li et al. 2005a). Other analyses demonstrated significant associations with exposures in the first year of life to wood

Table 5.6. Estimates of the effects of radon and smoking in the Colorado plateau uranium miner cohort by different sampling designs. (Reproduced with permission from Langholz and Goldstein 1996).

Model[a]	Full cohort	Random sampling		Counter-matching		Hybrid 1:1:1
		1:1	1:3	1:1	1:3	
Univariate models						
Radon (β_R)[b]	0.36 (0.10)	0.41 (0.19)	0.41 (0.15)	0.33 (0.11)	0.36 (0.11)	0.35 (0.11)
Smoking (β_S)[c]	0.16 (0.05)	0.18 (0.07)	0.20 (0.06)	0.37 (0.15)	0.23 (0.08)	0.19 (0.07)
Adjusted model[d]						
Radon (β_R)	0.38 (0.11)	0.42 (0.20)	0.43 (0.16)	0.39 (0.14)	0.41 (0.13)	0.44 (0.16)
Smoking (β_S)	0.17 (0.05)	0.23 (0.10)	0.20 (0.07)	0.25 (0.10)	0.19 (0.07)	0.23 (0.09)
Interaction model[e]						
Radon (β_R)	0.67 (0.27)	0.51 (0.29)	0.53 (0.24)	0.54 (0.28)	0.50 (0.21)	0.62 (0.30)
Smoking (β_S)	0.24 (0.08)	0.25 (0.12)	0.22 (0.08)	0.30 (0.13)	0.22 (0.079)	0.29 (0.11)
Interaction (β_{RS})	−0.68 (0.27)	−0.41 (0.70)	−0.41 (0.42)	−0.53 (0.46)	−0.31 (0.36)	−0.53 (0.42)
Number of distinct subjects	3,347	478	837	473	765	670

or oil smoke, cockroaches, herbicides, pesticides, farms and a negative association with sibship size (Salam et al. 2004).

Andrieu et al. (2001) discuss the possibility of counter-matching jointly on two interacting risk factors, such as a surrogate for a gene (say, family history) and a surrogate for exposure. In a 1:1:1:1 counter-matched design, each matched set would include one case and three controls, with each combination of surrogate exposure and family history strata represented. They showed that this design was more efficient for testing gene–environment interactions than either a standard 1:3 nested case-control designs or 1:3 counter-matched designs that used only one of the two factors for counter-matching.

Case-distribution studies

In this section, we consider several examples of a broad class of what Greenland (1999b) calls "case-distribution studies." All these designs are characterized by the exposure distribution of a case series being compared to a hypothetical exposure distribution representing a theoretical complete population or distribution. Greenland provides a unified likelihood framework for the analysis of such designs and a thorough discussion of the assumptions needed for its validity. Here, we introduce the basic idea through several examples of particular types of case-distribution studies.

Case-crossover design

The case-crossover design was developed to study acute responses to environmental triggers by using each subject as his own control in the spirit of a randomized crossover trial. The idea was first proposed by Maclure (1991) to study precipitating events for myocardial infarction. He proposed to assign to each observed event a comparison time at some predefined earlier time point that was otherwise comparable (e.g., same day of the week). The analysis would then compare the frequencies or means of various exposure factors at the case times versus the crossover times using a matched analysis, just like a standard matched case-control study.

Using this design, Peters et al. (2004) found a highly significant odds ratio of 2.9 (95% CI 2.2–3.8) for nonfatal myocardial infarctions in relation to exposure to traffic one hour prior to onset; the association was consistent for use of public transport, private cars, motorcycles, or bicycles and virtually unchanged by adjusting for level of exercise. Schwartz (2005) used a similar approach to study ozone exposures, controlling for weather. As will be discussed in Chapter 8, time series analyses of the relation between short-term fluctuations in air pollution and corresponding fluctuations in mortality can be quite sensitive to how one adjusts for weather; associations with ozone are particularly sensitive because high ozone days tend to be quite hot. Schwartz compared two case-crossover analyses, both matching on season and individual factors, but one also requiring the control day to have the same temperature as the case day. The two point estimates for all-cause mortality were quite similar: 0.19% per 10 ppb change in ozone (95% CI 0.03–0.35) without matching on temperature but controlling for it using regression splines; 0.23% (0.01–0.44) matching on temperature, although the latter led to somewhat wider confidence limits.

This design entails the implicit assumption that there is no consistent trend in exposure prevalence or intensity. In some circumstances, this can be overcome by the use of a "bidirectional case-crossover" design (Navidi 1998), in which two crossover times are chosen, one before and one after the time of event. For an event like death, this might seem strange, as exposures occurring after death could not possibly be causally related and would be meaningless for factors specific to the individual (e.g., activities)! In other contexts, such as ambient air pollution, however, the exposure intensities at the two crossover times could reasonably be interpreted as provided an estimate of the population distribution of exposures expected at the time of event under the null hypothesis. Nevertheless, the bidirectional design is not immune to bias, particularly for cyclical exposures, unless one analyzes the deviations from some smooth curve rather than the absolute exposure intensities (Janes et al. 2005).

Poisson process model for recurrent events

It can be shown that the case-crossover design estimates the same parameter as that estimated by time-series studies described in Chapter 8. To see this, let $\lambda(t, Z(t))$ denote the usual hazard rate at time t in relation to exposure at time t. The likelihood is obtained by conditioning on the fact that an event occurred at time t_i for subject i and did not occur at his or her crossover time s_i, hence

$$L(\beta) = \prod_i \frac{\lambda(t_i, Z_i(t_i))}{\lambda(t_i, Z_i(t_i)) + \lambda(s_i, Z_i(s_i))}$$

If we assume $\lambda(t, Z_i(t)) = \lambda_{0i} \exp(\beta Z_i(t))$, then the baseline hazard terms cancel out, without any requirement of homogeneity across subjects, and we are left with the usual matched case-control comparison of relative risk terms evaluated at the different times for the same subject. For this cancellation to occur, however, it is necessary to assume constancy of baseline rates over time. If instead one were to assume $\lambda(t, Z_i(t)) = \lambda_0(t)$ $X_i \exp(\beta Z_i(t))$, where X_i represents an unobservable "frailty" for each subject—his relative risk, relative to other subjects with the same covariate values at any given time—then the X_is still cancel out of the previous likelihood but the $\lambda_0(t)$ terms do not. Nor can the baseline hazard function be estimated from case-crossover data alone, but it could be estimated if supplemented by the time-series of event counts, assuming constancy of the population size and risk factors other than $Z(t)$. One would also need to assume that the exposures at case's crossover times could be used to estimate the distribution of population exposures at these times, at least relative to other times. Since this could be a questionable assumption, one might also want to include an appropriate control group to estimate the distribution of $Z(t)$ in the population at risk. The full likelihood is then a combination of a Poisson likelihood for the number of events at each point in time, multiplied by a case-control likelihood for which subject was the case, and also multiplied by the conditional likelihood given above for the timing of the case's event, given that he or she experienced an event:

$$L(\beta, \lambda_0(t)) = \prod_i \Pr(Y(t)|\bar{\lambda}(t))$$

$$\times \prod_{i|Y_i=1} \Pr\left(Y_i = 1|\tilde{R}_i, \{\bar{Z}_j\}_{j \in \tilde{R}_i}\right)$$

$$\times \prod_{i|Y_i=1} \Pr\left(Y_i(t_i) = 1|Y_i = 1, \{Z_i(t)\}_{t \in \tilde{T}_i}\right)$$

where $\bar{\lambda}(t) = \sum_{j \in \tilde{R}(t)|Y_j(t)=0} \lambda_0(t) \exp(\beta Z_i(t))/n_{\tilde{R}(t)}$, $\tilde{R}_i$ is the set compri case i and matched controls, $\tilde{R}(t)$ the set of all subjects at risk at time t and

$\tilde{T}_i$ the case and sampled crossover times for subject i. We might thus call this the "case-crossover-control-crossover (C^4)" design!

This idea has been more formally developed by Dewanji and Moolgavkar (2000; 2002) as a Poisson process model for recurrent event data with person-specific baseline rates. By conditioning on the total number of events for each person, the baseline hazards cancel out, leaving a likelihood that depends only on the time-dependent ambient exposure data (which would be the same for all individuals at any given time). The theoretical connection with case-crossover and time-series methods, along with applications to air pollution data, will be described further in Chapter 8.

Case-specular design

The case-specular design is similar to the case-crossover one, except that the comparison is spatial rather than temporal. It was first proposed by Zaffanella et al. (1998) for the purpose of testing the hypothesis that childhood leukemia was related to residential exposure to magnetic fields predicted by the configuration of electric power lines near the home. Rather than comparing the wiring configurations or magnetic field measurements of the case's home with that of a real control, a "specular" home was imagined as having the corresponding location on the opposite side of the street and the magnetic fields predicted by wiring configurations at that location is used for comparison. (Unlike most of the other designs, the specular home need not contain an eligible control person, or even exist!) When applied to two earlier studies of childhood cancers, the results supported the original conclusions of an effect of wire codes, rather than confounding by socio-economic status or other neighborhood characteristics (Ebi et al. 1999).

A very similar idea is being used to investigate radio frequency emissions from cellular phones as a risk factor for brain cancers in an international case-control study being coordinated by the International Agency for Research on Cancer (Cardis et al. 2007). Here, the comparison is within the individual, specifically between the estimated dose at the actual location of the tumor in the brain to the "specular" dose at the mirror image location on the opposite side of the brain. Because most people habitually use cell phones on one side or the other, these estimated doses are generally quite different at the two locations. This would provide a valid comparison, provided handedness is unrelated to the baseline risk of cancer on the two sides of the brain.

Langholz et al. (2007) have extended this basic idea to a comparison of doses at multiple locations across an organ, specifically scatter radiation doses to the contralateral breast from radiotherapy for breast cancer in the WECARE study. For this purpose, doses were estimated to each of eight quadrants plus the central nipple area and showed a clear gradient between inner and outer quadrants. There is also considerable variability in the baseline frequency of cancer (the frequency in unirradiated women) across the nine sectors, which would appear to be a recipe for confounding. However, it can be shown that a conditional likelihood like that for a conventional 1:8 matched case-control study with separate intercepts for each quadrant yields a consistent estimator of the relative risk parameter. Let Z_{is} denote the dose to sector s in case i and s_i denote the sector in which the tumor actually appeared, then the appropriate likelihood is

$$L(\beta, \boldsymbol{\alpha}) = \prod_i \frac{\exp(\alpha_{s_i} + Z_{is_i}\beta)}{\sum_{s=1}^{9} \exp(\alpha_s + Z_{is}\beta)}$$

setting $\alpha_1 = 0$ for identifiability. Note that no control individuals are used in this form of analysis. Langholz et al. compared the efficiency of this analysis to a conventional 1:2 matched case-control comparison of average doses to the entire breast between case and the two control individuals, a 1:2 matched case-control comparison of the doses Z_{isi} and Z_{jsi} to the actual tumor location in the case i and the same location for her matched controls j, and a 1:26 matched comparison of Z_{isi} and all remaining locations Z_{is} ($s \neq s_i$) for the case and all Z_{js} for the controls (this last, also requiring sector-specific intercepts). The results of this comparison are shown in Table 5.7. The first of these analyses is generally not valid unless an appropriately weighted dose (weighted by the baseline risk in the different sectors) is used. Even if the correct weighted average is used, however, this analysis does not exploit the actual tumor locations, so is less efficient than the 1:2 or 1:26 matched comparisons of the case's dose at the tumor location. The relative efficiency of the remaining analyses depends upon the ratio of variances in doses between subjects within

Table 5.7. Simulated bias and efficiency of the alternative case-specular comparisons for the WECARE study

Design	Bias (%)		Asymptotic relative efficiency	
	$\alpha_3/\alpha_1=5$	$\sigma_\psi=1$	$\alpha_3/\alpha_1=5$	$\sigma_\psi=1$
Case-control (1:2)				
crude average dose	+20	−30	116	63
weighted average dose	0	−20	89	83
dose to tumor location	0	−20	100*	100*
Case-specular (1:8)	+5	+5	27	46
Case-specular/control-specular (1:26)	0	−15	124	140

* Reference design for relative efficiency comparisons.

matched sets and between locations within cases. In the WECARE study, the within-subject variance is small compared to the between-subject variance (in part because the counter-matched design ensured a high degree of discordance), so the case-only design was quite inefficient, compared with either of the other two. Of the two, the 1:2 matched comparison of doses at the actual tumor location is less efficient, as less information is used, but is not sensitive to assumptions about the adjustment for baseline risk needed in the 1:26 matched comparison. On the other hand, both these analyses are potentially confounded by subject-specific risk factors, whereas the case-only analysis is not.

Case-only and case-parent-triad designs

In Chapter 12, we will consider various approaches to studying interactions, including gene–environment interactions. Before concluding this section on case-distribution studies, it is worth mentioning two other designs that are specifically intended for studying gene–environment interactions. The simplest is the *case-only* design (sometimes called a *case–case* design). This design relies on the basic result that if two factors are independently distributed in the population at risk, they will only be independently distributed among cases if their joint effects on the risk of disease are multiplicative. Thus, if we are prepared to believe the two factors are independent in the source population, an observation that they are associated among cases would be tantamount to a departure from a multiplicative risk model. Now for most environmental agents, there is no particular reason why we would expect them to be independent in the population. However, for genes we can rely on the random assortment of alleles at meiosis—which is not going to be influenced by subsequent environmental exposures, except possibly by differential survival—so that we might indeed be willing to assume that most genes and most environmental factors are independent. The case-only design (Piegorsch et al. 1994; Umbach and Weinberg 1997; Weinberg and Umbach 2000; Albert et al. 2001) therefore simply tests for association between genes and exposures among cases, without reference to any controls, as an indirect test of gene–environment interaction on a scale of risk. Of course, it is also possible that the gene under study has some effect on an individual's propensity for exposure: one is reminded of Fisher's (1958) famous conjecture that the association between smoking and lung cancer might be due to a gene that confers a propensity to smoke or to become addicted and independently an increased risk of lung cancer. [See Stolley (1991) for further discussion of this historical controversy.] In such a situation, the assumption of gene–environment independence would be suspect.

The *case-parent-triad design* is widely used in genetic epidemiology for studying the main effects of genes, but is also relevant in environmental epidemiology for studying gene–environment interactions. Like the case-crossover and case-specular designs, this design does not use any real control subjects but instead compares the "exposure" status (here, genotype) of the case to the distribution of hypothetical exposures, here the set of genotypes the case could have inherited from his or her parents. For study of gene–environment interactions, only the case's environmental exposures are used, not those of the parents (which are unlikely to be comparable because of generational and other time-related differences). Instead, the analysis compares the effect of the genes under study between exposed cases and unexposed cases (Schaid 1999). We defer the details of the analysis of such designs to Chapter 12, where we consider gene–environment interactions. It is worth noting that this design entails a similar assumption of gene–environment independence as the case-only design, but somewhat weaker because all comparisons are made within families, not between individuals, so it is sufficient to assume gene–environment independence conditional on parental genotypes.

Modeling exposure–time–response relationships

In Chapter 4, we introduced the general linear model (GLM), and special cases of it that are of particular importance in epidemiologic analysis of binary disease data—logistic, Poisson, and Cox regression. In that section, we confined attention to a generic statistical model, the canonical form for the link function $g[\mathrm{E}(Y|\mathbf{Z})] = \mathbf{Z}'\beta$ appropriate for the particular data structure. Although an extremely flexible modeling framework, it is not based on any biological concepts, but rather chosen to yield some desirable statistical properties, for example, that predicted probabilities always be in the range [0, 1] and that the estimated model parameters β be asymptotically normally distributed.

In environmental epidemiology, we frequently will have occasion to want to incorporate concepts from biology into the form of our mathematical models. For example, radiobiological theory suggests that cancer dose–response from ionizing radiation is expected to be a linear-quadratic function of dose, with an additional negative exponential factor at very high doses due to cell killing (NAS 1990; UNSCEAR 2000). The theory of multistage carcinogenesis further suggests specific forms for how dose–response is expected to be modified by age and latency (Chapter 13). Additional complexities come into play when analyzing extended time-varying exposure histories. In other circumstances, we might not want to be confined by the strictures of either a statistical or a biological model, but simply wish to let the "data speak for themselves" in a very flexible manner, making only minimal assumptions such as a certain degree of smoothness. In this chapter, we develop a broader modeling framework that will address such aims.

Although there are subtle distinctions between the ways one might want to model individual binary, aggregate count, and survival data, it generally suffices to frame this discussion in terms of models for the hazard rate $\lambda(t, \mathbf{Z}(\cdot))$ where $\mathbf{Z}(\cdot)$ denotes the history of exposure (and any other covariates) up to time t. From this, one can then express the likelihood for any particular data structure following the general principles outlined in Chapter 4.

Basic dose–response models

In this chapter, we will generally use the terms "exposure" and "dose" interchangeably. From a statistical point of view, the distinction is of no significance, but the semantic difference is important. More precisely, the word dose refers to the amount of some hazardous agent delivered to the target organ, whereas exposure refers the environmental concentration to which a person is subjected. Various steps—activity, inhalation or ingestion, internal deposition, metabolism, delivery to the target organ, uptake and retention—can intervene between exposure and dose, all of which might be modeled mathematically, as discussed in Chapter 13. Therefore, we use the letter Z (or in Chapter 11 the letter X to distinguish true and measured dose) to denote either quantity.

We begin by considering a broad class of parametric dose–response models—those that follow some specific mathematical form involving one or more parameters (e.g., slope and shape coefficients) that we wish to estimate. The greater the number of parameters we allow, the more flexible the model and the better we expect it to fit the data, but the less precisely we expect to be able to estimate them. In the limit as the number parameters approaches the effective number of data points, a parametric model would do no more than reproduce the data, without providing any insight into its patterns. Somewhere in between these two extremes lies what we will call "nonparametric" or "flexible" models. Or course, those are not really without assumptions, but here we aim to make as few mathematical restrictions as possible to allow the patterns and the data to reveal themselves.

We begin by restricting Z to be a time-constant quantity—for example, average exposure intensity—and will extend the models to time-varying and extended exposure histories later. We also limit attention in this chapter to a single exposure factor, deferring discussion of unscrambling multiple exposures and interaction effects to Chapter 12.

Parametric models

The basic hazard rate model we discussed in Chapter 4 took the parametric proportional hazards form

$$\lambda(t, Z) = \lambda_0(t) \exp(Z'\beta),$$

the exponential form for the relative risk $r(Z) = \lambda(t, Z)/\lambda_0(t) = \exp(Z'\beta)$ being chosen for mathematical convenience so that the hazard rate could never be negative. However this function could, in principle, be replaced by any mathematical function that made biological sense. More generally, we distinguish two basic forms of hazard rate models that are commonly used:

Relative risk model: $\lambda(t, Z) = \lambda_0(t) r(Z'\beta)$

Excess risk model: $\lambda(t, Z) = \lambda_0(t) + e(Z'\beta)$

where $r(\cdot)$ and $e(\cdot)$ can be any specific function, subject to the constraint that $\lambda(t, \mathbf{Z}) \geq 0$ for all observed values of t and $\mathbf{Z}$. Probably the most widely used "general relative risk model" (Thomas,1981) is the linear model $r(\mathbf{Z}'\boldsymbol{\beta}) = 1 + \mathbf{Z}'\boldsymbol{\beta}$, but of course there is nothing to stop us from adding terms, for example, a linear quadratic form $r(Z; \boldsymbol{\beta}) = 1 + \beta_1 Z + \beta_2 Z^2$. Radiobiological theory suggests for ionizing radiation a general form

$$r(Z; \boldsymbol{\beta}) = 1 + (\beta_1 Z + \beta_2 Z^2) \exp(-\beta_3 Z)$$

known as the linear-quadratic cell-killing model (NAS, 1980). The linear term represents the probability of double-strand break being caused by a single hit to a strand of DNA by a quantum of energy (e.g., a gamma ray or alpha particle). The quadratic term represents the probability of two independent traversals, each causing single-strand damage at homologous loci. The exponential term represents the probability of radiation-induced cell death before it can become malignant. This model is generally relatively well behaved provided all the regression coefficients are positive. However, if some are not significantly positive, their lower likelihood-based limits can end up being determined by the single observation with the largest Z, since the overall log likelihood goes to minus infinity as $r(Z; \beta) \to 0$.

The background rates of most chronic diseases like cancer rise dramatically with age. Before discussing models for $\lambda_0(t)$, it is worth considering the behavior of the relative risk and excess risk models in terms of age–exposure interactions. Note that in their present forms, neither model explicitly includes any age interaction terms. If we consider their simplest linear forms $\lambda(t, \mathbf{Z}) = \lambda_0(t)(1 + \mathbf{Z}'\boldsymbol{\beta})$ and $\lambda(t, \mathbf{Z}) = \lambda_0(t) + \mathbf{Z}'\boldsymbol{\beta}$, however, we see that each can be rewritten in the other's form as $\lambda(t, \mathbf{Z}) = \lambda_0(t) + \mathbf{Z}'\boldsymbol{\beta}\lambda_0(t)$ and $\lambda(t, \mathbf{Z}) = \lambda(t)[1 + \mathbf{Z}'\boldsymbol{\beta}/\lambda_0(t)]$, respectively. Thus, it follows that if the relative risk were constant with age, the excess risk must increase with age in proportion to the background rate. Conversely, if the absolute risk were constant with age, the relative risk must decline, inversely proportionally to the background rate. Of course, it is highly likely that neither parameter is truly constant with age, but the pattern is intermediate, absolute excess rates increasing and relative risks decreasing with age. This is clearly shown in Figure 6.1 for solid cancer mortality among the atomic bomb survivors. In such a case, explicit modeling of the exposure–age modification is needed, as discussed below.

In addition to specifying the form of the relative risk $r(\mathbf{Z}; \boldsymbol{\beta})$ or excess rate $e(\mathbf{Z}; \boldsymbol{\beta})$, one must also specify the baseline rate $\lambda_0(t)$. This too can be done parametrically or nonparametrically, the latter essentially eliminating them using partial likelihood as described in Chapter 4. Since in environmental epidemiology we are generally concerned with excess risks that are small in comparison with background rates, and since the latter

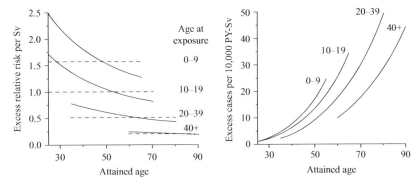

Figure 6.1. Excess relative risk $r(Z)$ (left panel) and excess absolute risk $e(Z)$ at 1 Sv by age at exposure groups for solid cancers in the atomic bomb survivors (Reproduced with permission from Preston et al. 2004.)

may have a very strong and complex dependency on age and other factors that may be correlated with exposure, it is essential that baseline rates be modeled carefully if a parametric approach is to be used. A sufficiently rich parametric model may suffice for this purpose, subject to the constraint that it be positive, or more precisely that the total rate (background plus excess) be positive. For example, the EPICURE package (Preston et al. 1993) that has been widely used in modeling the atomic bomb survivor data allows the following general form for the baseline hazard:

$$\lambda_0(U) = (1 + U'\alpha)\exp(U'\gamma)$$

where U is a vector of baseline risk factors such as age, sex, calendar year, or year of birth, and possibly additional risk factors (other than exposure Z) and interactions between them. (Note that some variables like age can be time varying, while others like sex are fixed.) Some of these covariates could appear in the linear component, some in the loglinear component, or even in both.

Sometimes an investigator may wish to use a set of external rates $\lambda^*(t)$ for comparison with an exposed cohort. However, there could be some question about the appropriateness of this comparison (recall the discussion of the "healthy worker effect" in Chapter 2, for example). If one thought that such a bias was similar across the entire cohort, one might take $\lambda_0(t) = \alpha\lambda^*(t)$, where the population/baseline rate ratio α would be estimated jointly with the relative risk coefficients β that are of primary interest. More generally, one might wish to allow this baseline rate ratio to be modified by such factors U as time since start of employment in an occupational cohort and adopt a model of the form $\lambda_0(t, U) = \lambda^*(t)\exp(U'\gamma)$.

The EPICURE package supports both relative and excess risk models, where both the excess relative risk terms $r(\mathbf{Z}, \mathbf{U})$ and the excess absolute risk terms $e(\mathbf{Z}, \mathbf{U})$ can include linear and/or loglinear components. Thus, the full model can be written in the form.

$$\lambda(\mathbf{U}, \mathbf{Z}) = \lambda_0(\mathbf{U})[1 + \Sigma_k r_k(\mathbf{Z}, \mathbf{U})] + \Sigma_\ell \, e_\ell(\mathbf{Z}, \mathbf{U})$$
$$= e^{\alpha_0' \mathbf{U}}[1 + \Sigma_k \beta_k' \mathbf{Z} \, e^{\gamma_k' \mathbf{U}}] + \Sigma_\ell \, \eta_\ell' \mathbf{Z} \, e^{\omega_\ell' \mathbf{U}})$$

The full generality is seldom needed, but its flexibility can be very useful in studying interactions, as we shall see in Chapter 12. For example, one might want to allow radiation (Z) and smoking (S) to interact additively with each other, but both multiplicatively with age:

$$\lambda_0(\mathbf{U}, Z, S) = \exp(\mathbf{U}'\boldsymbol{\alpha})[1 + \beta_1 Z + \beta_2 S]$$
$$= \lambda_0(\mathbf{U})[1 + r_1(Z) + r_2(S)]$$

Rich as this general form of parametric model for baseline rates is, it is still using relatively few parameters to describe what could be a very complex dependency on several temporal dimensions and other factors, so one might wish to adopt some less restrictive form. For example, in Poisson regression of data cross-tabulated by categories of U and Z, one might estimate a separate parameter λ_s for each stratum s of the factors in U. Similarly, in Cox regression of individual data, one could allow $\lambda_0(t)$ to have a separate value of each of the observed event times. Such models are called "semi-parametric" because the modeling of $\lambda_0(t)$ is nonparametric, while the modeling of $r(Z)$ or $e(Z)$—and the combination of baseline and excess risks—is parametric. This has the great advantage that no assumptions are made about the form of the baseline risk, which if violated could have biased inferences about the excess risks that are of primary interest. For relative risk models, this approach works very well, since one can use a form of conditional likelihood to estimate the parameters in $r(Z)$ directly without actually having to estimate the baseline parameters λ_s or $\lambda_0(t)$. Excess risk models are more difficult to treat semiparametrically because of the constraint that $\lambda(t) + e(Z) \geq 0$ for all t and Z, not just at the observed event times. Furthermore, lack of orthogonality can also cause problems with convergence to asymptotic normality if many parameters are being estimated in $\lambda_0(t)$. See Borgan and Langholz (1997) for further discussion of this problem.

However one specifies the model for $\lambda(t, Z)$, fitting is done by maximum likelihood, following the general principles for likelihood formation discussed in Chapter 4. For example, in Poisson regression, the likelihood

Table 6.1. Estimates of mortality dose–time–response relationships for all solid cancers combined in the atomic bomb survivors β is the slope per Sv; γ_1 the percent change per decade of age at exposure; γ_2 is the change per year of attained age. (Data from Preston et al. 2004.)

Model	Dose effect (β)	Age at exposure (γ_1)	Attained age (γ_2)
Relative risk	0.47 (0.37, 0.57)	−31% (−42%, −20%)	−0.70 (−1.4, 0.08)
Excess risk	30 (24, 36)	−23% (−34%, −12%)	+3.6 (3.0, 4.4)

is still written as

$$L(\alpha, \beta) = \prod_{s,z} \lambda(s, z)^{D_{sz}} \exp[-Y_{sz}\lambda(s, z)]$$

where s and z denote the levels of the stratification of U and Z, and D_{sz} and Y_{sz} the number of observed cases (or deaths) and person-years at risk in the corresponding cells of the cross-tabulation. Table 6.1 compares the parameter estimates for linear relative and excess risk models for all solid tumors combined from Poisson regression of the atomic bomb survivor data (Preston et al. 2004). The modifying effect of age at exposure is similar on relative and absolute risk scales, declining about 20-30% per decade. On the other hand, relative risks decline with advancing attained age (although this decline is not significant if age at exposure is in the model), whereas excess absolute risks increase strongly.

For leukemia, the best-fitting model is an age-at-exposure stratified excess absolute risk model of the form

$$\lambda(t, Z, E, S) = \lambda_0(t, S) + \left[\beta_{1E}Z + \left(\frac{\beta_2}{\beta_{1E}}\right)Z^2\right] \exp\left[\gamma_E \ln(t/\overline{E}) + \gamma_S\right]$$

(Preston et al. 2004), where Z denotes radiation dose, E denotes three strata of age at exposure (with mean age $\overline{E}$), and S denotes sex. Thus the dose–response is linear-quadratic with a slope that increases with age at exposure and declines with time since exposure; the decrease with latency is steepest in the age group 0–19 and nonsignificant thereafter.

Nonparametric models

We now turn to even more flexible ways of handling the excess risk part of the model, for example, for visualizing the fit of the data to some parametric dose–response model or asking whether there is any evidence of a threshold or saturation effects.

Of course, one could always address such questions parametrically and with greater statistical power by reserving a single parameter to test the specific hypotheses of interest, but then any such inferences would be model dependent and one could never be certain that the test or estimate did not reflect some lack of fit somewhere else in the data. There is no substitute for "seeing" the data itself, but in high dimensional data sets, this is easier said than done. In general, some combination of parametric and nonparametric modeling is the best approach.

Consider, for example, the question of testing for the existence of a threshold. Parametrically, one might test for an absolute threshold by comparing a linear relative (or excess) risk model $r(z) = 1 + Z\beta$ against a two-parameter linear spline of the form.

$$r(Z) = 1 + (Z - \tau)I(Z - \tau)\beta \equiv 1 + (Z - \tau)_+\beta \qquad (6.1)$$

where $I(u)$ is an indicator function taking the value 1 when the condition u is true, 0 otherwise, and $(u)_+ = u$ if $u > 0$, otherwise 0. Here, τ represents the threshold below which there is absolutely no increase in risk. One can estimate τ by maximizing the likelihood jointly with respect to τ and β, test the null hypothesis $H_0{:}\tau = 0$ with a likelihood ratio test, $G^2 = 2\ln[L(\hat\beta, \overline\tau)/L(\hat\beta, \tau = 0)]$, where $\hat\beta$ denotes the restricted MLE of β at $\tau = 0$. Perhaps more useful, the likelihood-based confidence limits on τ are given by the solution to

$$2\ln[L(\hat\beta, \overline\tau)/L(\hat\beta(\tau), \tau)] = \chi^2_{1-\alpha/2}$$

where again $\hat\beta(\tau)$ represents the MLE of β as a function of τ.

(*Technical aside*: Because this threshold model does not have continuous derivatives, the conditions for asymptotics (Cox and Hinkley 1974) do not strictly hold. Thus, for example, a plot of the likelihood as a function of τ will have an abrupt change of slope at each observed value of Z. In large enough sets of individual data, this is of no real practical significance, as the distribution of Z will tend to a continuum, but in smaller or aggregate data, hypothesis tests may not have exactly their nominal size or confidence limits their nominal coverage. A second technical point is that the null hypothesis $\tau = 0$ lies on the boundary of the space of allowable values, since negative thresholds would be meaningless. Maximum likelihood inference in such situations has been considered by Self and Liang (1987). In this two-parameter model with only a single constraint, the solution is quite straight-forward: one simply uses one-tailed tests and confidence limits, hence the $\alpha/2$ in the expression for the upper confidence limit (that is, unless $\tau \gg 0$ and one is interested in finding both its upper and lower limits).)

Some might be uncomfortable with the notion of an absolute threshold and believe that nature varies continuously. Furthermore, one might believe that inter-individual variability is a universal feature of biology: if an absolute threshold were to exist, it would not be the same for everyone. These ideas have led some to propose a more flexible class of models with the property that the slope of the dose–response is zero at $Z = 0$ and the transition to a positive linear relationship is continuous. One way to accomplish this is to propose that the threshold τ is an unobserved, latent variable for each individual, having some distribution $f_\theta(\tau)$. Then the observable, marginal dose–response is obtained by integrating over the unobserved person-specific thresholds:

$$R(Z; \beta, \theta) = \int_0^\infty r(Z; \beta, \tau) f_\theta(\tau) \, d\tau = 1 + \beta \left(Z F_\theta(Z) - \int_0^Z \tau f_\theta(\tau) \, d\tau \right)$$

where $F_\theta(\tau)$ denotes the cumulative distribution function of $f_\theta(\tau)$. Taking the normal or gamma distribution for $f_\theta(\tau)$, it is possible to derive closed-form expressions for this integral (see Chapter 15). As above, θ can be estimated jointly with β by maximum likelihood.

Estimates and tests of τ (or μ) are seldom sufficient by themselves to provide compelling evidence about the existence or not of a threshold. It is therefore helpful to superimpose the fitted models on some kind of plot of the observed data. A simple way to do this is just to form a fine stratification of dose and estimate the relative risk and confidence limits for each category. Figure 6.2 illustrates the categorical estimates for all solid concerns combined among the atomic bomb survivors, together with a smoothed dose–response curve (with confidence bands) and linear and linear-threshold models. Visually, there is no evidence for departure from low-dose linearity, and this is supported by the likelihood-based tests and confidence limits from a linear threshold models. This analysis shows that any threshold greater than 0.06 Sy can be rejected, even though none of the individual data points in this region are statistically significant. This illustrates the trade-off between resolution on the dose axis

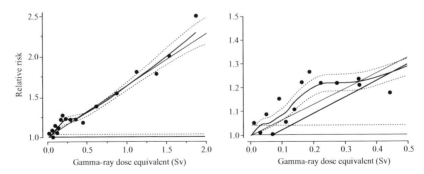

Figure 6.2. Dose response for all solid cancer mortality combined in the atomic bomb survivors, 1958–1994: left panel, entire dose–response relationship; right panel, enlargement of the region below 0.5 Sv, with upper confidence limit on a linear threshold model shown in gray. (Reproduced with permission from Pierce et al. 2000.)

and statistical precision on the risk axis: the more categories we form, the more detail we can see about the possible location of a threshold, but the more unstable each estimate becomes. It is rather like Heisenberg's Uncertainty Principle—we cannot simultaneously know both the location of a threshold and the magnitude of the risk in the immediately surrounding neighborhood!

The debate over risks from low-level exposures has been particularly intense in the radiation field. Despite the apparent linearity of the dose–response demonstrated above, several groups of scientists (Academie des Sciences 1997; Radiation Science and Health 1998) with strong connections to the nuclear industry have argued in support of the hypothesis that low-levels of ionizing radiation are actually protective ("hormesis"), based on a combination of evolutionary biology and mechanistic theories about DNA repair. Various authoritative bodies have reviewed the evidence in support of this hypothesis and concluded that "the assumption that any stimulatory hormetic effects from low doses of ionizing radiation will have a significant health benefit to humans that exceeds potential detrimental effects from the radiation exposure is unwarranted at this time" (NAS 2006). Doll (1998) provides a particularly thoughtful discussion of this commentary. See also (Little and Muirhead 1996; Hoel and Li 1998; Little and Muirhead 1998; Little 2000; 2002; Baker and Hoel 2003; Little 2004b) for further analyses of the evidence for threshold effects among the atomic bomb survivors.

To get around the "uncertainty principle" problem, a rich literature on flexible modeling strategies has evolved. A very simple strategy is to use some form of moving average, such as

$$\bar{r}(Z) = \sum_{|Z'-Z|<\delta} \hat{r}(Z')$$

a simple average of the estimated relative risks with some window of size δ around each value of Z, or better inversely weighted by their respective variances and/or by some function of $|z - z'|$. Instead, however, we will focus on a particular class of flexible models known as "splines." There are many variants of this approach, including natural splines, regression splines, basis or B-splines, and LOESS smoothers. Generalized additive models (GAMs) allow several such functions to be combined within the general framework of GLMs as $E[f(\mathbf{Z})] = \Sigma_j s_j(Z_j)$, where $s_j(Z_j)$ represents a flexible model (see (Hastie and Tibshirani 1990; Green and Silverman 1994) for comprehensive treatments). We illustrate the general approach here with cubic splines. We have already encountered one form of spline above, the linear spline (Figure 6.1) used to introduce the concept of a threshold. Cubic splines generalize this in two ways: first, by using cubic polynomials rather than linear functions, as it yields a function that is not only continuous but also has continuous first and second derivatives, conforming better to our intuitive notion of "smoothness"; second, rather then a single threshold τ, we allow as many change points (known as "knots") as needed to provide a good fit to the data. The general form of a cubic spline can be written as

$$r(Z) = 1 + \beta_1 Z + \beta_2 Z^2 + \beta_3 Z^3 + \sum_{k=1}^{K} \gamma_k (Z - \tau_k)_+^3$$

Since each of the terms $(Z - \tau_k)_+^3$ is zero at their respective knot τ_k, it is easy to see that the resulting function $r(Z)$ is continuous. Likewise, since their first and second derivatives are also zero at the corresponding knots, $r(Z)$ will also have continuous first and second derivatives. In practice, the number and location of the knots is specified in advance, depending upon the degree of flexibility or wiggliness desired, and then only the βs need to be estimated. Note that this is no more difficult than any other multiple regression problem, since $r(Z)$ is linear in the coefficients β and γ.

How wiggly is too wiggly? If, for argument sake, one were to put a knot at every data point, then the spline would fit the data perfectly but would be so wiggly that it would be impossible to detect any patterns. To overcome this problem, one can impose a smoothness penalty. If we define smoothness at any point as the degree of curvature, defined as the second derivative $r''(Z)$, then a natural definition of overall smoothness is its integral (or more precisely, the integral of the squared second derivative, since curvature can be positive or negative, but we care only about is magnitude),

$$S(\beta, \tau, K) = \int [r''(Z)]^2 \, dz$$

This integral is easily computed, since $r''(Z)$ is simply a piecewise linear function. Now if we let $\ell(\beta, \tau, K)$ denote the log likelihood function derived from $r(Z)$, we can form a "penalized" loglikelihood by subtracting the quantity $\sigma S(\beta, \tau, K)$ and maximize this function with respect to β, τ, and K for whatever degree of smoothness penalty σ we wish to impose. Furthermore, although at first blush it would appear that the penalized log likelihood is monotonically decreasing in σ, one must remember that $\hat{\beta}$, $\hat{\tau}$, and $\hat{K}$ themselves depend upon σ, so the penalized profile likelihood is a nonlinear function of σ. Thus, it is possible to estimate the optimal smoothness penalty by choosing the number and location of the knots to maximize the penalized likelihood. In this way, these choices are not entirely arbitrary. Recent work by Robins et al. (2007) on semiparametric regression using higher-order influence functions appears promising as a means of deciding how many degrees of freedom to allocate to complex,

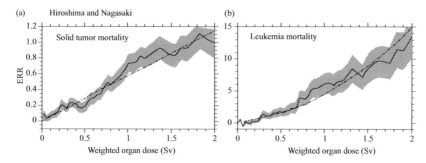

Figure 6.3. Nonparametric estimates of the dose–response relationships for solid cancers and leukemia in the atomic bomb survivors. (Reproduced with permission from Chomentowski et al. 2000.)

multidimensional confounders, while still yielding "honest" confidence limits on the main effect of interest.

Figure 6.3 illustrates nonparametric fits to the atomic bomb survivor data on leukemia and solid concerns using a sliding window approach, together with confidence bounds on the fitted curve. Note that the simple linear model for solid cancers and the linear-quadratic model for leukemia fit easily within these bounds all the way down to zero dose, although thresholds less than 0.25 Sv cannot be excluded for leukemia. A better sliding windows approach was illustrated in Figure 6.2, where each of the neighboring points was weighted by prior quadratic weights and the inverse of their variances. For large datasets like this, there is arguably little to choose between alternative fitting methods, which will generally give similar results. For smaller datasets, the use of LOESS or other spline methods may be preferable.

Splines have not been widely used in the radioepidemiologic literature, perhaps in part because of the strong radiobiological theory supporting the linear-quadratic cell-killing model. The approach has been very widely used the air pollution field, however, where there is no such theoretical basis for favoring any particular mathematical form for the dose–response relationship. Figure 6.4, for example, provides plots using the generalized additive model for daily mortality rates as a function of temperature and particulate pollution levels in the city of Birmingham, AL (Schwartz 1993). Whereas the relationship with temperature is highly nonlinear, with elevated death rates at very high and very low temperatures, the air pollution curve is remarkably linear, even at levels down to the lowest observed pollution levels, well below the current regulatory limit. In the following section, we will see how splines can also be used to describe the effects of time lags between peak exposure and peak mortality, as well as other temporal modifiers.

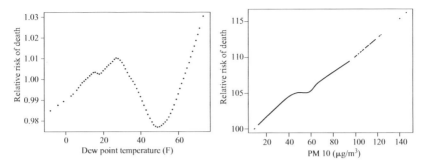

Figure 6.4. Fit of generalized additive models for dew point temperature (left) and PM10 to daily mortality data from Birmingham, AL. (Reproduced with permission from Schwartz, 1993.)

A joint analysis of time-series data on daily particulate pollution levels and mortality from the 20 largest U.S. cities in the NMMAPS project (Daniels et al. 2000) used a hierarchical model with cubic splines to describe the shape of the dose–response relationship (see Chapter 8 for a general description of time series methods). For cardiovascular and respiratory causes of death combined, there was no evidence of departure from a simple linear relationship, whereas for all other causes combined, a model with a threshold at 65 $\mu g/m^3$ provided a better fit. Schwartz and Zanobetti (2000) introduced a simple approach to combining nonparametric dose–response curves by inverse-variance weighting and similarly concluded that a linear relationship provided a good fit for total mortality down to the lowest observed levels in 10 U.S. cities. These methods have also been applied to European data from the APHEA project ((Schwartz et al. 2001; Samoli et al. 2003; 2005)), again confirming the linearity of the particulate-mortality relationship, even after controlling for a nonparametric smooth function of SO_2 and between-city heterogeneity.

Another flexible class of models that has recently received some attention in the epidemiologic literature is "fractional polynomial models" (Greenland 1995; Royston et al. 1999; Bagnardi et al. 2004; Royston and Sauerbrei 2005; Faes et al. 2007; Sauerbrei et al. 2007). These are basically polynomial regression models, but supplemented with terms of the form Z^{-2}, Z^{-1}, $Z^{-1/2}$, $\ln(Z)$, and $Z^{1/2}$, together with product terms of the form $Z^p\ln(Z)$, for greater flexibility. The model is generally implemented with some form of stepwise variable selection or model averaging, although more than two such terms are seldom needed. Logarithmic or negative power terms cannot be included if there are subjects with zero or negative values of Z unless some suitable constant is added. A better solution in this case is often to include an indicator variable for no exposure and use the fractional polynomials only for exposed subjects.

As should be evident by now, even splines are not truly nonparametric, taking one of the several specific, albeit highly flexible, mathematical forms, none motivated by biological considerations beyond a general notion that nature should be smooth. As noted earlier, a truly nonparametric model would be useless as a estimate of a dose–response relationship since it would merely reproduce the data, although it could be used as a test of case-control differences or to avoid making assumptions about other parts of the model, like time in a Cox model. If, however, one were willing to make the minimal assumption that the dose response were monotonic—never decreasing as dose increased—then one could use the technique of "isotonic regression" (Barlow et al. 1972). Here one can show that the best fitting monotonic relationship is a step function, with jumps at some subset of the observed covariate values. For case-control data, Thomas (1983a) showed that such jumps can occur only at cases' exposures. The model is fitted using the "pool adjacent violators" algorithm: starting with completely unconstrained step function with intervals defined by the cases' Z values, each time the estimated dose–response goes down, one pools the two adjacent categories, and continues in this manner until there are no further violations. For hypothesis testing purposes, one can show that the chi square statistic for heterogeneity among the final set of categories has a distribution comprised of a weighted average of chi square distributions with various degrees of freedom, the weights corresponding to the probability of observing the corresponding numbers of categories under the null hypothesis of a completely flat dose–response. These probabilities are easily worked out by a combinatorial algorithm and have been tabulated (along with the critical values of the weighted chi square distribution) in Barlow et al.'s textbook. As an example, Figure 6.5 shows the estimated monotonic dose–response for asbestos and lung cancer, showing good agreement with the fit of a simple linear relationship ($p > 0.079$) but not with the loglinear model ($p > 0.0059$; note that these are lower bounds on the goodness-of-fit probability, for reasons explained in Thomas (1983a)).

Neither isotonic regression nor splines are limited to binary data, the theme on this chapter. Barlow et al., for example, discuss applications to continuous outcome data, for which the corresponding tests become mixtures of F-distributions. Of course, there are also situations where one might not wish to assume monotonicity, such as for high-dose radiation. Figure 6.6, for example, illustrates the dose–response for bone cancer among the radium dial painters. The best fitting model for absolute risk, constant after 5 years latency, is a quadratic cell-killing model, $\lambda(t) = (\alpha + \beta Z^2) \exp(-\gamma Z)$, which shows a marked down-turn in risk above about 20 Gy (2000 rad). This model fits significantly better than any model without the cell-killing term ($p = 0.05$ for a linear-quadratic model) or a linear cell-killing model ($p = 0.03$). Similar patterns have been observed for leukemia

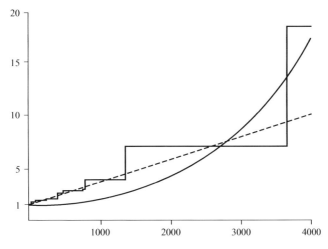

Figure 6.5. Dose–response for lung cancer and asbestos estimated by isotonic regression, compared with linear and loglinear parametric models. (Reproduced with permission from Thomas 1983a.)

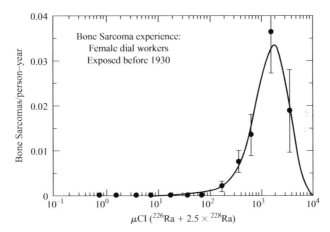

Figure 6.6. Bone sarcomas in the female radium dial painters. (Reproduced with permission from Rowland et al. 1978.)

among women irradiated for cervical cancer (Boice et al. 1987), as will be discussed below in the context of dose heterogeneity.

Both splines and isonomic regression also have multivariate generalizations. We will defer discussion of these approaches to the following section.

Incorporating temporal modifiers

Previously we noted that any statement about effect modification (inter-action) is necessarily scale-dependent: a factor that modifies the relative risk may not modify the excess absolute risk and vice versa, or may do so in opposite directions. Another way of saying this is that statistical interaction represents a departure from some model for the main effects of the factors, and the need for incorporating such statistical interactions will depend upon the form of the main effects model used. Thus, under the principle of parsimony, we might well prefer a main effects model that avoids the need to add interaction terms over some other model that requires them to obtain a decent fit. Here we consider a particular form of interaction—between a single exposure variable and one or more temporal modifying factors—and defer a general treatment of interactions between multiple exposures or between genes and exposure to Chapter 12.

There are a number of different temporal factors that could modify a dose–response relationship: age at exposure, age at risk of disease (here-after called "attained age"), time since exposure ("latency"), and calendar year of birth, exposure, or risk. Of course, for extended exposures, dura-tion, time at or since first or last exposure also become important, as discussed in the following section, but for now we focus on an instanta-neous exposure. Many exposures are more hazardous at younger ages, such as iodone-131 for thyroid cancer. Most chronic diseases have some minimum latent period following exposure before there is any increase in risk (often quite long in the case of solid cancers). The excess or relative risk may then rise to a peak, after which it may decline; it may go all the way back to background levels eventually or remain somewhat elevated indefinitely. Calendar year of exposure can be a useful surrogate for dose when it is impossible to assess it directly. For example, cancer among radi-ologists declined as a function of year first employed, reflecting increasing recognition of the hazards and improvements in technique (Court Brown and Doll 1958; Berrington et al. 2001).

What makes assessment of modification by temporal factors challenging is the interrelationships: attained age = age at exposure + latency; year of death = year of birth + attained age; time since first exposure = dura-tion of exposure + time since last exposure; and so on. Similar problems arise in age-period-cohort analysis in descriptive epidemiology (Holford 2006). Typically, one could consider any one of those variables alone, but their apparent modifying effects could be confounded by some other variable. Likewise, one could consider any pair of variables in one of these triplets, but not all three simultaneously because of their multicollinear-ity. Since our objective is generally to describe an exposure–time–response relationship rather than to draw inferences about any particular modifier, it suffices to choose some subset of the variables that fit the data well, but sometimes one subset may provide a more parsimonious description than another, for example latency alone rather than both age at exposure and attained age.

In most instances, it is reasonable to assume that at no point in time is the risk associated with a hazardous exposure ever protective. (An exception might be when exposure advances some events in time, leaving a trough in its wake, as in the "harvesting" of deaths among the most chronically ill by air pollution (Chapter 8). There was no evidence for this following the London Fog, where death rates remained elevated for weeks after the peak of pollution (Brunekreef and Hoek 2000), but it has been claimed to occur following other air pollution episodes (Martin 1964).) This assumption motivates a model for exposure–time–response relationships in which time modifies specifically the excess risk part of dose–response. The general relative risk form used in the EPICURE package discussed earlier, $\lambda(t, Z, W) = \lambda_0(t)[1 + \beta Z \exp(\alpha'W)]$, lends itself naturally to this type of assumption, where t represents attained age, Z exposure, and W one or more a temporal modifying factors. Thus, the slope of the dose–response relationship is $\beta \exp(\gamma'W)$, which is always positive for any value of W (provided $\beta > 0$).

Consider, for example, modeling latency. If one wished to allow risk to peak at some point, one could include a quadratic function in W. It might be convenient to parameterize the model as $\exp[-\gamma_1(W - \gamma_2)^2]$ or $\exp[-\gamma_1 \ln^2(W/\gamma_2)$, so that γ_2 represents the time of maximum relative risk after exposure, and γ_1, measures how quickly risk decreases before or after that peak. Of course, multiple temporal factors can be incorporated in this general regression framework, including interactions among them. For example, one might ask whether those exposed at younger ages or higher doses experienced shorter latencies. For leukemia, these questions were examined by the BEIR V committee (NAS, 1990) by adding interaction terms between latency and age at exposure or between latency and dose, and revealed strong modifications by age, but none by dose (Figure 6.7). The time to the maximum excess RR/Gy increases slightly with age at exposure from less than 5 years (the start of follow-up) among those exposed at age 5 to about 8 years among those exposed at age 40. Although the overall modification of level of relative risk with age is quite significant, this modification of the location of the peak is not ($p = 0.13$), and there is no trace of a modification of the peak latency by dose (i.e., higher doses having shorter latency). Lacking any data on the first five years after exposure (and having only data grouped into five-year intervals thereafter) severely handicaps any analyses of the time to peak risk, however.

The model shown in Figure 6.7 is fully parametric in dose and in latency. One could use flexible models for describing either or both. Consider the

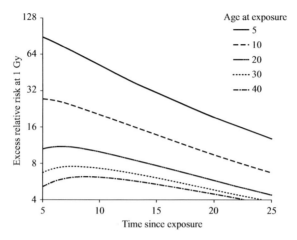

Figure 6.7. Fitted excess relative risks at 1 Gy for leukemia mortality among the atomic bomb survivors. (Based on data from LSS report 11.)

following special cases:

$$RR = 1 + \beta Z \exp[-\gamma_1(W - \gamma_2)^2] \qquad (6.2a)$$

$$RR = 1 + s_1(Z) \exp[-\gamma_1(W - \gamma_2)^2] \qquad (6.2b)$$

$$RR = 1 + \beta Z \exp[s_2(W)] \qquad (6.2c)$$

$$RR = 1 + s_1(Z) \exp[s_2(W)] \qquad (6.2d)$$

$$RR = 1 + \exp[s(Z, W)] \qquad (6.2e)$$

Equation (6.2a) is the fully parametric form shown in Figure 6.7. Equations (6.2b) and (6.2c) replace one part of the model or the other by a one-dimensional flexible model, such as a cubic spline. Equation (6.2d) treats both parts flexibly, but assumes they combine multiplicatively. Equation (6.2e) relaxes this last assumption to allow for any form of joint effect dose and latency.

These various models are compared in Figures 6.8 and 6.9. Both factors are highly nonlinear, although the downturn in dose–response occurs at doses over 2.5 Gy where there are few cases, and likewise the increasing risk in the 5–10 year latency is based on only a single stratum of time with relatively few cases. Modifying the form of the dose–response has little influence on the latency curve and vice versa, so that modeling both factors flexibly, as in Eq. (6.2d) yields estimates of $f(Z)$ and $g(W)$ that are very similar to those shown in for each factor separately (with the other modeled parametrically, as in Figure 6.8). Their joint relationship is shown in Figure 6.9. The fit of the multiplicative model [Eq. (6.2d)] and the two-dimensional spline [Eq. (6.2e)] are virtually identical, as indicated by the

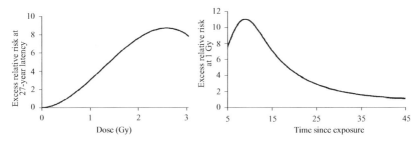

Figure 6.8. Flexible dose–response and latency curves for leukemia in the atomic bomb survivors. Left panel: cubic spline for dose–response with a single knot at 3 Gy, treating latency parametrically with terms for $\ln(E)$ and $\ln(T)$ in the loglinear modifier in Eq. (6.2b). Right panel: cubic spline for latency with a single knot at 15 years, treating dose–response as linear and a term for $\ln(E)$ in the loglinear modifier in Eq. (6.2c). (Based on data from LSS report 11.)

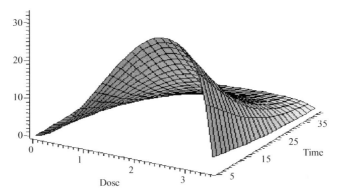

Figure 6.9. Dose–time–response relationship for leukemia in the atomic bomb survivors using cubic splines in both dose and latency, with $\ln(E)$ in the loglinear modifier, assuming a multiplicative effect of the two factors, Eq. (6.2d). The fit of Eq. (6.2e) using a two-dimensional spline is visually indistinguishable, as supported by the likelihood ratio tests in Table 6.2.

comparisons of deviances in Table 6.2: the last two lines corresponds to a change in deviance of only 0.11 on 2 df. Indeed, of the various possible comparisons of nested alternative models, only the addition of a quadratic term in the dose–response part of the model is statistically significant (e.g., 5.43 on 1 df in the models with a linear term for latency). Ulm (1999) used two-dimensional isotonic regression to describe the joint dependence of cancer risk on total exposure and time since first exposure to "particles not otherwise specified."

A particularly interesting temporal interaction involves duration of exposure or its inverse, dose rate: for the same total exposure, is a long, low dose–rate exposure more or less hazardous than a short, intense one?

Table 6.2. Deviances for alternative parametric, semi-parametric, and full two-dimensional spline models for leukemia in the atomic bomb survivors.

Dose–response model $f(Z)$	Latency model $g(W)$	Deviance (number of parameters)
Linear	Linear	1744.21 (8)
Linear-quadratic	Linear	1738.78 (9)
Cubic spline	Linear	1736.37 (11)
Linear	Linear-quadratic	1743.58 (9)
Linear-quadratic	Linear-quadratic	1738.17 (10)
Linear	Cubic-spline	1742.16 (11)
Linear-quadratic	Cubic-spline	1736.96 (12)
Cubic spline	Cubic spline	1735.14 (14)
Two-dimensional spline		1735.03 (16)

For ionizing radiation, it turns out that the answer is different for high- and low-LET radiation. Such an analysis needs to be carefully controlled for the potential confounding effects of the other temporal variables, however. Before we can address this question, however, we need to consider in general how to go about modeling risks from extended time-varying exposures, which we will do in the next section.

Before leaving this topic, we need to consider some fallacies of certain naïve analysis. Suppose, for example, that one tried to study latency by simply plotting the distribution of intervals between exposure and disease among cases. What this fails to acknowledge is that the distribution is "censored": as follow-up of the cohort increases, more and more long-latency cases will occur. Furthermore, if one wished to make comparisons between cases with different ages at exposure, differential censoring could lead to biased comparisons. This simply emphasizes the importance of using appropriate statistical methods for censored survival data for this purpose. Even so, confounding by other temporal dimensions can distort inferences about latency from either cohort or case control studies. These problems have been discussed in greater detail elsewhere (Thomas 1987; Thomas 1988).

Extended exposure histories

Few exposures in environmental epidemiology are even approximately instantaneous like the radiation from the atomic bombs or certain accidents like those discussed in the disasters section of the final chapter. Most are chronic exposures, often extending over one's entire lifetime.

Many also vary in intensity over time, although often it may be difficult, if not impossible, to quantify this temporal variation: at best one can estimate each individual's or group's average lifetime exposure rate. In other circumstances—tobacco smoking comes to mind—one may be able to determine the age at first and last exposure and the average level of exposure (e.g., cigarettes per day), but not all the variations in level or dates of intervening periods without exposure. Studies of residential exposures (e.g., radon, air pollution, electromagnetic fields) typically obtain complete residence histories (dates and addresses) and attempt to characterize the exposure levels of each home. Usually this is more easily done for the current residence (and possibly a few other recent or longest occupied); the remainder must be imputed in some way, perhaps with an average based on location or some kind of a spatial prediction model, as discussed in Chapter 9. In all these situations, let us denote the exposure history as $\mathcal{Z}(t) = \{z(u)\}_{u < t}$, where $z(u)$ denotes the exposure intensity at age u. Note that this notation, $\mathcal{Z}(t)$ represents the entire history, not just some summary exposure variable that will be used in the analysis.

Before proceeding further, we must confront the fundamental question of how the joint effects of intensity and duration of exposure combine. As we shall see, this question is intimately tied up with the choice of scale for excess risk. Consider first the case of a lifelong exposure at a constant rate z. Do we expect the excess due to exposure to increase with age or remain constant? That depends, amongst other things, on whether we are talking about the absolute or relative excess. Recall that in this situation, age and duration of exposure are the same. Thus, if we think of the natural increase in population cancer rates to be a response to a lifetime of exposure to a "sea of carcinogens" at some essentially constant rate z, then it would seem reasonable to assume that the excess due to an incremental exposure would be proportional to it that is, $\lambda(t, z) = \lambda_0(t) + \beta z \lambda_0(t) = \lambda_0(t)[1 + \beta z]$, that is, a constant relative risk, proportional to intensity z, not cumulative exposure $Z = zt$. Here, multiplication by t is not needed because our multiplication by $\lambda_0(t)$ serves the purpose of describing the biological response to a lifetime of exposure. If background disease rates increase rapidly with age (say, exponentially or as some power of age, like most solid cancers), then why should we expect the excess due to exposure to increase only linearly?

Of course, the biological mechanisms responsible for background cases might be different from those causing the exposure-induced cases: perhaps they are due to genetics, for example. If we think of the two as completely disjoint (no interaction between exposure and background causes), then it might make more sense to consider an excess risk model, of which the simplest form might be a excess proportional to cumulative exposure, $\lambda(t, z) = \lambda_0(t) + \beta zt$. Here, cumulative exposure zt is simply a stand-in for our ignorance about the basic form for the biological effect of exposure.

The excess could be proportional to some more complex function of age, say $zf(t)$; or it might be nonlinear in intensity but linear in age, say $g(z)t$, or nonlinear in both, say $g(z)f(t)$; or age and intensity might even combine in some non-multiplicative fashion, say $h(z,t)$. (Of course, the same could be said of relative risks.)

Which is the most plausible model also depends upon how we think the effects of each increment of exposure are modified by temporal factors, as discussed in the previous section, and how they affect each other. To consider an extreme situation, suppose we assume that the effect of exposure at each age is manifest after a latent period of exactly τ. Then the excess at attained age t would depend only at the exposure rate at age $t - \tau$ and would be independent of duration of exposure. Depending upon our choice of scale for excess risk, a constant lifelong exposure would then yield a dose–response either of the form $\lambda(t, z) = \lambda_0(t)(1 + \beta z)$ or $\lambda_0(t) + \beta z$.

More realistically, we might expect the effect of exposure at age u to be spread out over some range of later ages, say with probability density $f(\tau)$. Thus, the excess at attained age t would represent the cumulative effects of insults received at ages $t - \tau$ with weights $f(\tau)$. (For simplicity, we assume for now that this weight function is not modified by age at exposure or attained age.) But how exactly do the effects of past exposures combine? The simplest assumption would be independently, i.e., that the excess is simply a function of the (weighted) sum of the effects of all past exposures,

$$\beta z \int_0^t f(t - u) \, du = \beta z F(t)$$

where $F(t)$ denotes the cumulative distribution function of the latency distribution.

But what if the contributions of each prior increments of exposure were not independent? What if each exacerbated the damage done by prior exposures? Then we might think of the rate of change of risk at age t being proportional to the product of the accumulated risk and dose at that age, leading to the differential equation

$$\frac{d\lambda(t)}{dt} = \beta \lambda(t) z(t)$$

whose solution is

$$\lambda(t) = \exp\left(\beta \int_0^t z(u) \, du\right) = \exp(\beta z t)$$

in other words, an exponential function of cumulative exposures. (This derivation ignores latency and other temporal modifiers, but the basic result would be similar.)

With these preliminaries behind us, we can now turn to the general case of time-varying exposures. Suppose we adopt the basic assumption of additivity of effects of past exposures and assume that each effect is modified by some function of age at or time since exposure $f(t, u)$. Then we obtain an excess risk model of the form

$$\lambda(t) = \lambda_0(t) + \beta \int_0^t z(u) f(t, u)\, \mathrm{d}t$$
$$= \lambda_0(t) + \beta \overline{Z}(t)$$

where $\overline{Z}(t)$ represents the time-weighted cumulative exposure given by the integral. If $f(t, u)$ were constant, then this would be simply cumulative dose zt.

For the reasons discussed above, it is more appealing to model relative risks, under the assumption that the effect of exposure parallels that of background causes. Assuming that background causes have been essentially constant over time (without loss of generality, we can set them to unit intensity) and assuming that the same temporal modifying function $f(t, u)$ applies to both, we obtain

$$\lambda(t) = \int_0^t [1 + \beta z(u)]\, f(t, u)\, \mathrm{d}u$$

However, $\int_0^t f(t, u)\, \mathrm{d}u = \lambda_0(t)$, so we can rewrite this model in terms of $\lambda_0(t)$ as

$$\lambda(t) = \lambda_0(t) \left[1 + \beta \overline{Z}(t)/\lambda_0(t)\right]$$

where $\overline{Z}(t)$ is again the time-weighted cumulative exposure given by the integral in the excess risk model, but with the additional constraint that $f(t, u)$ must equal the derivative of $\lambda_0(t)$ with respect to t for all u.

Whatever form of model we adopt, it is fitted in the same way as described in the previous section by maximum likelihood. If we are prepared to specify the weight functions $f(t, u)$ exactly, say $f(t, u) \equiv 1$ yielding $\overline{Z}(t)$ as simple cumulative dose, then we are left with a model involving only one free parameter, β. Or $f(t, u)$ could be some parametric function involving parameters to be estimated, like those discussed in the previous section. Conceptually, this raises no new issues, but then fitting is more complicated as the likelihood requires integration of this function against $Z(t)$. This is easily done numerically, however.

Finally, it is also possible in principle to treat $f(t, u)$ or $g(z)$ nonparametrically in the spirit of Eqs. (6.2b–e), for example,

$$\lambda(t, z(t)) = \int_0^t s_1(t - u)[1 + s_2(z(u))]\, \mathrm{d}u$$

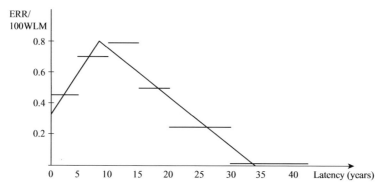

Figure 6.10. Bilinear model for latency in the Colorado Plateau uranium miner data. (Reproduced with permission from Langholz et al. 1999.)

where s_1 and s_2 represent one-dimensional spline functions in latency and exposure intensity respectively. For example, Langholz et al. (1999) described a simple approach to modeling latency effects for extended exposures using linear splines, as shown in Figure 6.10. Here, there are three free parameters to be estimated: (τ_0, τ_1, τ_2), the minimum, peak, and maximum latency respectively. Fitting is done by evaluating the likelihood over a grid of these three parameters, maximizing each with respect to the slope parameter β for "effective" dose $Z(\tau_0, \tau_1, \tau_2)$. In their application to the data from the Colorado Plateau uranium miners cohort, they found the best fit with $\tau_0 = 0$ years, $\tau_1 = 8.5$ years, and $\tau_2 = 34$ years. Subsequently. Hauptman et al. (2001) reanalyzed these data using splines and found that the smoothed latency curve was consistent with elevated risks between 9 and 32 years after exposure; these estimates were not significantly modified by attained age, duration of exposure, dose–rate, or smoking.

Before concluding this section, let us revisit the question of intensity and duration effects for ionizing radiation. It has long been known that the same total dose of low-LET radiation (e.g., X-rays) delivered in multiple fractions has a lower risk than if delivered all at once. This phenomenon is generally ascribed to the possibility of repair of single-strand breaks induced by one fraction before the later fractions are delivered. In classical radiobiological theory, if the dose–response is a linear-quadratic function representing the probability of a double-strand break being induced by a single or two concurrent independent traversals respectively, then as exposure duration becomes small, the quadratic term will disappear, since the probability of two traversals occurring close enough in time for repair not to have occurred in the interim will vanish. This is the basic principle underlying the use of fractionated doses in radiotherapy.

Curiously, exactly the reverse phenomenon is seen with high-LET radiation in both human data and various *in vitro* systems. Among the uranium miner cohorts, for example, a long low-intensity exposure appears to be more, not less, hazardous than a short intense one for the same total exposure. (There is some inconsistency in this finding across studies, however; for example, Grosche et al. (2006) found this effect only at high exposures.) The biological mechanism for this phenomenon is not well understood, but "bystander" effects (Curtis et al. 2001) are thought to offer a possible explanation (see Chapter 13).

From the statistical modeling perspective, one could in principle think of both phenomena as forms of effect modifications by either intensity or duration. However, it is more appealing to model this in terms of intensity modifications, as the alternative would entail describing the instantaneous effect of an increment of exposure as depending on past or future events, which gets more complicated. It is not biologically implausible, however, as the earlier discussion of potentiation of one exposure by later exposures revealed.) Thus, a simple way to incorporate intensity and duration effects would be by allowing a nonlinear function of intensity, for example $\int_0^t g[Z(u)]\,\mathrm{d}u$ (possibly modified by $f(t, u)$ as above). For example, if we take $g(z) = z^\alpha$, then if $\alpha > 1$, we get the sparing affect of fractionation seen with low-LET radiation, and conversely if $\alpha < 1$, the increased risk for protracted exposures seen with high-LET radiation. Alternatively, if we wished to model the phenomenon in terms of potentiation, we might consider a model involving in terms of the form $\int_0^t \int_0^u z(u)z(v)\,\mathrm{d}v\,\mathrm{d}u$, again possibly involving weights depending on u, v, and t. See Thomas (1988) for further discussion of such models. Finally, it is worth noting that the apparent modifying effects of intensity and duration could also be accounted for by exposure measurements errors having a bigger effect on intensity estimates than duration. This possibility will be discussed further in Chapter 11, but briefly Stram et al. (1999), reanalyzing the U.S. uranium miner data showed that the effect of low dose-rate exposures was relatively unchanged by measurement error correction, while high dose–rate effects increased, thereby somewhat reducing the observed dose–rate effect.

Modeling temporal modifiers for tobacco smoking

Without doubt, tobacco smoking is the strongest risk factor for many chronic diseases, above all for lung cancer. It has been well established by numerous epidemiologic cohort and case-control studies that the risk increases with intensity and duration of exposure and is strongly modified by such factors as age at starting and time since quitting. Given this wealth of data, it is not surprising that there is a large literature on methods of modeling the risk of lung cancer in relation to lifetime history of smoking (Hammond 1966; Doll and Peto 1978; Peto 1986; Whittemore 1988; Moolgavkar et al. 1989; Freedman and Navidi 1990; McKnight et al. 1999; Peto et al. 2000; Leffondre et al. 2002; Flanders et al. 2003; Dietrich and Hoffmann 2004; Knoke et al. 2004; Rachet et al. 2004; Vineis et al. 2004; Hazelton et al. 2005; Thurston et al. 2005; Leffondre et al. 2006; Lubin and Caporaso 2006; Lubin et al. 2007a; 2007b). Whether or not one views tobacco smoking as an "environmental" exposure like the various external exposures that this book is primarily concerned with, much can be learned about methods for modeling exposure–time–response relationships for extended exposures from this experience. Smoking has also been shown to interact strongly with other risk factors like asbestos (Selikoff et al. 1968), radiation (Prentice et al. 1983; Thomas et al. 1994; Hornung et al. 1998), air pollution (Xu and Wang 1998), and arsenic (Hazelton et al. 2001; Chen et al. 2004). Given its potential as a strong confounder or modifier of associations with other external agents, it is thus essential that smoking effects be modeled carefully even in studies where other factors are of primary interest.

Using data from the British doctors study, Doll and Peto (1978) showed that the lung cancer rate was proportional to the square of the number of cigarettes smoked per day ($+6$) and years of smoking (-3.5) raised to the 4.5 power. The addition of 6 to smoking intensity was needed to account for the rate in nonsmokers due to other factors and the subtraction of 3.5 years was intended to allow for the lag between the appearance of the first fully malignant cell and death. A recent modification of the model (Knoke et al. 2004), fitted to the American Cancer Society cohort study added age at starting or attained age as an additional multiplicative factor:

$$\lambda(t, Z, t_0) = 2.21 \times 10^{-13}(Z + 6)^{1.02}(t - t_0 - 3.5)^{2.35} t^{2.68}$$

where Z denotes cigarettes per day, t_0 age at starting, and t attained age (alternative models replace duration in the second factor or attained age in the last factor by age at starting, yielding similar fits). Although such results have been widely interpreted as duration having a much more important effect than intensity, it must be remembered that this is a model for *absolute* risk, not *relative* risk. The baseline risk in nonsmokers in the ACS cohort is best fitted by $\lambda_0(t) = 5.29 \times 10^{-13} (t - 3.5)^{4.83}$ (Knoke et al. 2004; Thun et al. 2006). The resulting model for excess relative risk per pack-year (PY) can thus be expressed as

$$\frac{ERR}{PY} = \left(\frac{\lambda(t, Z, t_0) - \lambda_0(t)}{\lambda_0(t)[Z(t - t_0)]}\right) \cong c \frac{(t - t_0 - 3.5)^{2.35} t^{2.68}}{(t - 3.5)^{4.83}(t - t_0)}$$

This rather complex function is plotted in Figure 6.11, showing that the risk per pack-year increases in a nonlinear fashion, eventually declining after an amount that is roughly double the age at starting.

Lubin et al. (2006; 2007a, b) fitted models of the form $ERR = \beta Z \cdot (t - t_0) \cdot \exp[g(Z, t-t_0)]$ to a variety of other datasets, where $g(Z, t-t_0)$ included

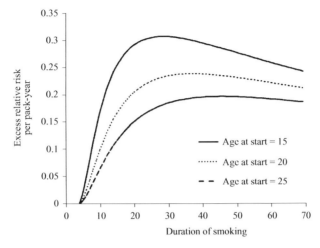

Figure 6.11. Predicted excess relative risk per pack year for continuing smokers from the Knoke et al. (2004) model as a function of duration of smoking at age at starting.

various transformations of intensity or duration, the best-fitting model being a linear-quadratic function of $\ln(Z)$. Thus, they concluded that the *ERR per pack-year* increased with increasing intensity up to about 20 cigarettes per day, then declined with further increases in intensity above that amount.

Most of these models were fitted to data only on continuing smokers. Patterns of absolute risk for ex-smokers are more complex, generally starting to level off or even declining somewhat shortly after quitting, gradually approaching but never attaining the risk for never smokers (Freedman and Navidi 1990; Peto et al. 2000), so that the *relative* risk declines with time since quitting. While it may be tempting simply to include various temporal modifying factors in a logistic or Cox regression model, this approach can lead to strange results. For example, a relative risk model of the form $\exp[\beta_1 Z + \beta_2 (t - t_0) + \beta_3 Z \cdot (t - t_0) + \ldots]$ (possibly including additional terms for time since quitting) would imply an increasing risk with intensity (β_1) even for individuals with miniscule duration and likewise an increasing risk with duration (β_2) even for individuals with miniscule intensity. Furthermore, while a relative risk model for pack-years of the form $\exp[\beta_1 PY + \beta_2 PY \cdot (t - t_1)]$ with β_2 negative does imply that risk declines with time since quitting, it will continue declining indefinitely, so that once $t - t_1 > \beta_1/\beta_2$, prior smoking will actually appear beneficial! Addition of a main effect of time since quitting does not really help, as it would imply a reduction in risk independent of the intensity of smoking, so that the point when ex-smokers would attain a risk lower than never smokers simply occurs sooner for lighter than for heavier smokers. Finally,

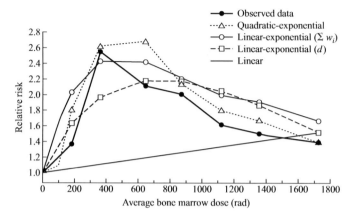

Figure 6.12. Dose–response relationships for leukemia following radiotherapy for cervical cancer. (Reprinted with permission from Boice et al. 1987.)

Table 6.3. Fit of linear cell-killing models with age and latency terms to the cervical irradiation data: cohort data includes only indicator variable for treated or not by radiotherapy; case-control data includes doses to 14 compartments of the bone marrow, but for radiotherapy subjects only; the combined analysis uses both data sets. (Adapted with permission from Thomas et al. 1992b.)

Data set	Dose	Cell-killing	Latency	Age at diagnosis	LR χ^2 (df)
Cohort	0.22	—	−1.17	−1.71	12.15
	(0.22)		(0.75)	(1.76)	(3)
Case-control	1.27	−5.11	−1.56	−10.86	9.83
	(2.89)	(3.44)	(1.37)	(7.63)	(4)
Combined	1.07	−3.88	−1.00	−2.02	19.28
	(1.34)	(3.90)	(0.42)	(1.35)	(4)

the interpretation of the modifying effects of age at starting, duration, time since quitting, and attained age is complicated by the fact that they are linearly related. For an age-matched case-control study, for example, the sum of the first three variables must be the same for any matched set, so any two of these three will lead to identical fits.

The fits of this model and various submodels are compared in Figure 6.12 (risks plotted against the average dose $\overline{Z} = \Sigma w_k Z_k$). Since risk is a highly nonlinear function of dose, the use of the average dose in $r(\overline{Z}, \beta)$ yields a much poorer fit to the data, although the likelihood ratio tests comparing the various models using only the case-control data are not significant.

Thomas et al. (1992b) later reanalyzed these data combining the detailed dose information from the nested case-control study with the radiotherapy present/absent comparison from the original cohort study. Since the case-control component included only radiotherapy patients, the dose range is restricted, and adding in the unirradiated patients considerably improved the power to distinguished alternative dose–response models (Table 6.3).

Modeling short-term fluctuations in exposure

While most of this chapter is concerned with the effects of long-term changes in exposure, short-term fluctuations—on the scale of biological processes governing tissue doses (hours, days)—can also be important if these processes are nonlinear. Chapter 13 will develop mechanistic models for such metabolic processes, but here we sketch out an empirical approach to this problem. Suppose the risk of disease as a function of exposure is, as described above $\int r[Z(t)]dt$, and we approximate $r(Z)$ by a Taylor series $r(\overline{Z}) + (Z - \overline{Z})r'(\overline{Z}) + (Z - \overline{Z})^2 r''(\overline{Z})/2 + \cdots$ Then substituting this expression into the integral over time yields an approximation $\beta_1 E(Z) + \beta_2 \mathrm{var}(Z)$. In other words, without having to specify the form of the dose-response in detail, one can test for the possibility of short-term nonlinearity simply by adding the temporal variance of exposure to a simple linear risk model as an additional covariate. One could further investigate the temporal scale of such nonlinear effects by decomposing the temporal variance into different time scales (e.g., hourly, daily, weekly, seasonal, annual) and adding each of these components to the model. Instantaneous risk models of the form $r \propto Z^\alpha(t)$ can generate "threshold-like" ($\alpha > 1$) or "saturation-like" ($\alpha < 1$) behavior for short-term exposure rates; Taylor series approximation yields metrics of the form $\int z(t) \ln[Z(t)]dt$ which can be added to a model containing $\overline{Z}$ and $\overline{Z}^2$ without having to create a whole family of metrics with specific thresholds. Indices of the form $\int Z(t)Z'(t)dt$ or $\int [Z''(t)]^2 dt$ can be useful for investigating the effect of rapidity of temporal fluctuations. In a similar manner, short-term interaction effects could be investigated by adding the temporal covariance between two exposure variables.

Analyses such as these of the CHS data revealed a significant effect of hourly, daily, and weekly variation in ambient ozone concentrations on FVC, where the mean level showed no assocation, whereas none of the variances in NO_2 had a significant effect beyond the mean (unpublished data). No threshold-like effects on dose-rates, lags, or autocorrelation effects for ozone were found, however.

Dose heterogeneity

So far, we have acted as if the dose to a target organ were homogenous. This may be a reasonable approximation in many instances, such as for whole-body irradiation from the atomic bombs or for lung doses from gaseous air pollutants. In other circumstances, however, there can be marked variation in local dose across a target organ. For example, women irradiated for cervical cancer receive extensive scatter radiation to the bone marrow, leading to increased leukemia risk. An international cohort study comparing cervical cancer patients who did and did not receive radiotherapy found 77 leukemias among the exposed group compared with 65.83 expected, with the highest risk of acute and myeloid leukemia occurring 1–4 years after diagnosis of cervical cancer (28 observed vs. 13.41 expected) (Day and Boice 1983). However, different portions of the bone marrow receive very different doses, some relatively low on the linear-quadratic portion of the dose–response curve, some quite high, well within the cell-killing range. As a result, a simple average dose across the entire bone marrow does not well describe the average leukemia risk. Boice et al. (1987) subsequently conducted a nested case-control study within the earlier international cohort (adding some additional centers) in order to retrieve radiotherapy records and estimate the doses Z to each of 14 compartments containing proportions w_k of the bone marrow. The overall leukemia risk was then computed as a weighted average of the risk to each compartment,

$$\lambda(t, Z) = \lambda_0(t) \sum_{k=1}^{14} w_k r(Z_k, \beta)$$

where

$$r(Z, \beta) = 1 + (\beta_1 Z + \beta_2 Z^2) \exp(-\beta Z)$$

Similar issues have arisen in other studies of second breast cancers in relation to radiotherapy for cancer in the contralateral breast (Boice et al. 1992; Bernstein et al. 2004) and brain cancers in relation to cell phone use (Cardis et al. 2007), as described at the end of the previous chapter.

7 Longitudinal models

Up this point, we have focused on disease incidence or mortality data—discrete events, which by definition can only occur once. Not all traits of interest to the epidemiologist are like that. Underlying any binary trait is likely to be a complex system of continuous or discrete changes developing over time. Diseases come and go, and their pattern of recurrence, particularly in relation to changes in the environment, may be of interest. Often these subtler changes in intermediate traits can shed light on biological mechanisms or, by being closer to the immediate environmental cause, prove to be a more sensitive endpoint. In this chapter, we discuss the analysis of biomarkers, physiological measurements, clinical symptoms, and other traits that vary over time, either as endpoints of interest in themselves or as surrogates for the development of some unobservable disease process.

We begin with the regression analysis of a continuous, normally distributed trait, measured at a single time point and its rate of change between two observations, and then generalize to repeated observations. For this purpose, the general linear model (GLM) framework developed at the end of Chapter 4 will prove helpful, and we extend it to the general linear mixed model (GLMM). This will provide a natural way to apply the framework to categorical outcomes and multivariate outcomes, including combinations of continuous and discrete data. We can then address questions like which change occurs first, which trait affects which, and how this helps predict the risk of some ultimate disease incidence or death. A final technical detail we will have to deal with is the problem of missing data introduced in Chapter 4. This can be simply a nuisance if it occurs randomly, but more difficult if the probability of missingness depends upon other measured or unmeasured variables, including the true state of the variable under study.

To illustrate these various approaches, we turn away from the examples of cancer incidence and mortality, which we have relied on in previous chapters, and use the data from the Children's Health Study (CHS), particularly the annual measurements of lung function, incidence of new diagnoses of asthma, and exacerbations of various bronchitic symptoms. One aspect of the CHS that we will defer to the following chapter is school absences. Although conceptually similar to other forms of binary longitudinal data, we shall see that such data on rapid fluctuations over time are better analyzed by the techniques of time-series analysis. Accordingly, in this chapter we are less concerned with deviations from a general temporal trend than with the evolution of the trend itself, what we shall call the "growth curve."

Continuous outcomes

Let Y now denote a continuous, normally distributed outcome variable. (More precisely, it is not the trait itself but the residuals—the deviations from some model for their mean—that will be assumed to be normally distributed.) In $\mathbf{Z} = (Z_1, \ldots, Z_p)$, we include all the variables used to predict Y, including the exposure variable(s) of primary interest and their various confounders and modifiers. As before, we let $i = 1, \ldots, n$ index the study subjects and introduce a second level of subscripts $k = 1, \ldots, K$ to denote the times t_k of observation. But we are getting ahead of ourselves: first let us review the standard linear regression model for single observations on each subject, say the baseline measurements in a follow-up study, as introduced in Chapter 4, to motivate the various extensions leading to the GLMM. (For this purpose, we can omit the second level of subscripts.)

Single observations

As noted in Chapter 4, the standard linear regression model requires four basic assumptions:

1. The dependence of the mean of Y and $\mathbf{Z}$ is correctly specified by a linear additive function.
2. The residuals $e_i = Y_i - \mathbf{Z}_i'\boldsymbol{\beta}$ are independent.
3. The residuals are normally distributed.
4. The residuals have constant variance σ^2.

The first of these assumptions can sometimes be relaxed by appropriate transformations of either Y or $\mathbf{Z}$ or both. For example, if the relationship were quadratic rather than linear, one could simply include a Z^2 term as an additional covariate; if multiplicative rather than additive, a log transformation of Y (and possibly $\mathbf{Z}$ if needed) would accomplish this. In both these cases, the dependence of $E[f(Y)]$ on $g(\mathbf{Z})'\boldsymbol{\beta}$ remains linear in $\boldsymbol{\beta}$, so the basic form of the estimator, its variance, and significance tests remain the same. Nonlinear regression models (meaning nonlinear in $\boldsymbol{\beta}$) are also possible and can be fitted by correctly specified maximum likelihood or least squares in a manner similar to the general relative risk models discussed in Chapter 6.

The last assumption—homoscedasticity of the residual variance—might be addressed by modeling the variance and the mean jointly, using either maximum likelihood or weighted least squares. Suppose one had a specific way of predicting the variance of each observation, say $\text{var}(Y_i|Z_i) = s_i^2$ without requiring any free parameters. Then the weighted sum of squares would take the form $\text{WSS}(\beta) = \Sigma_i (Y_i - Z_i'\beta)^2/s_i^2$ and the likelihood $L(\beta) \propto \exp(-\text{WSS}(\beta)/2)$. Minimizing $\text{WSS}(\beta)$ or maximizing $L(\beta)$ again leads to the same closed form solution,

$$\hat{\beta} = \left(\sum_i Y_i Z_i/s_i^2\right) \left(\sum_i Z_i' Z_i/s_i^2\right)^{-1}$$

or in matrix notation,

$$\hat{\beta} = (\mathbf{Y}'\mathbf{W}\mathbf{Z})(\mathbf{Z}'\mathbf{W}\mathbf{Z})^{-1} \quad \text{and} \quad \text{var}(\hat{\beta}) = n(\mathbf{Z}'\mathbf{W}\mathbf{Z})^{-1}$$

where $\mathbf{Y} = (Y_1, \ldots, Y_n)$, $\mathbf{W} = \text{diag}(s_1^2, \ldots, s_n^2)$, and $\mathbf{Z} = (Z_1, \ldots, Z_n)$, an $n \times p$ matrix.

Seldom, however, can one specify the residual variance exactly. One may know some components, specific to each observation, but there remains some residual variance due to imperfect specification of the means. Thus one might choose $\text{var}(Y_i|Z_i) = s_i^2 + \sigma^2$, where σ^2 is the unknown model misspecification variance. One might have no knowledge specific to each subject but only a general belief that the depends in some way on the predicted mean, for example, $\text{var}(Y_i|Z_i) = \exp[\alpha_0 + \alpha_1 \text{E}(Y_i|Z_i)]$. (The exponential form is convenient to constrain the variance to be positive.) Or one could adopt some general form that depends on each covariate separately, for example, $\text{var}(Y_i|Z_i) = \exp(Z_i'\alpha)$. In general, let us denote any of these choices by $\text{var}(Y_i|Z_i) = v(Z_i, \alpha, \beta)$ where $v(\cdot)$ is some specific function of involving additional variance parameters α and possibly also the means parameters β. The joint likelihood now takes the form

$$L(\beta, \alpha) = \exp\left(-\sum_{i=1}^{n} \frac{(Y_i - Z_i'\beta)^2}{v(Z_i, \alpha, \beta)} - \frac{1}{2} \ln v(Z_i, \alpha, \beta)\right)$$

which in general requires numerical methods to maximize. However, if one ignored the dependence of the variance on β and maximized the likelihood (or equivalently minimized WSS) with respect to β for any given α, then the weighted least squares estimator given above would result. One could then compute the residuals e_i for each observation and fit them to $\text{E}(e_i^2) = v(Z_i, \alpha, \beta)$ to estimate α, holding β fixed, and repeat this process with the new α until convergence. This procedure in general would not lead to the joint maximum likelihood estimate (MLE) of β and α, but nevertheless can still be shown to yield an estimator of β that is asymptotically normal with variance given by the "sandwich estimator", Eq. (4.6).

Although we used the score function derived from the likelihood (which involves the assumption of normality), we do not really require that assumption to posit the GEE as an estimator of $\boldsymbol{\beta}$. The mere fact that it comprises a sum of independent random variables is sufficient to establish the asymptotic normality of $\hat{\boldsymbol{\beta}}$ and derive its variance. Furthermore, it is not really even necessary that the residual variances v_i be correctly specified: an incorrect specification will lead to inflated residuals and hence a larger sandwich estimator, which will correctly reflect the full variance of $\hat{\boldsymbol{\beta}}$ under the misspecified model. That, in fact, is ultimately the justification for being able to ignore the possible dependency of v_i on $\boldsymbol{\beta}$. However if the residual variance is correctly specified, the GEE estimator will be nearly fully efficient, relative to the optimal MLE, and furthermore, it is possible to compute the optimal weights using a second estimating equation of the form

$$\sum_{i=1}^{n} \left[e_i^2 - v(Z_i, \alpha, \beta) \right] \left[\mathrm{var}(e_i^2) \right]^{-1} \frac{\partial v_i}{\partial \alpha} = 0$$

The asymptotic normality and sandwich estimates of the variance still require the assumption of independence, however. This can be overcome if the data can be organized into blocks of dependent observations that are mutually independent. For example, in a family study, one might treat each family as a vector of observations, the members of which are dependent, but different families are independent. More relevant to this chapter, we could take repeated observations on each subject as a vector of correlated values, but independent between subjects. The same basic form of GEE is still used, the only difference being that the Y_i are now vectors and W_i becomes the inverse of the matrix of covariances $\mathrm{cov}(Y_{ij}, Y_{ik} | Z_{ij}, Z_{ik})$ among observations within a block. Of course, this assumes that the entire data set is not just a single block of correlated observations, in which case the effective sample size is just one and asymptotic theory would not apply. We will return to this approach for treating longitudinal data in Chapter 8.

Paired observations

Use of multiple observations over time accomplishes several things:

- It shifts the focus of the analysis from *levels* of the outcome to *rates* of change.
- As a consequence, inferences are unconfounded by factors that affect between-individual differences in level, only by those that affect rates of change.
- The residual variance against which effects are tested is that *within* rather than *between* individuals, and will generally be smaller, leading to a more powerful test, depending upon how widely spaced the observation are; and

- By serving as replicates, the precision of estimates of level is also improved.

To develop these ideas more formally, consider the following linear model for pairs of observations on each subject:

$$Y_{ik} = \alpha_0 + \alpha_1 Z_i + (\beta_0 + \beta_1 Z_i + e_i^T)t_k + e_i^B + e_{ik}^W \qquad (7.1)$$

where $k = 1, 2$, $\text{var}(e_i^B) = \sigma_B^2$, $\text{var}(e_i^T) = \sigma_T^2$, and $\text{var}(e_{ik}^W) = \sigma_W^2$. Here, $\alpha_0 + \alpha_1 Z_i$ describes the mean of the trait at baseline $t = 0$ and $\beta_0 + \beta_i Z_i$ the mean rate of change per unit t. (For now, we assume Z is constant over time.) Then the paired differences are given by model

$$Y_{i2} - Y_{i1} = (\beta_0 + \beta_1 Z_i + e_i^T)(t_2 - t_1) + e_{i2}^W - e_{i1}^W$$

Thus, the residual variance of the differences is

$$\text{var}(Y_{i1} - Y_{i1}) = (t_2 - t_1)^2 \sigma_T^2 + 2\sigma_W^2$$

compared with the variance of a single measurement

$$\text{var}(Y_{ik}) = t_k^2 \sigma_T^2 + \sigma_B^2 + \sigma_W^2$$

or of the average of two measurements

$$\text{var}(\overline{Y}_i) = \overline{t}^2 \sigma_T^2 + \sigma_B^2 + \sigma_W^2/2$$

Now suppose we code the time scale so that $\overline{t} = 0$, that is, $t_1 = -t_2 = \Delta t/2$. Then $\text{var}(Y_{i2} - Y_{i1}) = (\Delta t)^2 \sigma_T^2 + 2\sigma_W^2$ and $\text{var}(Y_{ik}) = \sigma_B^2 + \sigma_W^2$. Thus, the variance of the difference will be smaller than the variance of a single measurement if $\sigma_W^2 + (\Delta t)^2 \sigma_T^2 < \sigma_B^2$.

Now the variance of $\hat{\beta}_1$ is given by $\text{var}(Y_{i2} - Y_{i1})/n(\Delta t)^2 \text{var}(Z_i)$ whereas the variance of $\hat{\alpha}_1$ estimated from a single observation per subject is $\text{var}(Y_{ik})/n \text{var}(Z_i)$. Thus $\text{var}(\hat{\alpha}_1) > \text{var}(\hat{\beta}_1)$ if

$$\frac{(\Delta t)^2 \sigma_T^2 + 2\sigma_W^2}{(\Delta t)^2} < \sigma_B^2 + \sigma_W^2$$

that is, if $(\Delta t)^2 < 2\sigma_W^2/(\sigma_B^2 + \sigma_W^2 - \sigma_T^2)$.

What might one expect about the relative sizes of α_1 and β_1? Suppose the rate of change and its dependence on Z remained constant over the lifetime and at birth there was no relationship between $Y(0)$ and Z, that is, $\alpha_1 = 0$. Then at time t_1, we would expect $Y_1 = \alpha_0 + \beta_0 t_1 + \beta_1 Z t_1 +$ error. If we had only this single observation, then we could not estimate β_1 directly without assuming $\alpha_1 = 0$, and so this would be equivalent to fitting $Y_1 = \alpha_0^* + \alpha_1^* Z + \text{error}$ where $\alpha_1^* = \beta_1 t$. With this equivalence, we find that the paired test is more powerful than the unpaired test if

$$\frac{t_1^2}{\sigma_B^2 + \sigma_W^2} < \frac{(\Delta t)^2}{(\Delta t)^2 \sigma_T^2 + 2\sigma_W^2}$$

Thus, it is clear that either test could be the more powerful, depending on the magnitude of the three variance components, the age of first observation, and especially, the interval between the two observations.

Repeated measurements

Now suppose we have K measurements per subject at times t_k. (For the time being, we assume complete data with all subjects observed at all times; later we will consider the more realistic situation where some observations are missing or some subjects drop out prematurely.) Let us again assume that the linear model given by Eq. (7.1) applies. Note first that this is equivalent to the following hierarchical model:

$$Y_{ij} = a_i + b_i t_j + e_{ij}^W \qquad (7.2a)$$

$$a_i = \alpha_0 + \alpha_1 Z_i + e_i^B \qquad (7.2b)$$

$$b_i = \beta_0 + \beta_1 Z_i + e_i^T \qquad (7.2c)$$

where a_i and b_i are random person-specific intercepts and slopes respectively. Thus, one could in principle proceed in a two-stage manner, first fitting a separate regression, Eq. (7.2a), to each subject separately to obtain estimates $(\hat{a}_i, \hat{b}_i)$ and their sampling variances $v_i^a = \text{var}(\hat{a}_i)$ and $v_i^b = \text{var}(\hat{b}_i)$, then perform weighted least squares to fit the $\hat{a}_i$ and $\hat{b}_i$ to Eqs. (7.2bc). Here the appropriate weights would be given by $1/(v_i^a + \sigma_B^2)$ and $1/(v_i^b + \sigma_T^2)$ respectively, which involve the unknown residual variances σ_B^2 and σ_T^2, thus requiring an iterative procedure (Stram 1996). Also note that unless the t_k are centered at their mean, the $\hat{a}_i$ and $\hat{b}_i$ will be correlated, requiring a bivariate fitting of Eqs. (7.2bc) with weights given by $(V_i^{-1} + \Sigma^{-1})^{-1}$ where $V_i = \text{cov}(\hat{a}_i, \hat{b}_i)$ and $\Sigma = \text{diag}(\sigma_B^2, \sigma_T^2)$. Fortunately, a number of programs are available to fit the full model in a single stage, such as ML_n (Kreft and de Leeuw 1998) or procedure MIXED in the SAS package (Littel et al. 1996). For example, using SAS, one would specify the following code:

```
INPUT ID TIME Z Y;
PROC MIXED;
    CLASS ID TIME;
    MODEL Y = Z TIME Z*TIME/SOLUTION;
    RANDOM ID ID*TIME;
```

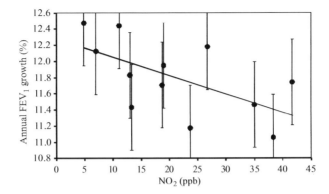

Figure 7.1. Regression of individually-adjusted community-mean 4-year FEV_1 growth rates on community ambient NO_2 levels in the CHS. (Adapted from data in Gauderman et al. 2000.)

This will yield estimates of the overall intercept α_0 and the coefficients of $Z(\alpha_1)$, TIME (β_0), and $Z*TIME = (\beta_1)$, as well as the three residual variances SUBJECT (σ_B^2), SUBJECT*TIME (σ_T^2), and RESIDUAL (σ_W^2). Figure 7.1 illustrates the regression of individually adjusted community-mean 4-year growth rates for FEV_1 on community ambient NO_2 concentrations in the CHS using this form of analysis (Gauderman et al., 2000), showing a strong inverse relation between the two $(R^2 = -0.61, p < 0.025)$.

In using the hierarchical mixed model (GLMM) in this way we are assuming that all three error terms e_i^B, e_i^T, and e_{ik}^W are normally distributed. In some circumstances, we may be primarily interested in the growth rate parameters β and do not wish to assume a specific model for the intercepts a_i, including linearity and normality. In this case, one could simply omit Eq. (7.2b) and treat the a_i as fixed effects—a vector of nuisance parameters, one per subject. Table 7.1 compares the estimated parameters from the fixed and random effects models from the CHS (Berhane et al. 2004). (Here we have included a third level of random effects for communities, as will be explained in Chapter 10.) The results for both slopes and intercepts are quite similar whether the other terms are treated as fixed or random.

Over relatively short periods of time, linearity of growth rates may be a reasonable approximation, but over periods where subjects are developing rapidly there can be marked nonlinearity. Furthermore, covariate values could also be varying substantially over time. For example, in the CHS, children are observed before and after their adolescent growth spurts (growth curves also differ markedly between boys and girls). Although the main focus of the analysis concerns long-term effects of air pollution, annual variation in pollution levels also influences annual growth rates. For these reasons, we need a more general framework.

Table 7.1. Comparison of fixed effects (columns 2 and 3) and random effects (columns 4 and 5) estimates of the intercepts and slopes for MMEF changes in the CHS (Reprinted with permission from Berhane et al. 2004.)

Pollutant	Cross-sectional intercepts (%)	Longitudinal slopes (%)	Full model	
			Intercepts (%)	Slopes (%)
O_3	0.80(0.93)	−0.20(0.26)	1.15(1.14)	−0.18(0.27)
PM_{10}	−1.54(0.93)	−0.49(0.20)	−1.65(0.80)	−0.45(0.21)
$PM_{2.5}$	−2.62(1.58)	−0.74(0.34)	−2.83(1.44)	−0.68(0.37)
NO_2	−1.69(1.15)	−0.47(0.25)	−1.97(1.07)	−0.46(0.27)
Acid	−0.44(1.15)	−0.43(0.22)	−0.78(1.08)	−0.41(0.24)

First consider the effects of time-dependent covariates, still assuming linear growth rates with respect to time. A simple way to accommodate time-dependent covariates would be to consider average levels $\overline{Z}_i$ and deviations $Z_{ik} - \overline{Z}_i$ at the different levels of the model:

$$Y_{ik} = a_i + b_i t_k + \beta_1 \left(Z_{ik} - \overline{Z}_i \right) + e_{ik}^W \tag{7.3a}$$

$$a_i = \alpha_0 + \alpha_1 \overline{Z}_i + e_i^B \tag{7.3b}$$

$$b_i = \beta_0 + \beta_2 \overline{Z}_i + e_i^T \tag{7.3c}$$

Thus the effect of covariates can be tested at both the temporal level [Eq. (7.3a)] and the subject level [Eqs. (7.3b,c)]. In addition to testing H_0: $\beta_1 = 0$ and H_1: $\beta_2 = 0$, one could also test H_2: $\beta_1 = \beta_2$. If H_2 is not rejected, one might set $\beta_1 = \beta_2$, thereby essentially estimating a weighted average of the two effects (Neuhaus and Kalbfleisch 1998).

This framework is conceptually a bit unsatisfying, however, as it postulates a dependence of intercepts a_i and annual observations Y_{ik} on contributions to $\overline{Z}_i$ that have yet to occur! More appealing, then, is to focus on the *annual*, rather than *average*, rates of change and postulate that

$$Y_{i1} = \alpha_0 + \alpha_1 Z_{i1} + e_i^B$$

$$Y_{ik} - Y_{i,k-1} = \beta_0 + \beta_1 Z_{ik} + e_i^T + e_{ik}^W, \quad k = 2, \ldots, J_i$$

which by successive summation yields:

$$Y_{ik} = \alpha_0 + \alpha_1 Z_{i0} + \beta_0 t_k + \beta_1 W_{ik} + e_i^B + k e_i^T + f_{ik}$$

where

$$W_{ik} = \sum_{\ell=1}^{k-1} Z_{i\ell} \quad \text{and} \quad f_{ik} = \sum_{\ell=1}^{k-1} e_{i\ell}$$

Thus, baseline levels are regressed on some index Z_{i0} of cumulative exposure before the first observation and the subsequent levels are regressed on another index W_{ik} of cumulative exposure during the ensuing interval. Also, the variance structure in this model is somewhat different from Eq. (7.3) since the residuals f_{ik} are no longer independent: $\text{var}(f_{ik}) = (k-1)\sigma_W^2$ and $\text{cov}(f_{ik}, f_{i\ell}) = (k-\ell)\sigma_W^2$. As a simple approximation, however, using standard GLMM software, one could simply adopt the same variance structure as previously, that is, three variance components e_i^B, e_i^T, and e_{ik}^W, assumed to be independent with constant variances σ_B^2, σ_T^2, and σ_W^2, respectively. In any event, use of GEE methods would avoid any problems of misspecification of the variance structure.

To address the problem of varying baseline growth rates with age, we can proceed parametrically or nonparametrically. Parametrically, one might simply add additional polynomial terms in age to Eq. (7.2a), for example,

$$Y_{ik} = a_i + b_i t_k + c_i t_k^2 + d_i t_k^3 + e_{ik}^W$$

and model each of the random effects (a_i, b_i, c_i, d_i) in the same manner as Eqs. (7.2b, c), with additional regression coefficients γ, δ, and so on for Z_i. This is a very flexible model, but the individual regression coefficients can be difficult to interpret. Supposing one were expecting a generally sigmoidal relationship, one could still retain the cubic polynomial form of dependence on age, but reparameterize it in terms of more interpretable parameters, such as the peak rate of growth, b_{max}, the age at peak growth rate, t_{max}, the baseline Y_0, and maximum attained change ΔY.

$$Y = Y_0 + b_{max}(t - t_{max}) - c(t - t_{max})^3$$

where c is the solution to a quadratic equation involving b_{max} and ΔY. Any or all of these four parameters could then be treated as subject-specific random effects, each possibly regressed on covariates. However, as some may be poorly estimated, one might prefer to treat only one or two of them as random or exposure-dependent and the remainder as single population parameters.

Alternatively, one might use some form of flexible modeling approach for the age effect, at either the individual or population level. For example, one might treat the shape of the growth curve $s(t)$ as a population function and model only individual deviations from it,

$$Y_{ik} = s(t_k) + \alpha Z_i + e_i^B + e_{ik}^W$$

but this would not allow for differences in rates of growth, only differences in level. To accommodate this, one could add linear deviations of the form

$$Y_{ik} = s(t_k) + \alpha Z_i + (\beta Z_i + e_i^T)t_k + e_i^B + e_{ik}^W$$

where $\beta Z_i t_k$ describes the regression of individual deviations in slope on Z_i and $e_i^T t_k$ describes the random deviations from the population average rate of change. Alternatively, one could model each individual's growth curve flexibly, say $s_i(t)$, and then treat certain functionals $F_p(s)$ of these curves—say, the maximum rate of growth or the maximum overall change as random effects to be modeled in terms of covariates.

$$Y_{ik} = s_i(t_k) + e_{ik}^W$$
$$F_p(s_i) = \beta_{p0} + \beta_{p1} Z_i + e_i^p, \quad p = 1, \dots, P$$

For this approach to be viable, however, one must take care to avoid overparameterizing the model by having more flexibility in the individual growth curves than the number of observations can support. Gauderman et al. (2004) used this approach to describe the maximum 8-year changes in lung function in the CHS subjects using linear splines with two fixed knots (at ages 12, 14, and 16, Figure 7.2). Berhane and Molitor (2007) used cubic splines to model instead the population growth curves $s(t)$ for each community separately and then regressed the community-average maximum attained lung volume (or other functionals like the peak growth rate) on air pollution.

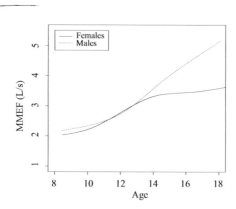

Figure 7.2. Cubic spline estimates of lung function growth in males and females from the CHS. (Reprinted with permission from Berhane and Molitor 2004.)

Binary and categorical outcomes: General linear mixed models

Now consider a trait that can be either present or absent at any time point, say some respiratory symptom. Rather than consider only the first transition from absence to presence in subjects with no previous history of the trait (incidence) as in previous chapters, suppose now we were interested in describing the entire time course of the trait. For a binary trait, this would be fully described by the probability of transitions from absence to presence and vice versa:

$$\text{logit } \Pr(Y_{ik} = 1 | Y_{i,k-1} = 0) = \alpha_k + a_i + \alpha Z_{ik}$$
$$\text{logit } \Pr(Y_{ik} = 0 | Y_{i,k-1} = 1) = \beta_k + b_i + \beta Z_{ik}$$

where a_i and b_i are random effects for subjects and α_k and β_k are fixed effects for time. Categorical response can be modeled in a similar manner with a fully saturated model given by a matrix of equations for all possible transitions $\Pr(Y_{ik} = r | Y_{i,k-1} = s)$, although one might prefer a more parsimonious model by imposing certain constraints on the parameters.

To frame the problem in a more general context, let us consider the GLMM:

$$\theta_{ik} = g(\mu_{ik}) = \alpha_k + a_i + \alpha Z_{ik}$$

where, as before, $\mu_{ik} = E(Y_{ik} | Z_{ik})$ and $g(\mu)$ is the canonical link appropriate to the distribution of Y, for example, the logit for binary data. What distinguishes the GLMM from the GLM discussed earlier is the addition of the random effect a_i, whose distribution is unspecified but assumed to heave zero mean and variance σ^2. This is thus a "marginal model," describing the mean and variance of Y at each time for each person, rather than the conditional model for the transition probabilities.

Unlike in linear models, the interpretation of the regression coefficients in nonlinear exposure–response relations differs between marginal and conditional models. Consider, for example, a logistic model of the form $E(Y_{ik} | Z_{ik}, X_i) = \text{expit}(\alpha + \beta Z_{ik} + X_i)$, where X_i denotes an unobserved random effect for subject i (a "frailty") having some distribution with variance σ^2, as discussed in the sections of Chapter 4 on overdispersion. Here, β has the interpretation of the log relative risk per unit Z *for individuals with the same frailty*. In the corresponding marginal model $\mu_{ik} = E(Y_{ik} | Z_{ik}) = \text{expit}(\alpha + \beta^* Z_{ik})$ with $\text{var}(Y_{ik} | Z_{ik}) = \mu_{ik}(1 - \mu_{ik})(1 + \sigma^2)$, β^* estimates instead a *population average* log relative risk. The average risk $E_X[E(Y | Z, X)]$ obtained from the conditional model is not exactly a logistic relationship (although it has a similar shape), and its slope coefficient is generally has a somewhat larger than that from the marginal model, particularly if the average response probability is near zero or one. A quantitative example of this phenomenon for the probit model (where an analytic solution is possible) is discussed in Chapter 15 (Figure 15.3).

Combined longitudinal and survival outcomes

In some instances, one may have longitudinal observations of some biomarker or other continuous outcome along with a censored failure-time disease endpoint. For example, in the CHS, each child's lung function is measured annually and one might want to ask whether the diagnosis of new-onset asthma can be predicted by the history of prior lung function measurements. Let Y_{ik} denote the longitudinal measurements, as before, and let D_i represent the binary disease endpoint, taking the value 0 until diagnosis at time T_i. The predictors, such as air pollution Z_{ik} could be either fixed or time-dependent. One way to relate the two outcomes might be to introduce a latent variable process, $X_i(t)$, representing the underlying propensity to disease at any point in time, and treat the longitudinal measurements as flawed indicators of that process. (This can be thought of as a longitudinal form of measurement error model that will be considered in greater detail in Chapter 11.) Thus, we might specify the model in terms of three submodels:

$$X_i(t) = a_i + b_i t + \alpha_2 [Z_i(t) - \overline{Z}_i], \qquad \text{latent biological process} \quad (7.4a)$$

where $a_i = \alpha_0 + \alpha_1 \overline{Z}_i + e_i^B$ and $b_i = \beta_0 + \beta_1 \overline{Z}_i + e_i^T$

$$Y_{ik} = X_i(t_{ik}) + e_{ik}^W, \qquad \text{biomarker measurement} \quad (7.4b)$$

$$\lambda(t) = \lambda_0(t) \exp[\gamma_1 X_i(t) + \gamma_2 Z_i(t)], \quad \text{disease risk} \qquad (7.4c)$$

Here, we are assuming a linear growth curve model for the unobserved biological process, but of course more complex parametric or nonparametric models could be considered. The person-specific intercepts a_i and slopes b_i are treated as random effects, regressed on covariates as before, with coefficients and error variances to be estimated. In addition to allowing inference on the effects of exposure on individual growth rates (α_1, β_1), the model also provides an estimate of the dependence of disease risk on the latent biological process (γ_1), possibly after adjusting for the direct effect of exposure on disease (γ_2).

Figure 7.3 illustrates a simple descriptive analysis showing how asthmatics have consistently lower MMEF than nonasthmatics, the earlier the onset, the lower the lung function. In this analysis, no incident asthmatics during the period of follow-up were included, so it cannot be determined whether cases diagnosed at older ages already had reduced lung function before their diagnoses. Such a question would require a more complex analysis like that described above.

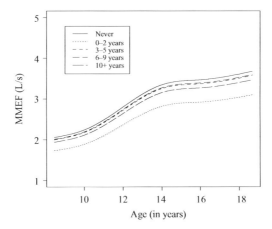

Figure 7.3. Fitted spline curves for MMEF growth in asthmatic children, by age at onset, and in nonasthmatic children. (Reprinted with permission from Berhane et al. 2000.)

Fitting of this model requires special purpose software. Faucett and Thomas (1996) described an MCMC approach in which the various random effects are sampled conditional on the data and the current estimates of the population parameters, then these parameters are sampled from their distributions given the sampled random effects, continuing in this manner for many iterations. Their application to data on the incidence of AIDS in relation to longitudinal observations of CD4 counts illustrates the problem of informative censoring, since counts below a certain level would mean death, leading to premature truncation of the longitudinal observations. Thus, joint estimation of the CD4 process and AIDS incidence is necessary to overcome the bias this would produce. Their estimate of the mean slope [β_0 in Eq. (7.4a), with no covariates Z in the model] was 7% higher from the joint analysis than that from an analysis of only the CD4 data. Likewise, the relative risk coefficient [γ_1 in Eq. (7.4c)] was 35% higher than that from the naïve regression of AIDS incidence on measured CD4 counts. The AIDS incidence data thus informs the longitudinal analysis of CD4 counts and vice versa. See Wulfsohn and Tsiatis (1997); Faucett et al. (1998); Hogan and Laird (1998); Henderson et al. (2000) for further discussion and applications.

Panel studies

In addition to the time-series methods discussed in the next chapter, which are generally based on aggregate data, acute effects of air pollution or other rapidly time-varying exposures can be studied at the individual level using a design known as a panel study. These are essentially the same as the longitudinal studies for chronic effects described earlier in this chapter, but on a shorter time scale. Typically, one or more relatively small panels of subjects (often highly sensitive individuals like asthmatics) will be enrolled and observed frequently (perhaps daily) over a period of a few weeks or months. Since the burden on participants is substantial, it is not realistic to expect the same participants to continue for long and the burden on investigators is also heavy enough to preclude studying panels of more than a few dozen at a time, but one might study several panels over different time periods. What this design accomplishes, however, is to allow quite intensive measurement of many acute effects at fine temporal resolution.

The typical data structure might thus be represented as (X_{pt}, Y_{pit}), where X_{pt} denotes the air pollution exposure at time t for panel p, and Y_{pit} denotes the outcome variable measured for subject i at time t. A natural linear model might then take the form

$$Y_{pit} = \beta_0 + \alpha_p + \gamma_{t(p)} + \beta X_{pt} + e_{pi} + e_{pit}$$

where α and γ represent fixed effects for panel and time, respectively and e_{pi} and e_{pit} are random effects for between-subject and within-subject deviations with some assumed distribution (typically Gaussian). Conditioning on the between-subject deviations by subtracting the subject means of the Ys and the panel-time means from the Xs eliminates the need for a parametric assumption about the distribution of subject-level random effects and yields the equivalent simple regression model

$$Y_{pit} - \overline{Y}_{pi} = \beta_0 + \alpha_p + \beta\left(X_{pt} - \overline{X}_p\right) + e_{pit}$$

The following SAS code will readily fit the full model:

```
INPUT PANEL ID DAY X Y;
PROC MIXED;
    CLASS PANEL ID(PANEL) DAY(PANEL);
    MODEL Y = X PANEL DAY/SOLUTION;
    RANDOM SUBJECT = ID(PANEL);
```

Of course, this assumes that every subject is observed for the entire time period, so that $\overline{X}_p$ is the same for every subject. If this is not the case, then one should create a person-specific mean and regress the daily outcomes on both the person-specific means and on the daily deviations from the person-specific means. The former constitutes a between-person, the latter a within-person estimator of the air pollution effect.

An alternative model is the "transition" model in which each response is regressed on the residual from previous observation,

$$Y_{pit} = \beta_0 + \alpha_p + \gamma_{t(p)} + \beta X_{pt} + \eta(Y_{pi,t-1} - \beta X_{pi,t-1}) + e_{pit}, \quad t = 2, \ldots, T$$

Forming the differences between successive observations for a linear model yields the equivalent form (with some reparameterization),

$$Y_{pit} - Y_{pi,t-1} = \gamma^*_{t(p)} + \beta^*(X_{pt} - X_{pi,t-1}) + \eta^* Y_{pi,t-1} + e_{pit}, \quad t = 2, \ldots, T$$

SAS Proc Mixed can fit this model by replacing the RANDOM command by the following line:

> REPEATED DAY(PANEL)/TYPE = AR(1)
> SUBJECT = ID(PANEL);

which assumes a first-order autocorrelation process for the residuals. A broad menu of more complex alternative specifications is also available. For binary outcomes, the GEE procedure GENMOD should be used instead, which has slightly different syntax. GENMOD also produces GEE estimates of the parameters and their variances (whatever the form of the outcome variable). Rather than modeling the correlation structure directly, one might also perform the analysis on the residuals from LOESS smooths of both X and Y on date to remove spurious associations due to confounding by season.

Although we have focused on linear models for continuous outcomes, the general framework of GLMMs can be applied to other types of outcomes, such as a logistic model for binary variables. This is clearly preferable to the traditional approach to analyzing panel studies of, say, asthma exacerbations by aggregating over subjects and treating the panel's daily attack rate as a continuous or Poisson variable, to be regressed on daily pollution levels and temporal confounders like weather. That approach fails to properly account for the dependence of events over time, assumes a constant variance, and cannot deal with dropouts or intermittent events, amongst other problems (Korn and Whittemore 1979).

For example, Dr. Xiao-chuan Pan and his colleagues at Beijing Medical University (unpublished data) looked at the effects of particulate pollution in three cities of China and South Korea resulting from the annual springtime dust storms in the Gobi Desert using a panel study design. Three panels (one in each region), each comprising about a hundred children, were enrolled and observed over a two-month period surrounding the anticipated peak pollution period. Lung function tests were performed twice daily (morning and evening). For illustration, we compare the effects on evening peak flow (PEFR) in relation to daily PM_{10} levels using the data from the two Chinese sites. Fixed covariates include age, height, and gender. Presence of cough can be treated as a time-dependent covariate, but could also be an intermediate variable on a causal pathway, so its inclusion might constitute overadjustment. Table 7.2 compares the fits of the

Table 7.2. Illustrative results from the panel study of the effects of particulate pollution resulting from the Gobi desert dust storms on one panel of 106 children in China (unpublished data from Pan et al.)

Outcome variable	Adjusted for cough?	Repeated measures model			Random effects model		
		PM_{10}	Cough	Corr(RE)	PM_{10}	Cough	var(RE)
PEFR[1]	Yes	−12.11	−15.51	0.67	−4.67	−14.21	1,848
		(1.96)	(1.68)		(1.77)	(1.51)	
PEFR[1]	No	−13.24	—	0.67	−5.09	—	1,817
		(1.96)			(1.77)		
Cough[2]	N.R.	−0.025	—	0.50	−0.031	—	0.036
		(0.008)			(0.011)		

[1] Fitted using Proc Mixed with an AR(1) correlation structure (for the repeated measures analysis).
[2] Logistic model fitted using Proc Genmod with an AR(1) correlation structure for the repeated measures analysis and an exchangeable correlation structure for the random effects model.

two models for PEFR with and without adjustment and treating cough as a binary endpoint. Although cough is strongly related to both air pollution and PEFR, adjustment for it reduces the effect of PM_{10} on PEFR only slightly in either the random effects or repeated measures model.

See Sheppard (2005) and Dominici et al. (2003b) for more extensive treatments of panel studies and their relationships with time-series, case-crossover, and cohort studies.

Missing longitudinal data

In Chapter 4, we considered the problem of missing data on fixed exposure variables in the context of case-control and cohort studies. We now turn to the more complex problem of missing data on outcomes in longitudinal studies. Here, the "complete case" analysis discussed in Chapter 4 would require restriction to those subjects with data at *all* observation times; aside from possibly leading to an unacceptable loss of sample size, this strategy is not immune to bias as we saw earlier. What is often done is simply to omit the missing observations for each subject, but this too can lead to bias. Although the basic principles of missing data models and techniques for dealing with them are similar, it is helpful to review these first and discuss different missingness patterns that can arise in the longitudinal context. Of course, data on time-dependent exposures can also be missing, but for simplicity, we assume the problem concerns only the relationship between a single fixed (not time-dependent) exposure variable Z and a sequence $\mathbf{Y} = (Y_1, \ldots, Y_K)$ of observations of a continuous outcome variable, some of which could be missing.

Recall that we earlier defined the concepts of missing completely at random (MCAR), missing at random (MAR), ignorably missing, and non-ignorably missing. In longitudinal studies, we would define these concepts as follows. Again, we introduce a missing value indicator, now as a vector $\mathbf{M} = (M_1, \ldots, M_K)$ corresponding to whether the outcome variable is missing at each time point. Here, we define MCAR as missingness $\mathbf{M}$ being independent of $\mathbf{Y}$ at *all* observation times, although it could depend upon Z or other measured variables. MAR is defined as missingness M_k at a given time being independent of Y_k at that time, although it could depend on $Y_{(-k)}$ at other times. Ignorable missingness is MAR, plus there being no functional relation between the parameters in $\Pr(\mathbf{Y}|Z; \beta)$ and $\Pr(\mathbf{M}|\mathbf{Y}, Z; \alpha)$.

Outcomes may be missing at some observation times either because planned visits were missed or because of drop-outs from the study. We call the former "intermittent missingness." If the only missing values result from drop-outs (i.e., $M_k = 1$ implies $M_{k+1} = M_{k+2} = \cdots = M_K = 1$), we say the data have a "monotone missingness" pattern. Intermittent missingness is likely to be ignorable in situations where the probability of a subject missing a particular observation is unrelated to the true value at that time (although it could be related to other measurable characteristics). However, it can be awkward to deal with by some of the standard multiple imputation strategies: there may be no easy way to model the dependence of outcomes at a given observation time on previous times if there are many different subsets of subjects with different subsets of the data available. Truncated observations, on the other hand, can have more serious implications for bias if the reason for dropping out is related to the outcome variable, for example, if children in polluted communities who develop asthma are more likely to move to cleaner air than nonasthmatics in the same communities. Fortunately, however, the missing data imputation strategies are relatively easy to apply in this circumstance.

Under the MAR assumption, simply ignoring the missing Ys still yields unbiased estimators of the intercept α and slope β in the mixed model Eq. (7.2) if maximum likelihood is used (this may not be true for GEE methods, however). In the general model involving random effects for both a_i and b_i, even subjects with only a single Y_k will contribute some information towards estimation of α and β, although their individual a_i or b_i, cannot be estimated. In the model with intercepts treated as fixed effects, only subjects with at least two observations are informative about β. These contributions result from the information about the relevant variance components from each deviation from the model prediction. In the two-stage analysis described earlier, such subjects do not contribute, but for each of those with three or more observations, the first stage yields estimates $\hat{a}_i$ and $\hat{b}_i$ and their variances $v_i^a = \text{var}(\hat{a}_i)$ and $v_i^b = \text{var}(\hat{b}_i)$. These are then used in weighted regressions with weights $1/(v_i^a + \sigma_B^2)$ and $1/(v_i^a + \sigma_B^2)$, respectively. The subject-specific components of these weights depend on the number and spacing of the observations, specifically inversely with $n_i \, \text{var}(t_{ik})$. Thus, the subjects who receive the heaviest weight are those with the most complete and most widely spaced observations.

Given that all subjects contribute appropriately in the mixed model, what is the validity of analyzing only the available observations, and is there any need to impute the missing observations? If the data are indeed MAR, then this analysis is unbiased and nothing further can be gained by imputing the missing observations.

Now suppose the missingness is nonignorable, that is, that the probability that a particular observation is missing depends upon the true value of Y_k conditional on Z and the available observations of the other Ys. The parametric (regression) and semiparametric (propensity score) methods discussed in Chapter 4 are readily applied to longitudinal data with monotone missingness patterns, simply by building separate models for each time based on the fixed covariates and the observations from previous times. For example, in the propensity score method, one might fit a separate a logistic model $\Pr(M_k = 1 | \mathbf{Y}_{(-k)}, \mathbf{M}_{(-k)}, Z)$ at each time point k, and use this model to stratify the subjects and impute values for the missing Y_k observations from the nonmissing ones at that time (Lavori et al. 1995). Fitzmaurice et al. (1994) discuss parametric regression approaches to missing longitudinal binary outcomes assuming that $\mathbf{Y}$ is a member of the exponential family; since the Ys are binary, it becomes feasible to evaluate the full likelihood, summing over all possible combinations of the missing data, without resorting to multiple imputation.

If the missingness pattern is not monotone, some form of likelihood-based analysis (or multiple imputation approximation to it) is required. Such methods are based on one of the two following factorizations of the full likelihood:

$$\Pr(\mathbf{Y},\mathbf{M}|Z) = \begin{cases} \Pr(\mathbf{Y}|\mathbf{M}, Z) \Pr(\mathbf{M}|Z) & \text{``Pattern mixture'' models} \\ \Pr(\mathbf{M}|\mathbf{Y}, Z) \Pr(\mathbf{Y}|Z) & \text{``Selection'' models} \end{cases}$$

Pattern–mixture models (Little 1993; 1994; 1995b; Little and Wang 1996; Demirtas 2005) are particularly convenient for data with nonmonotone missingness patterns, since one can reconstruct the marginal model by averaging over all observed missingness patterns:

$$\Pr(\mathbf{Y}|Z) = \sum_{\mathbf{M}} \Pr(\mathbf{Y},\mathbf{M}|Z) = \sum_{\mathbf{M}} \Pr(\mathbf{Y}|\mathbf{M}, Z) \Pr(\mathbf{M}|Z)$$

Essentially, one stratifies the data by the pattern of missingness, fits a separate model for each stratum, then takes a weighted average of these stratum-specific models. The weights $\Pr(M|Z) = \prod_k \Pr(M_k | M_1, \ldots, M_{k-1}, Z)$, are readily estimated by, say, logistic regression since the M and Z data are completely observed. Estimation of $P(Y|Z, M)$ is more difficult, since by definition the full Y vector is not observed, requiring some additional assumptions for identifiability. For example, suppose Y follows a growth curve model like Eq. (7.2), so that the individual a_i and b_i are estimable even if the Y_i vector was incomplete. Then it might be reasonable to assume that the means and covariances $(\alpha_m, \beta_m, \Sigma_m)$ of the distributions of (a_i, b_i) for subjects with $M_i = m$ were "similar" for subjects with similar missingness patterns. One can write the likelihood contribution for a particular subject as

$$\Pr(Y_{obs}, M, (a, b)|Z] = \Pr(Y_{obs}|(a, b), Z, M) \Pr((a, b)|Z, M) \Pr(M|Z)$$

The first factor is a conventional growth curve model like Eq. (7.2a) using only the observed data for subjects with a given missingness pattern. The second factor is a model for the distribution of random coefficients as a function of the completely observed covariate and missingness patterns. The third might be a logistic model for the missingness probabilities as a function of covariates. Demirtas (2005) describes a Bayesian approach to smoothing across missingness patterns.

Selection models (Little and Rubin 1989b) model instead the marginal probability of Y given Z directly, but supplement the likelihood with a model for the missingness process as a function of the (potentially missing) Y and complete Z data. The two approaches are equivalent, of course, if the parameters of $\Pr(Y|Z)$ and $\Pr(Y|Z, M)$ are the same and the parameters of $P(M|Y, Z)$ and $\Pr(M|Z)$ are the same. Little (1995b) compared the two approaches in the context of modeling the drop-out process in a longitudinal study. The two approaches can also be combined, using one for part of the data and another for the rest, as described in Little (1993). Hogan and Laird (1997) extend the pattern-mixture models approach to joint analysis of a longitudinal continuous variable and a censored survival endpoint that depends upon it.

8 Time-series models for acute effects

Although most of this book is concerned with the long-term health effects of environmental exposures, there are also short-term effects for some outcomes. Obviously not for cancer, which is the result of a long accumulation of carcinogenic exposures (combined with underlying genetic susceptibility), nor for such outcomes as asthma incidence or chronic obstructive pulmonary diseases. Short-term elevations in disease rates following major population exposure events, like the London Fog that led to a doubling of all-cause mortality (Figure 8.1) (Logan 1953; Waller et al. 1973; Bell and Davis 2001), may be obvious and do not need sophisticated statistical methods to detect them. (Although perhaps the best known, it was not the first such epidemic of air pollution-related health effects to be recognized. Earlier events occurred in the Meuse Valley of Belgium in 1930 (Firket 1931; Nemery et al. 2001) and in Donora, PA in 1948 (Ciocco and Thompson 1961).) However, very small relative risks that are consistently manifest over many such episodes and many locations can also be highly informative about acute effects, but require sensitive statistical methods to overcome potential confounding and other problems. These methods are the subject of this chapter.

A great advantage of time-series approaches is that comparisons across time are unlikely to be sensitive to the types of confounders that would affect comparisons between places or people, although they are subject to other types of time-varying confounders. We begin by considering

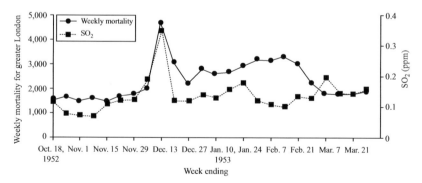

Figure 8.1. SO_2 concentrations and mortality during the London fog episode of 1952. (Reproduced with permission from Bell and Davis 2001.)

time-series methods for looking at the relationship between daily variation in the rates of discrete disease events and daily variation in exposure
levels, allowing for variability in latency, and then extend the approaches
to multiple time-series across different locations and ultimately to repeated
events within individuals.

Methods for a single time series

Suppose we have data from routine surveillance of a population of (essentially) constant size and structure over some period of time and observe
$Y(t)$ events at time t. Here, the time scale could be measured in any discrete intervals, but for argument sake, suppose it is measured in days. Also,
suppose we have a vector of explanatory variables $Z(t)$ measured on the
same time scale; this might include the exposure of interest (say, concentration of some air pollutant), as well as other potential confounders
like weather and the prevalence of influenza infections. Because the events
may be rare, many of the values of $Y(t)$ could be zero, so rather than
using standard linear regression techniques, a more natural choice would
be Poisson regression, as discussed in Chapter 4 . Thus, we might assume
$Y(t) \sim \text{Poisson}[\lambda(t)N]$ where N is the population size and $\lambda(t)$ the rate,
which we might model as $\lambda(t) = \exp(Z(t)'\beta)$. Implicit in this model is an
intercept term β_0 for the constant covariate $Z_0(t) \equiv 1$, so that the baseline
expected number of events is $Ne^{\beta_0} = \exp(\ln(N) + \beta_0)$; thus, assuming it is
relatively constant over the duration of the study, the population size can
simply be included as part of the intercept term. If we could assume that
the daily counts of events were independent, then the likelihood would
be formed simply by multiplying these time-specific Poisson probabilities
together,

$$L(\beta) = \prod_{t=1}^{T} \exp\left(Y(t)Z(t)'\beta - e^{Z(t)'\beta}\right) \Big/ Y(t)!$$

There are several problems with this naïve formulation, however. First,
daily rates are probably not independent, so this multiplication would be
inappropriate. Second, the daily rates are doubtless not perfectly explained
by the measured covariates, so there will be additional sources of daily
variation not explained by the model. Thus, the counts would not have a
Poisson distribution but would be "overdispersed," that is, their variance
is larger than that predicted by the Poisson distribution (equal to the mean).
Third, there are likely to be other unmeasured risk factors that are correlated with exposure, leading to temporal confounding. Amongst these
temporal confounders could be long-term trends and seasonal variation

in disease rates and exposures. Simply observing that disease rates tend to be highest in winter when some pollutant is also high would not be a sufficient basis for inferring a causal connection between the two—one would be more interested in demonstrating a short-term correlation within season. Thus, these long-term and seasonal trends need to be eliminated from the data before proceeding to a correlation analysis. Finally, the effect of exposure on disease is unlikely to be instantaneous, but only after some latent period that itself could be variable.

Rather than adopt the full likelihood approach described above, requiring proper specification of the entire multivariate distribution of **Y**, let us recast the problem in the form of a GLM by specifying only the mean and covariance structure. The naïve Poisson regression model given above can be written in the form

$$E[Y(t)]) = \exp(\mathbf{Z}(t)'\beta) \equiv \mu(t)$$

$$\text{var}[Y(t)] = \mu(t)$$

$$\text{cov}[Y(t), Y(u)] = 0 \text{ for } t \neq u$$

suggesting ways the model could be extended to address some of these problems. First, the overdispersion problem could be solved by adopting some more general dependence of the variance upon the mean, say $v(t) \equiv \text{var}[Y(t)] = f[\mu(t); \sigma^2]$, where σ^2 is an overdispersion parameter to be estimated along with β. Likewise, the serial correlation could be addressed by adopting some autoregressive model, perhaps a simple first-order autocorrelation model with decay parameter ρ, also to be estimated. The time trends and seasonal variation could be addressed by adding some smooth function of time, $s(t)$, to the means model. Finally, the latency problem might be addressed by subtracting a lag parameter δ from time in the means model. (Later we consider an extension of this known as the "distributed lag model" to address variability in the lag.) Putting it all together, the full model now might take the form

$$E[Y(t)] = \exp[\mathbf{Z}(t - \delta)'\beta + s(t)] \equiv \mu(t)$$

$$\text{var}[Y(t)] = \mu(t)[1 + \sigma^2\mu(t)] \equiv v(t)$$

$$\text{cov}[Y(t), Y(u)] = \sqrt{v(t)v(u)}e^{-\rho|t-u|} \equiv S(t, u)$$

The generalized estimating equations (GEEs) machinery introduced in Chapter 4 (Liang and Zeger 1986) can be used to fit such models. Suppose we have a number of independent time series $i = 1, \ldots, n$, say in different locations or by taking a single long time series and breaking it up into shorter series that would be virtually independent of each other (say, a 20-year observation period being treated as 20 independent 1-year time series). Then the estimating equation for β would be

$$U(\boldsymbol{\beta}) = \sum_{i=1}^{n} (\mathbf{Y}_i - \boldsymbol{\mu}_i(\boldsymbol{\beta}))' \mathbf{V}_i^{-1} \left(\frac{\partial \boldsymbol{\mu}_i(\boldsymbol{\beta})}{\partial \boldsymbol{\beta}} \right) = 0$$

where $\mathbf{Y}_i = (Y_i(1), \ldots, Y_i(T_i))$ is the vector of all the correlated event counts in city or year i, $\boldsymbol{\mu}_i$ the corresponding vector of means, and $\mathbf{V}_i$ the matrix of covariances given above. The solution to this equation is the estimator of $\boldsymbol{\beta}$ (which could include the lag parameter δ and any parameters in the smoothing function $s(t)$). The asymptotic variance of $\boldsymbol{\beta}$ is then given by the "sandwich estimator," Eq. (4.6). Parameters in the variances can in turn be estimated using GEE-2, by forming the empirical cross-products

$$C(t, u) = [Y(t) - \mu(t)][Y(u) - \mu(u)]$$

and regressing these pairs of observations on the model predictions $S(t, u)$ given above. Thus, σ^2 could be approximately estimated as the mean of $C(t, t)/\mu(t) - 1$, and ρ by the regression of the log of the standardized residuals $\ln\left[C(t, u)/\sqrt{v(t)v(u)}\right]$ on $t - u$. This approach has been termed "Generalized Iteratively Re-weighted Least Squares" (Zeger 1988; Zeger and Qaqish 1988), since in practice one iterates between an estimation of the parameters in the means model using the current estimates of the variance/covariance model, then computes these residuals and their cross-products and re-estimates the parameters in the variance/covariance model, until convergence.

The sandwich estimator requires multiple independent vectors of correlated observations to estimate the "meat" of the sandwich, $n^{-1} \sum_i \text{var}[U_i(\beta_0)]$. In time series, we have essentially only a single very long vector comprising the entire time series of correlated observations. If β_0 were known, one could estimate this by $n^{-1} \sum \sum_{ij} U_i(\beta_0) U_j(\beta_0)$, but the corresponding expression evaluated at $\hat{\beta}$ is simply zero. An *ad hoc* solution to this problem is to subdivide the time series into intervals such as years that are at least approximately uncorrelated and take the summation only over the within-interval pairs, but this is unsatisfying, since there is some correlation between the ends of the adjacent intervals and virtually no correlation between most of the within-interval pairs. Lumley and Heagerty (1999) described a broad class of weighted empirical adaptive variance estimators of the form

$$n^{-1} w_n^* \sum_{i=1}^{n} \sum_{j=1}^{n} w_{ij} U_i(\hat{\beta}) U_j(\hat{\beta})'$$

where w_{ij} are weights are some declining function of the distance between observations, which can be chosen adaptively so as to minimize the mean squared error.

In practice, this approach can be cumbersome because of the need to deal with the large matrices of covariances V_i. Rather than dealing with the serial correlation in the covariance model, a simpler approach is to remove the source of that correlation in the means model by filtering out the long- and medium-term variation through flexible models for $Y(t)$ and $Z(t)$. There are a variety of ways of doing this, the simplest being to filter using a simple moving average, say

$$\overline{Y}(t) = \sum_{u=t-h}^{t+h} \frac{Y(u)}{2h+1}$$

for some suitable choice of averaging time h, and similarly for $Z(t)$. More commonly what is done is to use some form of cubic spline model like those introduced in the previous chapter, using a large number of uniformly spaced knots, say one every 15 or 30 days. The assumption is then that the residuals $Y(t) - s(t)$ are serially uncorrelated, so ordinary weighted least squares can be used. Since this still involves the parameter σ^2 in the variance model, however, an iterative procedure like the one described above is still required. Dominici et al. (2002b) call this "Iteratively Reweighted Filtered Least Squares."

The choice of time scale for these filters is critical, since if one allowed them to vary too fast, they would essentially remove all the signal of the short-term causal effect one was looking for! On the other hand, taking too coarse a time scale might fail to remove some temporal confounding. We will revisit this question under the heading of "harvesting" later. Choice of the appropriate degree of "wiggliness" to allow for the background temporal variation $s(t)$, as well as adjustments for such time-dependent covariates as weather, poses one of the biggest challenges in time-series studies of air pollution. Recent work on semiparametric regression (Robins et al. (2007)) may offer a way forward to this problem.

Cubic splines are essentially parametric and, without some modifications (e.g., "natural" splines), can behave poorly near the ends of a series. Recently, nonparametric "smoothing" splines have become popular for their flexibility, the most widely used being the LOESS smoothers ("locally estimated polynomial regression" (Cleveland and Devlin 1988)), which have been incorporated into many statistical packages such as S-plus. When there are multiple covariates under consideration (e.g., several pollution variables along with seasonal and weather confounders), these smoothers can be strung together in what are known as "generalized additive models" (GAMs) (Hastie and Tibshirani 1990), which take the form

$$\mu(t) = \mathbf{Z}(t)'\boldsymbol{\beta} + \sum_j s_j[X_j(t)]$$

where $s_j(X_j)$ denotes any smooth function of covariate j. (Some of these terms could be joint functions of two or more variables, such as geographic coordinates.) An addition to nonparametric LOESS smoothers, alternatives include smoothing splines (Green and Silverman 1994) and parametric regression (natural- or B-splines) approaches (de Boor 1978; Cheney and Kincaid 1999). The use of GAMs with LOESS smoothers has become standard in the literature on the acute effects of air pollution since they were introduced in the mid-1990s (Schwartz 1994b). However, in about 2000, it was discovered that the default criteria used in the S-plus package were inadequate to ensure convergence when the range of relative risks is small (as is typical in air pollution applications) and multiple smoothers are included in the same model (also typical), and can lead to biased estimates. This prompted several simulation studies (Dominici et al. 2002b; Ramsay et al. 2003a, b) and re-analyses of certain major studies like NMMAPS (Dominici et al. 2005b). These generally found that there was some overestimation of the magnitude of the estimates (e.g., for total nonaccidental mortality, the original method yielded an estimate of 0.41% per 10 $\mu g/m^3$, which declined to 0.27% with a more stringent convergence criterion), but the basic qualitative conclusions about the relative contributions of different pollutants and their spatial heterogeneity were unchanged.

Although GAMs provide a flexible way of dealing with multiple confounders, concerns about residual confounding by risk factors like season and weather, which can have a much more powerful effect on mortality than does air pollution, continue to be raised (Moolgavkar 2005). One recurring criticism is that the health effects of weather are more than the sum of its parts, that is, temperature (particularly extremes and abrupt changes), humidity, wind, barometric pressure, sunlight, and so on. A suggested alternative to simply including these different factors in a GAM is to classify them into 10–20 discrete "synoptic" patterns based on climatological theory using various combinations of these factors. Pope and Kalkstein (1996) compared the two approaches using data from the Utah Valley and found that the synoptic method was as good or better than GAMs at describing the effect of weather, but there was little difference between the pollution effects estimated by the different methods. In detailed analyses of the NMMAPS data from Philadelphia, Samet et al. (1997) compared the synoptic approach with various other methods used to control for the effects of weather and found little difference in the estimated pollution effects.

Distributed lag models

Finally, we must address the problem of latency. Obviously, not every member of the population will be affected by an exposure event over the same time period, so the effect on population rates can be expected to be distributed over some range of times following the event. Equivalently, the disease rate at a particular time can be thought of as the result of exposures received over a distribution of times in the recent past. Since we are modeling outcomes, not exposures, the latter is conceptually the easier way to think about it. Thus, we might propose a model of the form

$$\ln(\lambda(t)) = \sum_{\delta=0}^{\Delta} b(\delta) Z(t - \delta)$$

where Δ is the maximum lag to be considered and $b(\delta)$ represents an array of regression coefficients to be estimated for the effect size at each lag δ. In practice, however, the Zs tend to be so highly correlated that direct estimates of b become very unstable. To overcome this difficulty, one could assume the $b(\delta)$ represents some smooth function in δ, say a polynomial, which rises to a peak and then declines. Thus, if we were to write $b(\delta) = \Sigma_{k=0}^{K} \beta_k \delta^k$, where the degree K of the polynomial is reasonably low, then one could rewrite the previous equation as

$$\ln(\lambda(t)) = \sum_{\delta=0}^{\Delta} \sum_{k=0}^{K} \beta_k \delta^k Z(t - \delta) = \sum_{k=0}^{K} \beta_k \sum_{\delta=0}^{\Delta} \delta^k Z(t - \delta) = \sum_{k=0}^{K} \beta_k \widetilde{Z}_k(t)$$

where $\widetilde{Z}_k(t) = \Sigma_{\delta=0}^{\Delta} \delta^k Z(t - \delta)$ is a vector of covariates than can be computed in advance.

This method was introduced by Schwartz (2000b) in an application to data on deaths over age 65 in 10 cities that had daily PM_{10} measurements. A quadratic distributed lag model showed that the excess risk was distributed over several days, gradually declining to zero by the fifth. He also showed that the estimate of the cumulative effect over that period was more than double that obtained by constraining the entire effect to occur in a single day. Applying similar methods to cause-specific mortality in these same cities, Braga et al. (2001) showed that respiratory deaths were generally associated with air pollution over the previous week, whereas cardiovascular deaths were more affected by the same day's exposure. Zanobetti et al. (2000; 2002; 2003) described further applications to the APHEA data.

Gilliland et al. (2001) fitted the distributed lag model to six-month time series of school absences in relation to daily ozone concentrations in each of the 12 communities. Figure 8.2 shows the resulting curve, averaged across communities using a hierarchical model described in Berhane

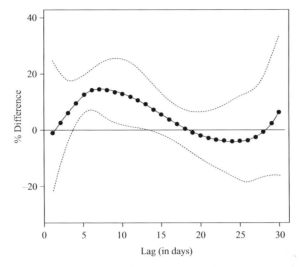

Figure 8.2. Fit of the distributed lag model for respiratory illness-related school absences in relation to daily 10AM–6PM ozone concentrations in the CHS. (Reprinted with permission from Gilliland et al. 2001.)

and Thomas (2002). Over the 30-day window ($\Delta = 30$) considered in this analysis, a cubic lag function ($K = 3$) showed a positive association of absences with ozone concentrations up to about 17 days earlier, significantly so over the previous 3–13-day interval.

The harvesting hypothesis

One interpretation of the association between daily fluctuations in death rates and air pollution is that it merely reflects the advancement in the time of death of individuals who were about to die anyway by a few days (Schimmel and Murawski 1976). If that were the case, the public health impact of the association would be much less important than if air pollution were causing deaths among members of the general population who, but for that exposure, might have lived much longer to die eventually of some unrelated cause. This hypothesis—called "harvesting" by Schimmel and Murawski or "mortality displacement" by Zeger et al. (1999)—can be developed more formally as follows. Suppose the population contains a very large number of healthy individuals and a small number $N(t)$ of "frail" individuals on a given day t. Suppose further that some number $I(t)$ from the healthy population become newly frail on day t. Finally, suppose that only frail individuals die, at a rate $\mu(t)$ that may depend upon air pollution levels $Z(t)$ on that day or some preceding day. (They may also recover at rate $v(t)$ that probably does not depend on air pollution.)

Then the size of the frail population will vary in response to air pollution according to the simple difference equation

$$N(t+1) = N(t)[1 - \mu(t) - \nu(t)] + I(t)$$

Zeger et al. suggest a model in which $I(t)$ has a Poisson distribution with a mean process that can allow for some serial correlation (say, due to influenza epidemics or weather patterns). By simulation, they showed that this model—where air pollution does not *cause* any frailty, only deaths among the frail subpopulation—can produce associations between mortality and air pollution on time scales only up to a maximum of twice the average time individuals spend in the frail pool, $1/(\overline{\mu} + \overline{\nu})$. A similar argument was made by Kunzli et al. (2001), who distinguished four possibilities: that air pollution increases the risk of frailty, the risk of deaths among the frail subpopulation, both, or neither (more on their discussion in the following section).

Schwartz (2000c) used Poisson regression with a range of averaging times, focusing on the mid-range as most likely to be unconfounded by long-term factors or short-term harvesting. He found that for all cause mortality and heart attacks, the association increased with increasing window size, whereas for chronic obstructive lung disease, the reverse occurred (consistent with the harvesting hypothesis). Pneumonia showed some evidence of harvesting in the first two weeks, followed by an increase in the strength of association for periods longer than a month. See Schwartz (2001) for a similar analysis of hospitalizations. Zanobetti et al. (2000; 2002; 2003) applied Poisson regression methods combining generalized additive and distributed lag models to data from multiple cities in the APHEA project and again found that harvesting could not explain the entire association and inclusion of longer-term effects more than doubled the overall effect of air pollution.

Zeger et al. introduced a novel "frequency domain" time-series method (Kelsall et al. 1999) to describe the magnitude of the association at different time scales. Suppose we represent the time series of mortality counts by $Y(t)$ and decompose it into a series of residuals $Y_k(t) = Y(t) - Y_{k-1}(t)$, representing the average mortality counts over successively finer intervals (e.g., years, seasons, months, weeks, days). We do the same for the air pollution series $X(t)$. Then the correlation of these residuals $Y_k(t)$ and $X_k(t)$ describes the strength of the association at each time scale. To do this on a continuous basis, Zeger et al. use Fourier transforms (see Box 8.1) to decompose the series into $1, 2, \ldots, T/2$ day cycles (where T is the total length of the series). Regression analysis of the amplitudes of the time series at the different frequencies for air pollution on the corresponding amplitudes for the air pollution series yields what they called a "frequency domain" analysis. They then provide a "harvesting resistant" summary

8.1 *Fourier series*

The Fourier series $y(k)$ of a given time series, say $Y(t)$, is a decomposition into sine and cosine waves of different frequencies $k = 1, \ldots, T/2$:

$$y(k) = \sum_{t=1}^{T} Y(t)[\sin(2\pi kt/T) + i \cos(2\pi kt/T)]$$

where $y(k)$ denotes a complex number $y_r(k) + i y_i(k)$. The original time series can then be reconstructed as

$$Y(t) = \sum_{k=1}^{T/2} y_r(k) \sin(2\pi kt/T) - \sum_{k=1}^{T/2} y_i(k) \cos(2\pi kt/T)$$

Thus if we had a linear model for the original time series

$$Y(t) = \beta_0 + \beta X(t) + e(t)$$

then on the transformed frequency scale, the same model can be written as

$$y(k) = \beta_0 + \beta x(k) + e'(k)$$

The frequency domain regression of Zeger et al. (1999) essentially regresses $y(k)$ on $x(k)$ to estimate β, and estimates the contributions from the different parts of the frequency spectrum by a weighted regression using Gaussian kernel weights centered around the time scale of interest. The time-domain regression of Dominici et al. (2003a) regresses $Y(t)$ on $X_k(t)$, the individual sine and cosine waves comprising the $X(t)$ series.

estimator based on accumulating the evidence, successively throwing away the longer-term cycles until the estimates stabilize.

The results for a 14-year time series in Philadelphia are shown in Figure 8.3. The left-hand panel shows no association at time scales of 2–3 days, indicating no harvesting, so excluding these cycles will lead to a larger effect estimate in these data. The right-hand panel shows the cumulative estimates, successively excluding more of the long-term cycle information as one moves towards the left (thus widening the confidence bands). The harvesting-resistant estimate (the vertical bars at the left-hand side of the panel) are 0.022 (95% CI 0.012 − 0.032) and 0.024

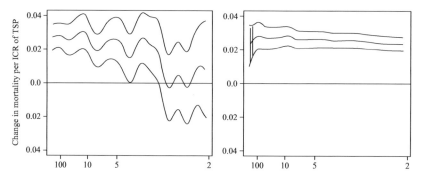

Figure 8.3. Time-scale specific (left) and cumulative (right) estimates of the log-relative risk for mortality in Philadelphia 1974–88 with current-day TSP as a function of the number of days in the cycle. (Reproduced with permission from the Zeger et al. 1999.)

(0.015 – 0.033) assuming mean times in the frail pool of 2 and 4 days respectively.

Dominici et al. (2003a) introduced a "time-domain" variant of this approach, again using a Fourier series decomposition of the air pollution time series to extract different time scales of variation. These were then taken as the independent variable in a Poisson regression of the mortality series, essentially a more sophisticated version of the Schwartz analysis described earlier. Their application to the four NMMAPS cities that had daily PM_{10} data available yielded similar results to the earlier Zeger et al. Philadelphia analysis. For a critique of their findings, see the accompanying editorial (Smith 2003) and the authors' response. Fung et al. (2005a,b) provide some simulation studies of the performance of the method under different frailty models.

Multiple time series

The air pollution time series literature has evolved gradually from the earliest papers, typically applying Poisson regression techniques to data from single cities, gradually incorporating various methodological advances described above and applying them to more and more such locations— hundreds of such publications (see Schwartz 1994a for a relatively early review of the literature). As a result, the findings tended to be conflicting and difficult to summarize, leading the editor of *Epidemiology* to call for a moratorium on publication of single-city time-series studies (Samet 2002).

Some authors had conducted multi-center studies and attempted to account for patterns that were seen across cities by various forms of meta-analysis. For example, Schwartz (2000a) analyzed data from 10 cities using a two-stage regression approach, treating the PM_{10} coefficients from the first stage (city-specific time-series analyses) as the dependent variable in the second stage, regressing these slope estimates on such city characteristics as their average levels of various copollutants. While the overall estimate was 0.67 per 10 $\mu g/m^3$ (95% CI $0.52 - 0.81$), adjustment for SO_2 lowered it slightly to 0.57 ($0.25 - 0.90$), adjustment for CO raised it (0.90, CI $0.42 - 0.97$), and adjustment for O_3 changed it very little (0.69, $0.53 - 1.26$), although all the adjusted CIs were substantially wider.

In general, however, the choice of locations for single city analyses have tended to be somewhat haphazard, depending upon the availability of suitable pollution, weather, and outcome data and upon knowledge of unique characteristics of the pollution profiles in these locations. Indeed, some critics have argued that meta-analyses of such data could be biased by the selection of cities that tended to show stronger relationships. The need for a more systematic approach was met by the NMMAPS led by Dr. Jon Samet from Johns Hopkins University and Dr. Joel Schwartz from Harvard University (Zeger et al. 1999; Daniels et al. 2000; Dominici et al. 2000; Samet et al. 2000a,b; Bell et al. 2004a; Dominici et al. 2004; 2006). In particular, this analysis, by being based on *all* of the largest cities, avoids the potential bias of city selection in favor of those showing associations. A parallel effort in Europe, the APHEA study was led by Dr. Klea Katsouyanni from the University of Athens (Katsouyanni et al. 1995; 1997; Touloumi et al. 1997; Atkinson et al. 2001; Samoli et al. 2001; 2005).

The NMMAPS study surveyed mortality and hospital admissions in the 100 largest U.S. cities using a common exposure and outcome assessment protocol (because of data limitations for some cities, the number included in the final analyses was reduced to 88). A hierarchical modeling strategy was developed to synthesize the results across cities and explain patterns of differences. Given the computational intensity of the single-city analyses, the various levels of the hierarchical model could not be fitted simultaneously, so each city was analyzed separately in the first stage of the analysis to yield a vector of regression coefficients b_c, which were then modeled in the second stage (Dominici et al. 2000; Samet et al. 2000a) in relation to a vector of potential explanatory variables X_c such as geographic region and mix of copollutants. Dominici et al. describe a sophisticated bivariate analysis of the coefficients for PM_{10} and O_3 in the 20 largest U.S. cities. At a one-day lag, the overall univariate PM_{10} effect (Figure 8.4) was estimated at 0.48% per 10 $\mu g/m^3$ (95% CI $0.05 - 0.92$). Using the proportion of the population below the poverty line, the proportion over age 65, and the long-term mean PM_{10} level as covariates in the second stage did not substantially change this estimate (0.52%, CI $0.06 - 0.98$), and none of the covariate effects was significantly different from zero. In a bivariate analysis, the standard deviations of both the PM_{10} and O_3

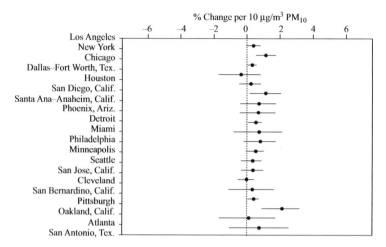

Figure 8.4. City-specific and combined relative risk estimates for the association of all-cause mortality with particulate pollution. (Adapted with permission from Samet et al. 2000b.)

coefficients were much smaller than in separate univariate analyses (0.36 cf. 0.76 for PM_{10}, 0.91 cf. 1.28 for O_3, with the correlation between them being -0.09). They also fitted spatial correlation models that we will defer to following chapter.

Dominici et al. (2002a) later extended this approach to 88 cities and a much broader set of second-stage covariates. Only the coefficient for average levels of PM_{10} was a significant modifier of the daily PM_{10} coefficient, indicating a stronger effect in the cities with lower average pollution levels (perhaps a saturation effect). The $PM_{2.5}/PM_{10}$ ratio was included as an indirect test of the hypothesis that fine particles were more important (data on $PM_{2.5}$ not being as widely available on a daily basis). However, when included with other modifiers, its estimated coefficient was negative (but with a large variance)—the reverse of what was expected—but it vanished when included alone in the second stage model.

Schwartz and Coull (2003) compared two different two-stage analysis approaches for multi-pollutant models in the presence of measurement error. Both entail estimating city-specific regression coefficients β_c for Y on X_1 and γ_c for X_2 on X_1 in the first stage, then in the second stage regressing the β_c on γ_c. One of the two possible estimators derived from this approach was shown to be unbiased with moderate loss of power (relative to the case of no measurement error), the other to be attenuated but with somewhat less loss of power. We will describe these two estimators more formally in Chapter 11.

An interesting variant of the multi-city time-series approach decomposes the regression of city-specific outcomes Y_{ct} on air pollution into two terms, a national average trend $\overline{X}_t$ and the local deviation from the national trend $X_{ct} - \overline{X}_t$ (Janes et al. 2007). The authors argue that the two should estimate the same quantity, so that differences would be symptomatic of uncontrolled confounding. In an accompanying editorial, however, Pope and Burnett (2007) point out that their analysis exploits neither between-city differences in levels of pollution and mortality (because of the inclusion of city-specific deviations in the model) nor short-term variability (because it is focused on longer-term trends using monthly averages rather than daily variability).

Time series for individual data

All the analyses described above were based on aggregate population rates and average population exposure data. If data were available at an individual level, would there be any advantage to analyzing the raw data at that level, so as to exploit between-individual as well as temporal comparisons? Arguably there would be little to be gained from such an analysis for a censoring event like mortality, particularly absent any data on individual variation in exposures, but perhaps an individual-level analysis would be more informative for recurrent events like hospitalizations, asthma attacks, or absenteeism. This problem was considered by Rondeau et al. (2005) in the context of school absences over a six-month period in the Children's Health Study cohort. Here, the data can be represented in terms of a pair of binary time series, $Y_{ci}(t)$, indicating whether child i in city c was absent or not on day t, and $R_{ci}(t)$, indicating whether that child was "at risk" of an absence (i.e., whether school was in session that day and, for incidence of new absences, whether he or she was in school the previous day). The model then takes the form

$$\text{logit } \Pr(Y_{ci}(t) = 1 | R_{ci}(t) = 1) = B_{0c} + b_c[Z_c(t) - Z_c] + \beta_2 Z_c + \alpha' W_{ci} + e_{ci}$$

where B_{0c} represents a baseline absence rate for children in community c, b_c the acute effect of air pollution in community c, β_2 the chronic effect, α the effect of subject-specific covariates W_{ci}, and e_{ci} a random effect for children representing their "absence proneness." In turn, the b_c are regressed on the long-term average exposures to the same or different pollutants in a second-level model

$$b_c = \beta_1 + \gamma' \left(Z_c - \overline{Z}\right) + e_c$$

to assess the overall acute effect β_1 and its modification γ by the overall level of pollution (or copollutants) in the community. There are three random effects variances in this model: $s_c^2 = \text{var}(e_{ci})$, the unexplained variability in absence rates between children within community c; $\sigma^2 = \text{var}(B_{0c})$, the between-communities variance in baseline rates; and $\tau^2 = \text{var}(e_c)$, the between-communities variance in acute effect slopes.

Several personal characteristics, such as race, income, asthma, and smoking, were found to be associated with illness rates. The risk on any given day was also found to depend on the previous two days' outcome by including autoregressive terms in the model. The only acute effect (β_1) of air pollution was an association of all absences with PM_{10} using a 30-day cubic distributed lag model, yielding a nonsignificant positive association in the first three days and a significant positive association between days 18–28 previously. Curiously, a significant negative association was seen with NO_2 at 5- and 15-day distributed lags. This could reflect a positive effect of O_3 with which it tends to be inversely correlated, but the association of absences with O_3 was not significant. There was also a significant *positive* chronic associations (β_2) of long-term individual absence rates with long-term average O_3 concentrations for total absences, as well as all illness-related and respiratory absences. In contrast, there was no chronic effect of PM_{10}, after adjusting for its acute effect, nor was there any modification (γ) of acute effects by chronic exposure to the same or different pollutants.

Comparing acute and chronic effect estimates

The previous distributed lag and harvesting discussions raise an important question about whether the overall public health effects of air pollution should be assessed by time series or cohort studies (Kunzli et al. 2001). Before delving into this question theoretically, it is worth briefly summarizing the findings on mortality from the few available cohort studies: the Adventist Health Study of Smog (AHSMOG) (Abbey et al. 1999), the American Cancer Society (ACS) study (Pope et al. 1995; 2002; Krewski et al. 2005a), and the Harvard Six Cities study (Dockery et al. 1993). Briefly, the AHSMOG studied mortality from 1977 to 1992 in 6638 nonsmoking Seventh Day Adventists living in California and reported significant associations of all cause mortality with PM_{10} (in males and in both sexes combined) and of lung cancer with both PM_{10} and O_3 in males only. The Six Cities study considered mortality over a 14- to 16-year period in 8111 residents of six cities with diverse air pollution levels, and found a 26% increase (95% CI 1.08–1.47) in all-cause mortality in the most compared to the least polluted city (Figure 8.5, left), and especially strong for lung cancer and cardiopulmonary disease. The ACS cohort comprised half a million adults from across the country and found significant associations of $PM_{2.5}$ and SO_2 with all-cause, cardiopulmonary, and lung cancer mortality (Figure 8.5, right). The magnitudes of the associations reported by these various studies were considerably larger that those estimates from the various time-series studies discussed earlier in this chapter (Table 8.1). Kunzli et al. (2001) and several other authors have pointed out that time series studies capture only the *number* of excess deaths attributable to changes in air pollution, and tell us nothing about the

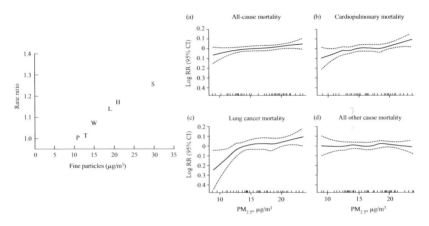

Figure 8.5. Association of total mortality with PM2.5 in the Harvard Six Cities Study (left) and of total and cause-specific mortality with PM2.5 in the ACS cohort (right). (Reproduced with permission from Dockery et al. 1993 and Pope et al. 2002.)

Table 8.1. Comparison of acute and chronic estimates of relative risk for particulates on mortality

	Total mortality	Cardiopulmonary	Lung Cancer
Time series studies			
NMMAPS[1]	1.005	1.007	N.R.
APHEA[2]	1.005	1.007	N.R.
Cohort studies			
AHSMOG[3]	1.12 (males)		2.38 (males)
	0.94 (females)	1.18	1.08 (females)
Six-Cities[4]	1.26	1.37	1.37
ACS[5]	1.04	1.06	1.08

[1] PM_{10} per 10 $\mu g/m^3$ (Samet et al. 2000a) (N.R. = not reported).
[2] PM_{10} per 10 $\mu g/m^3$ (Samoli et al. 2005) (N.R. = not reported).
[3] PM_{10} per 10 $\mu g/m^3$ (Abbey et al. 1999).
[4] Comparing most to the least polluted city (Dockery et al. 1993).
[5] $PM_{2.5}$ per 10 $\mu g/m^3$ (Pope et al. 2002).

Table 8.2. Major design features of time-series studies and cohort studies in air pollution epidemiology. (Reproduced with permission from Kunzli et al. 2001.)

Study design issue	Time-series studies	Cohort studies
Outcome	Counts	Person-time
Exposure variance	Temporal	Spatial
Time from exposure to outcome	Short (days or weeks)	Cumulative (years, lifetime)
Duration of exposure considered	Short term	Can be long term, in the past, etc.
Frailty assessment (underlying condition)	Indirect, by restriction, stratification, case-crossover design	May be investigated as the outcome
Morbidity history of "pollution victims"	Unknown	Known
Years of life lost	Assessable only for the (short) time explained by "harvesting" (or "mortality displacement")	Measured (person-time)

length of life lost as a result of these deaths (Table 8.2). Furthermore, they reflect only the acute effects, not the cumulative effect of a lifetime of exposure. On the other hand, cohort studies can estimate the effect of pollution on loss of life expectancy, but cannot distinguish whether a few individuals have lost a lot of lifetime or many individuals have lost only little (Rabl 2003). In the frailty framework described earlier, air pollution could in principle cause either an increase in the number of healthy people entering the frail pool or the death rate among frail individuals (or for that matter, the death rate in the otherwise "healthy" population). Time-series studies are best at capturing the association of mortality among the frail subpopulation (although as we have seen, the duration of the effect depends upon the average length of time frail individuals spend in that state). They do not capture as well the association between the risk of becoming frail and air pollution, as that may well depend much more on a lifetime history of damage than on the immediate exposure. See also Burnett et al. (2003); Rabl (2003); Thomas (2005a); Rabl (2005; 2006) for further discussion of these issues. We will revisit these issues in the context of risk assessment and individual compensation in Chapters 15 and 16, where we discuss the estimability of excess death rates and loss of life expectancy in the presence of unmeasurable variation in individual frailty.

A particularly promising approach to resolving this discrepancy entails a novel design for long-term cohort studies of recurrent outcomes that combines individual data on risk factors with ecologic-level time-dependent data on exposures (Dewanji and Moolgavkar 2000; 2002), introduced in Chapter 5. Goldberg et al. (2005) have undertaken such a study using Canadian Medicare files as the basis of sampling individuals with a various chronic diseases and, using their histories of doctor's visits, hospitalizations, and medication prescriptions, will be able to assess the effects of air pollution on these intermediate endpoints as well as mortality in relation to their latent time-dependent frailty status. This sort of approach to integrating acute and chronic effects offers the prospect of bringing some real insight into the biological mechanisms relating the two (Kunzli et al. 2008).

9 Spatial models

> Everything is related to everything else,
> but near things are more related than distant things.
> —(Tobler 1970)

Spatial models—and geographic maps in particular—have always played a central role in the "person, place, and time" triad of descriptive epidemiology. Nowhere is this better illustrated than in the pioneering work of John Snow (1855) on cholera (Figure 9.1) that led to the identification of the Broad Street pump as the source of the epidemic and his removal of the pump handle as arguably the most famous public health intervention of all time.

A nearly universal feature of environmental epidemiology is that the data are spatially distributed. Individuals' exposures are determined in part by where they live, work, play, or otherwise spend time. Thus, those who are at similar locations are likely to have similar exposures. They are also likely to have similar values of other unmeasured risk factors, so their outcomes are also likely to be correlated. In addition to violating the basic assumptions of independence underlying the standard statistical methods discussed in previous chapters, this spatial dependency can itself be a source of information about true exposures or unobserved shared confounders.

In this chapter, we consider methods that allow for spatial dependencies among subjects to obtain valid statistical tests and confidence limits and to understand the nature of these spatial relationships. These methods will also prove central to some of the methods of exposure modeling to be discussed in Chapter 11. Implementation of these statistical methods has been revolutionized by the development of sophisticated geographic information system (GIS) software for managing and analyzing spatial data.

A related aspect of discrete spatial data is sparseness. If the scale on which spatial dependencies operate is fine, then within any grid fine enough to capture this phenomenon, one might expect to see only a few cases in any cell, with great variability between cells. But over a fine grid, there are many cells, so some could be expected to show marked excesses of disease just by chance. The nonuniform distribution of the population at risk, not to mention spatial variability in baseline risk factors, will tend to produce the appearance of disease clustering, particularly if attention in focused

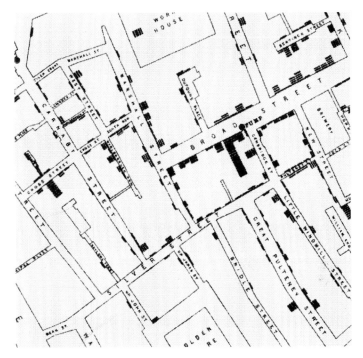

Figure 9.1. A portion of John Snow's (1855) map on the distribution of cholera cases in the vicinity of the Broad Street pump in London in 1854.

(say, by the public or the media) on a particular cluster in isolation (the so-called "Texas sharpshooter" effect explained below). Sorting out whether the *general pattern* of clustering is greater than one might expect by chance or can be explained by known risk factors—or whether *any particular cluster* may have a common cause or simply be a coincidence—are major statistical challenges, requiring specialized techniques collectively known as "small area statistics" or "disease clustering."

Of course, environmental epidemiology data have both spatial and temporal dimensions. In the two preceding chapters, we have considered the temporal dimension in detail. In this chapter, we also discuss a particularly interesting situation of dependency between the two dimensions known as space-time clustering, which can provide evidence of person-to-person transmission of an infectious agent or temporally restricted point-source exposures.

Spatial statistics is a complex field with applications in many different disciplines. For a more detailed treatment of the topic, see such textbooks as Cressie (1993); Lawson et al. (1999); Elliott et al. (2001); Waller and Gotway (2004); Lawson (2006). Additional references specific to GIS methods are provided below.

Mapping exposures and disease

Spatial models have been used in epidemiology to map either exposures or disease patterns or to correlate the two. Although at first glance, it would seem immaterial whether the data to be mapped are exposures or diseases, the nature of the two data sources tends to be quite different, requiring different methods. For example, exposure data might consist of a relatively sparse, irregularly spaced set of measurements, possibly complemented by information on sources and dispersion patterns. Disease data, on the other hand, could take the form of a set of rates (or Poisson counts of cases and numbers at risk or age-standardized expected counts) over an array of geographic areas ("aereal" data) or as the coordinates of a set of cases and suitable controls ("point process" data). Because the two are likely to be organized in different ways, relating exposure to disease can require rather sophisticated statistical methods. We begin by introducing a general model for spatial correlations in normally distributed data and then its extensions to other structures, such as Poisson counts on geographically defined areas or case-control locations.

Continuous spatial data

Spatial models have been particularly useful in environmental epidemiology for smoothing sparse exposure measurements so as to be able to assign exposures to individuals at locations where no direct measurements are available. These can also be combined with information on sources (e.g., point sources like toxic waste disposal sites or line sources like highways) and sophisticated dispersion models, possibly taking account of hydrology or meteorology.

Suppose for each observation of some variable X_i, $i = 1, \ldots, n$, we have a vector of associated covariates Z_i and a location p_i, generally comprising a pair of geographical coordinates. Depending upon the context, X could be some health outcome and Z some risk factors, or X could represent exposure and Z a vector of predictors of exposure like source information.

To begin with, we ignore the predictors Z and consider the problem of estimating the value of $X(p)$ at some unmeasured point p, using only a finite set of irregularly spaced measurements $X(p_i)$ at other locations. We then turn attention to how to incorporate covariates into such predictions. Whereas some environmental epidemiology studies have relied exclusively upon measurements (e.g., air pollution or magnetic field measurements at subjects' homes) and others exclusively on model predictions (e.g., distance from major highways or high-tension power lines, or physical models for predicted exposures), in general, exposure assessment is ideally based upon a combination of the two, taking account of their respective spatial covariance and measurement error structures (Beyea 1999).

Although one might simply use a two-dimensional generalized additive model (GAM, see Chapters 6 and 8), the standard method in spatial statistics has been "kriging." This is based on a simple weighted average of the available measurements in the neighborhood of any given point, with the optimal weights being derived from the observed covariance structure of the measurements. Essentially, the prediction of the field at some unknown location $\mathbf{p}$ is given by

$$\hat{X}(\mathbf{p}) = \sum_i X(\mathbf{p}_i) w(\mathbf{p} - \mathbf{p}_i)$$

for some suitable choice of weights $w(\Delta \mathbf{p})$. Now suppose one assumed the observed measurements were independently normally distributed around some unknown mean $\mu(\mathbf{p})$ with variance σ^2 and this mean was a stationary Gaussian random field with covariance $\tau^2 \rho(\Delta)$ for pairs of points separated by a vector Δ. Then it can be shown that the kriging estimator with the vector of weights given by

$$w(\mathbf{p} - \mathbf{p}_i) = \mathbf{r}(\mathbf{p})'[\sigma^2 \mathbf{I} + \tau^2 \mathbf{R}]^{-1}$$

where $\mathbf{r}(\mathbf{p})$ is the vector of $\rho(\mathbf{p} - \mathbf{p}_i)$ values and $\mathbf{R}$ is the matrix of $\rho(\mathbf{p}_i - \mathbf{p}_j)$ values, is the best linear predictor of $X(\mathbf{p})$, that is, that which minimizes the prediction mean squared error, $E\{[\hat{X}(\mathbf{p}) - X(\mathbf{p})]^2\}$. More generally, the weights are derived from the "variogram"

$$
\begin{aligned}
C(\Delta) &= \frac{1}{2}\mathrm{var}[X(\mathbf{p}) - X(\mathbf{p} + \Delta)] \\
&= \mathrm{var}[X(\mathbf{p})] - \mathrm{cov}[X(\mathbf{p}), X(\mathbf{p} + \Delta)] \\
&= \sigma^2 I(\Delta = 0) + \tau^2[1 - \rho(\Delta)]
\end{aligned}
$$

which can be solved for σ^2, τ^2, and $\rho(\Delta)$. $C(\Delta)$ might be estimated nonparametrically by a simple average over pairs of points within some window of distance Δ from each other

$$\hat{C}(\Delta) = \frac{1}{2|N(\Delta)|} \sum_{i,j \in N(\Delta)} [X(\mathbf{p}_i) - X(\mathbf{p}_i)]^2$$

where $N(\Delta)$ denotes the set of pairs of observations within that window and $|N(\Delta)|$ the number of such pairs. Traditionally, the variogram is presented as a function of both distance and direction, but if one assumed that the field is isotropic (independent of direction), it might be estimated simply as a function of distance. Parametric models, such as $\rho(\Delta) = \exp(-\rho\|\Delta\|)$, or more complex ones taking direction into account—an ellipsoidal function or atmospheric-physics-based dispersion models (see Box 9.1) taking meteorological, topographic, or other features into account—can be fitted by maximum likelihood.

9.1 *Dispersion modeling*

Atmospheric point and line source dispersion modeling plays an important role in air pollution exposure assessment. The basic idea is to compute the steady-state solution to the dispersion equation

$$\frac{\partial Z}{\partial t} = -\nabla^2 Z + \mathbf{v} \cdot \nabla Z$$

where $Z(\mathbf{p}, t)$ is the concentration of a pollutant at location $\mathbf{p}$ and time t, ∇Z denotes the curl operator (the sum of second partial derivatives with respect to space), and $\mathbf{v} \cdot \nabla Z$ the dot product of the wind speed/direction vector and the spatial gradient (the vector of first spatial derivatives) of pollutant concentration. In the absence of wind, the steady state solution reduces to a simple normal density with distance from the point source, hence the popularity of simple Gaussian plume models. In the presence of wind, a more complex ellisoidal shape downwind of the source results. Lawson (1993) describes a family of exponential distributions that can be used in an empirical manner to fit measured concentrations to various transformations of distance and direction from a point source.

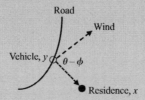

For line sources like roads, one must then compute the line integral along the length of each road segment y (see figure) and sum over all roads h, V_h, to obtain the predicted concentration C at a receptor site x. With variable winds, one must also integrate these results over the probability distribution P of wind speed v and direction θ:

$$\hat{C}(x) = \sum_h V_h \iint_{dv, d\theta} dP(v, \theta) \oint_{dy} f\left[||x - y||, v, \cos(\theta_v - \varphi_{x \to y})\right]$$

where $f(d, v, \theta - \varphi)$ is the equilibrium solution to the dispersion equation in terms of distance, wind velocity, and the angle between the two vectors. Turbulence, mixing height, irregular topography (e.g., canyons created by buildings in urban areas or mountainous terrain), atmospheric chemistry, deposition, and other factors can make these theoretical calculations more complex and in practice, they are generally evaluated using numerical methods by various software packages, such as the California Line Source Dispersion Model (CALINE4) model (Benson 1989) from the California Department of Transportation.

To incorporate spatial predictors of exposure such as traffic into the smoother, suppose we now assume that

$$X(\mathbf{p}_i) = \mathbf{Z}_i'\boldsymbol{\beta} + \mu(\mathbf{p}_i) + e_i$$

where $\mu(\mathbf{p})$ is a Gaussian random field with mean zero and covariance $\tau^2 \rho(\Delta)$ and $e_i \sim N(0, \sigma^2)$. Suppose further that we take $\rho(\Delta) = \exp(-\delta \|\Delta\|)$. Then it follows that the conditional distribution of any particular observation, given all the others is

$$\hat{X}(\mathbf{p}_i) = E(X(\mathbf{p}_i)|\mathbf{Z}, \mathbf{X}_{(-i)}) = \mathbf{Z}_i'\boldsymbol{\beta} + \frac{\tau}{\sqrt{\sigma^2 + \tau^2}}\overline{X}_{(-i)}$$

where

$$\overline{X}_{(-i)} = \frac{\sum_{j \neq i}(X_j - \mathbf{Z}_j'\boldsymbol{\beta})\exp(-\delta D_{ij})}{\sum_{j \neq i}\exp(-\delta D_{ij})}$$

is a weighted average of the deviations of all the other observations from their predicted values, weighted by the exponential decay function of their distances from $\mathbf{p}_i$. Its prediction uncertainty is

$$s_i^2 = \mathrm{var}\left(\hat{X}(\mathbf{p}_i)|\mathbf{Z}, \mathbf{X}_{(-i)}\right) = \sigma^2 - \mathbf{r}(\mathbf{p}_i)'[\sigma^2\mathbf{I} + \tau^2\mathbf{R}]^{-1}\mathbf{r}(\mathbf{p}_i) \quad (9.1)$$

In many applications, X_i (e.g., local air pollution measurements) will be available only on a subset of study subjects while Z_i (e.g., indicators of local traffic density or predictions from dispersion models) will be available for the entire cohort. In situations where Z_i are also available only for the same or different subset of subjects, then co-kriging methods (Sun 1998) can be used. These entail smoothing both the X and the Z fields simultaneously, exploiting the spatial correlations within each field as well as the correlation at the same location between the two fields.

Conceptually, the model might be fitted by the following iterative scheme. Initially assume observations are independent and use ordinary least squares regression to estimate $\boldsymbol{\beta}$ and σ^2. Now for each pair of observations, compute the cross-product of their deviations, $C_{ij} = (X_i - \mathbf{Z}_i'\boldsymbol{\beta})(X_j - \mathbf{Z}_j'\boldsymbol{\beta})$. Omitting the C_{ii} terms, regress $\ln(C_{ij})$ on $\ln \tau^2 - \delta D_{ij}$. Compute the $\overline{Y}_{(-i)}$ using this estimate of δ and regress $Y_i - \overline{Y}_{(-i)}$ on Z_i to obtain new estimates of $\boldsymbol{\beta}$ and σ^2, and repeat this process until convergence. A single-step procedure is implemented in the SAS procedure MIXED, using the following code.

```
proc mixed;
    model X = Z1–Zn/solution;
    repeated/local type = sp(exp) (Px Py);
```

This returns estimates of the regression coefficients for each Z, the residual variance τ^2, the spatial variance σ^2, and the distance decay parameter δ.

"Land use regression" (Briggs et al. 2000) is an example of this approach, incorporating predictors like buffers from point or line sources, indicators for zoning types, and so on. Briggs et al. illustrate this approach with an application to NO_2 measurements in the Huddersfield portion of the SAVIAH project.

Spatial smoothing and prediction models can also be combined, using the predictions of a sophisticated atmospheric dispersion model as one of the covariates in a spatial regression model. In Chapter 11, we will discuss a measurement survey of 259 homes selected at random from 10 of the 12 communities of the Children's Health Study (Gauderman et al. 2005). A multi-level spatial autocorrelation model for NO_2 concentrations was fitted, allowing for spatial dependencies both within and between communities and incorporating several traffic-related covariates. The single strongest predictor of measured NO_2 concentrations was the CO concentrations from freeway sources predicted by the CALINE4 model, but traffic-related indices (simple distance-weighted traffic counts) significantly improved the fit of the model, presumably by correcting for some imperfections in the physical model. The estimated between-community spatial variance was $\tau_B^2 = 6.5$ (ppb^2) with decay parameter $\delta_B = 0.0096$/km, the within-community spatial variance was $\tau_W^2 = 44.3$ with $\delta_W = 0.39$/km, and residual (independent) variance $\sigma^2 = 14.1$. This model could then be used to assign exposures to all subjects in the cohort, whether or not direct measurements are available at their homes.

See Jarup (2004); Nuckols et al. (2004) for reviews of various other applications. Williams and Ogston (2002) compared various methods based on distance, spatially smoothed measurements, and dispersion modeling for arsenic and copper pollution from a waste incinerator and concluded that either of the two approaches yielded similar assessments of exposure.

The Gaussian random field regression methods just described are fully parametric, both in the mean and covariance structures. Bayesian maximum entropy methods (Christakos and Li 1998) and general model-based geostatistics (Diggle et al. 1998) provide greater flexibility in incorporating nonlinear predictors, nonstationarity, and non-Gaussian error distributions over simple smoothers like GAMs and kriging. For example, the regression part of the model $Z'\beta$ could be replaced by a flexible GAM or the Gaussian random field $\mu(\mathbf{p})$ by a two-dimensional GAM. This would involve the usual trade-off between robustness and efficiency.

Distance might be measured on various scales. Although Euclidean distance is generally used, one might want to adopt some scale that reflects topography or prevailing winds if modeling air pollution levels (Le and Zidek 1992), or social distance if modeling unmeasured risk factors. For example, Brown et al. (1994) used the temporal correlation in air pollution levels between areas to define a distance metric (Figure 9.2), performed the spatial analysis on that scale, and then transformed back to Cartesian coordinates to display the final results.

Of course, exposure can have both spatial and temporal dimensions, and it can be advantageous to model the two components of variation jointly. For example, Whitaker et al. (2005) built a Bayesian spatial-temporal model for chlorination byproducts in drinking water for use in an epidemiologic study of adverse reproductive outcomes (Toledano et al. 2005; Nieuwenhuijsen et al. 2008). This provided estimates of trimester-specific exposures to trihalomethanes for each woman in the study, in a situation where the available spatial measurements alone were too sparse to provide robust estimates of exposure, but additional strength could be borrowed from measurements at the same location at different times. Gryparis et al. (2007) describe a Bayesian semiparametric latent variables modeling approach for spatial–temporal smoothing of sparsely distributed air pollution measurements.

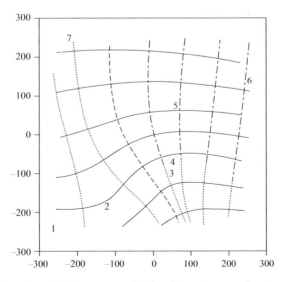

Figure 9.2. Transformation from geographical to dispersion space; locations of the available monitoring stations are indicated by the numbers, the "distance" between stations in this representation being proportional to the temporal correlations in their pollution measurements. (Reprinted with permission from Brown et al. 1994.)

Designing a measurement network

One use of spatial models is for the purpose of deciding where to locate a set of additional measurements S_2 based on an existing set of measurement locations S_1 so as to maximize the informativeness of the entire network $S = S_1 \cup S_2$ for predicting exposures at some even larger set N of unmeasured locations. This generally requires first an estimate of the prediction variance $V_S(\mathbf{p})$ given by Eq. (9.1) across the domain of points $\mathbf{p}$ of interest, using spatial model parameters $(\hat{\sigma}^2, \hat{\tau}^2, \hat{\delta})$ estimated from the available measurements S_1. This is then weighted by a "demand surface" $N(\mathbf{p})$ giving the population density at which future exposure estimates will be needed, to yield an estimate of the average variance $\overline{V}_S(N) = \int N(\mathbf{p}) V_{S_1 \cup S_2}(\mathbf{p}) \, d\mathbf{p}$. The challenge is then to choose the set S_2 that would minimize $\overline{V}_S(N)$ (Diggle and Lophaven 2005; Kanaroglou et al. 2005; Thomas 2007a). Intuitively, one would like to add measurement locations that would most improve the predictions at those points that have the largest prediction variance under the current network and are the most influential; thus, one might want to select points that are not too close to the existing measurement locations or to each other, but close to many subjects for whom predictions will be needed. For example, suppose the demand field is given by a discrete set of N locations of epidemiologic study subjects, then one would seek to minimize $\sum_{i \in N} V_{S_1 \cup S_2}(\mathbf{p})_i$, say by choosing the locations j from $N \backslash S_1$ one at a time that reduce the average variance the most until the target number of locations has been selected. Further improvements might then be possible by proposing to replace a randomly selected point in S_2 by a randomly selected point in $N \backslash S$ and accepting the swap if it results in a reduction in $\overline{V}_S(N)$. $V_S(P)$ requires a spatial model; estimation of the spatial decay parameter might be improved by including some proportion of close pairs in $S_1 \cup S_2$ (Diggle and Lophaven 2005).

Of course, one must first decide how many locations should be included in S_2, a decision that is generally constrained by cost, but one should consider the trade-off between the measurement costs and the costs of the main study N: there is a point of dimishing returns where further improvement in exposure prediction is less than the improvement in power that would result from increasing the sample size N for the main study for the same total cost. For example, rather than minimizing the average exposure prediction variance alone, what is really needed is to maximize the Fisher information for the relative risk parameter β from a model that will use these predictions. This may entail selecting measurement locations that will maximize the precision of the most influential points, such as those individuals predicted to be the most heavily exposed. See Thomas (2007a) for further discussion of these design trade-offs.

Poisson count data

Now suppose Y_c denotes the count of the observed number of cases in some geographical unit c, say a rectangular grid ("lattice") or some administrative unit like census tracts ("areal" data). These units typically differ in population size and composition, so that one would not expect similar counts across units anyway. Let E_c denote the expected number of events under the null hypothesis, computed in the usual way from the population distribution as described in Chapter 3 using age/sex/race standardized rates. Furthermore, suppose we have a vector Z_c of average covariate values for subjects in each geographic unit. Following Clayton and Kaldor (1987), we now assume that

$$Y_c \sim \text{Poisson}[E_c \exp(Z'_c \beta + \mu_c)] \tag{9.2}$$

where μ_c denotes a set of spatially correlated residuals not accounted for by the modeled covariates.

A simple way of modeling the residuals is by some form of local non-parametric process, say using a two-dimensional kernel density $K(D_{cd}/\sigma)$, where D represents the distance between the centriods of regions c and d and σ is a scale parameter that can be chosen by cross-validation. One then estimates μ_c by

$$\mu_c = \frac{\sum_{d \neq c} \mu_d K(D_{cd}/\sigma)}{\sum_{d \neq c} K(D_{cd}/\sigma)}$$

However, this deterministic model provides no measure of uncertainty.

The *intrinsic conditional autoregressive* (CAR) *process* (Besag 1974) puts this basic idea in a probabilistic framework by assuming the μs have a multivariate normal distribution with covariances specified by an adjacency matrix $\mathbf{A}$ consisting of binary indicators for whether a pair of regions i and j share a common boundary. The CAR model can be written in terms of a joint density function given by

$$\Pr(\mu) \propto \exp\left(-\frac{1}{2\sigma^2} \sum_{c \sim d} (\mu_c - \mu_d)^2\right)$$

where $c \sim d$ indicates a sum over all pairs of adjacent areas. Although this expression may not be a proper distribution (i.e., it does have a finite integral), its conditional distributions *are* proper and are simple normal densities

$$\Pr\left(\mu_c | \mu_{(-c)}\right) \sim N\left(\overline{\mu}_c, \sigma^2/n_c\right)$$

where $\overline{\mu}_c = \sum_{d \in \mathcal{N}_c} \mu_d / n_c$ and $\mathcal{N}_c$ denotes the set of n_c areas adjacent to area c. This model has only one free parameter, σ^2, so assumes that all the residual variation is spatially correlated. For greater flexibility, Besag et al. (1991) extended the model by allowing an additional independent deviate $\delta_c \sim N(0, \tau^2)$ to Eq. (9.2) corresponding to extra-Poisson variation that is spatially independent. Alternatives to assuming a normal distribution for these terms would be some heavier-tailed distribution such as a double-exponential (Besag et al. 1991), or a nonparametric (Clayton and Kaldor 1987; Kelsall and Diggle 1998) or mixture (Green and Richardson 2001) distribution. One could also allow the smoothing parameters σ^2 and τ^2 to vary spatially rather than be assumed to be constant across the region, so as to better capture discontinuities in the risk surface (Knorr-Held and Rasser 2000; Denison and Holmes 2001). More general weight matrices could be used in place of $\mathbf{A}$, say with continuous weights w_{ij} depending upon the length of the shared boundary S_{cd} or the distance D_{cd} between their centroids, so that

$$E(\mu_c | \mu_{(-c)}) = \alpha \sum_d w_{cd} \mu_d \Big/ \sum_d w_{cd}$$

and

$$\mathrm{var}(\mu_c | \mu_{(-c)}) = \sigma^2 \sum_d w_{cd}^2 \Big/ \left(\sum_d w_{cd} \right)^2$$

For example, one might choose $w_{cd} = \exp(-\rho D_{cd})$, but some restrictions may be needed for the resulting multivariate density to be proper, that is, for the inverse matrix $\mathrm{cov}(\mathbf{\mu}) = \sigma^2(\mathbf{I} - \alpha\mathbf{W})^{-1}$ to exist (Besag 1974). Hence, instead of starting with these conditional distributions, one could use the weights to specify the joint distribution of μ as a multivariate normal with mean zero and covariance matrix $\tau^2\mathbf{I} + \sigma^2\mathbf{W}(\rho)$. The main difference between the two models is that the intrinsic CAR model is not stationary (i.e., its mean is not constant across the region) whereas the joint model assumes that the marginal means $E(\mu_i)$ are zero for all i. On the other hand, fitting the intrinsic CAR model is substantially simpler because it involves only univariate normal distributions, without the need to invert the covariance matrix in the joint model, except for the purpose of estimating the covariance parameters (σ^2, τ^2, ρ). Lawson et al. (2000b) describe simulations comparing these various approaches and conclude that the independent gamma-Poisson and the parametric spatial model of Besag et al. (1991) generally perform better than either the nonparametric smoothing or mixture model approaches. See Pascutto et al. (2000) for a general review of these approaches. Gotway and Wolfinger (2003) provide a comparison between marginal and conditional models for spatial dependencies among disease rates, and Wall (2004) discusses the forms of dependency that are implied by the different models.

Clayton and Kaldor (1987) described an Markov chain Monte Carlo (MCMC) approach to fitting the intrinsic conditional autocorrelation model by treating each of the true rates μ_c as random variables, to be sampled from their conditional distributions given their observed data (Y_c, E_c) and the current values of each μ_d for all adjacent points $d \in N_c$, and then sampling the model parameters β, σ^2, τ^2 given the current estimates of all the μ_c. Their application to lip cancer incidence data for 56 counties of Scotland produced empirical Bayes estimates that were generally shrunk towards the overall mean relative to the maximum likelihood ones, as might be expected. However, the spatially correlated estimates were shrunk instead towards local means, which in some cases were higher and in some cases lower than the overall mean or even the raw observed rates themselves.

In Chapter 8, we described the hierarchical model for multi-city time series data used by Dominici et al. (2000) to analyze the 20 largest U.S. cities in the NMMAPS study. Their analysis also compared independent and spatially correlated models for the second stage regression. Recall that for each study, the first stage yields an estimate $\hat{b}_c$ of the coefficient for the effect of daily variation in air pollution on mortality counts for each city c, along with their variances. In the second stage, these estimates are regressed on city-specific covariates $\mathbf{X}_c$, appropriately weighted by their variances as described earlier. In the spatial version of the model, the $\hat{b}_c$s are treated as a vector of correlated observations. Dominici et al. considered two variants of the spatial model, one using a spatial autocorrelation model with correlation between being a function $\exp(-\rho D_{cd})$ of the distance D_{cd} between cities c and d, the other by treating the correlation as the same for cities in the same region, zero between regions. There was much less variation between cities in the spatially smoothed estimates than in the city-specific maximum likelihood estimates, as much of the apparent variation in the city-specific estimates was simply random sampling error. Under the continuous distance model, the correlation between $\hat{b}_c$ estimates for cities separated by the median distance was 0.61 (95% CI $0.3 - 0.8$), stronger at the lower 25th percentile $(0.86, 0.68 - 0.93)$, declining to 0.3 $(0.05 - 0.58)$ at the 75th percentile. In the regional model, the within-region correlation was 0.68 $(0.42 - 0.86)$. Estimates of the intercepts and regression coefficients for second stage covariates and of the between-cities variances in the $\hat{b}_c$s were essentially unchanged from the independence model, however.

Maps of estimated disease rates, even geographically smoothed, do not convey any information about the uncertainties of these estimates. To overcome this difficulty, Jarup and Best (2003) suggest plotting posterior probabilities that the true RR is greater than some threshold. For example, Figure 9.3 compares prostate cancer relative risks and posterior probabilities that RR > 1 at the ward level for two districts of England (Jarup

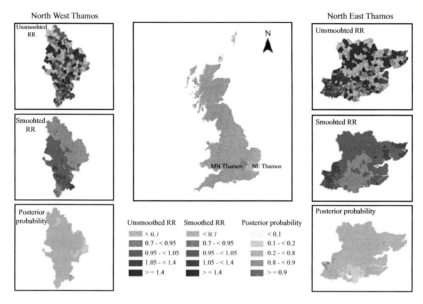

Figure 9.3. Comparison of unsmoothed and smoothed estimates of prostate cancer relative risk in two districts of England, along with the posterior probability that RR > 1. (Reproduced with permission from Jarup et al. 2002.)

et al. 2002). Two other districts (East Anglia and Yorkshire) that were also examined at this same level of detail showed virtually no variation in posterior *RR*s or posterior probabilities after smoothing, suggesting that environmental causes were unlikely to play a major role. Richardson et al. (2004) report an extensive simulation study comparing the performance of posterior relative risk and posterior probability estimates. To minimize false positive and false negative rates for calling areas "elevated" or not, they recommend as a criterion that Pr(RR > 1) be greater than 0.7 or 0.8. This yields good power for detecting true RRs of about 1.5–2 when the expected cell counts are about 20 or for detecting RRs of about 3 even when based on much smaller expected counts (in the latter case, however, the posterior *RR*s themselves tend to be underestimated by about 50%).

Most maps of disease rates have been based on administrative units rather than rectangular grids because that is the form in which aggregate data on rates are generally available. The sizes of the units have varied widely—countries, states, counties, census tracts, and so on—and the degree of smoothing that results and interpretation of associations depends upon the choice of areal unit. For example, Elliott and Wartenberg (2004) display maps of the percentage of housing before 1950 (a risk factor for childhood exposure to lead-based paint) by census block, zip code, and county showing how the more aggregated units tend to be dominated by denser population centers, obscuring information about prevalence in more rural local areas. This observation is known as the "modifiable areal unit problem" in geography (Openshaw 1984; Fotheringham and Wong 1991). The problem is particularly relevant to "contextual" variables used in ecologic analyses but also affects the degree of bias resulting from aggregation of individual data into groups, as discussed in the following chapter. Here, it is important that the unit of analysis be defined in a way that reflects the scale upon which these variables are hypothesized to act if the resulting associations are to be interpreted causally in terms of effects on individuals with these areas.

Individual case-control data

Now suppose we have conducted an unmatched case-control study, so that Y_i is an indicator variable for case-control status and, as before, each individual has an associated vector of covariates Z_i and a location p_i. Using the GLMM framework introduced earlier, we might assume

$$\text{logit } \Pr(Y_i = 1) = Z_i'\beta + \mu(p_i)$$

where $\mu(p) \sim CAR(C)$. The estimated spatial variance τ^2 from this model provides a test of whether cases tend to be closer to each other than controls, adjusted for their covariates.

With case-control data, interest might center either on a smoothed surface of residual risk $\mu(p)$ or on a test of clustering behavior. We defer the latter to a later section of this chapter. For the former, $\mu(p)$ is generally estimated by some form of kernel density estimator or GAM (Kelsall and Diggle 1998), as discussed earlier in this chapter. In this approach, the controls are essentially estimating the spatial density of the population at risk (which would not be expected to be uniformly distributed) and the cases would reflect this inhomogeneity *plus* any additional clustering due to spatial variation in disease risk. Key to the validity of this approach, however, is that controls be a random sample of the source population; thus, some modification is needed when this approach is applied to *matched* case-control data (Diggle et al. 2000), as described under "Tests of clustering behavior" below.

Spatial correlations of exposure and disease

Tests of correlation with a single point source

In the absence of detailed exposure measurements or model predictions, one might want simply to test the null hypothesis of no association between aggregated disease rates and proximity to some point source. For this purpose, Stone (1988) and Bithell and Stone (1989) introduced a simple nonparametric method where the regions could be ordered by distance and one tested H_0: $\lambda_c \equiv \lambda$ against H_1: $\lambda_1 \geq \lambda_2 \geq \cdots \geq \lambda_C$ (with at least one strict inequality) using isotonic regression (see Chapter 6). Morton-Jones et al. (1999) extend this to a semiparametric approach incorporating covariates and provide further discussion and references. Related methods for case-control data with individuals' locations relative to a specific point source have been developed by Diggle (1990). The original formulation was semiparametric, with a model of the form

$$\lambda(\mathbf{p}) = \lambda_0 g(\mathbf{p})[1 + \beta \exp(-\rho||\mathbf{p} - \mathbf{p}_0||)]$$

where $\mathbf{p}$ denotes any point in the region, $\mathbf{p}_0$ the location of the point source, $g(\mathbf{p})$ a smooth function of location for background risk, fitted using a nonparametric kernel density. Subsequent extensions include matched case-control data (Diggle et al. 2000) and flexible modeling of excess risk by isotonic regression (Diggle et al. 1999), amongst others.

Tests of association with spatially defined exposures

The problem of relating spatially correlated exposure data to spatially correlated disease outcomes requires careful consideration of the correlation structure in both variables. For aggregate data with both variables defined over the same lattice structure (e.g., counties), Cook and Pocock (1983) describe a maximum likelihood method for fitting a multiple regression model using the CAR model to allow for spatial dependence in the residuals. Clifford et al. (1989) address the problem by a simple calculation of the "effective sample size" (number of aggregate units) that takes the spatial correlation in both variables into account when evaluating the significance of the correlation between the two variables.

Richardson et al. (1995) performed one of the first analyses regressing spatially correlated disease data on spatially correlated environmental data, using 459 local authority districts of the United Kingdom as the observational units for incidence of childhood leukemia in relation to surveys of natural background radiation (gamma and radon). Although the results yielded no convincing associations with radiation, the paper broke new ground methodologically by Poisson regression using environmental covariates and a smooth spatial structure and by modeling spatial and nonspatial residuals explicitly. These analyses revealed that the majority of the extra-Poisson variation was due to the local spatial component and this structure was relatively stable over time. Issues of ecologic inference that arise in using aggregate data in such analyses will be addressed in the following chapter.

In practice, a major problem with spatial regression analyses is different scales of observation for exposure and disease ("spatially misaligned data"). Disease data may consist of rates by administrative units with irregular boundaries, while exposures may comprise measurements at discrete points or predictions from some model over a regular grid. Thus, Jarup (2004) warns that

Attempts to assess risk by overlaying maps of exposure and disease, given the (in)accuracy of the exposure estimates, latency periods, and migration problems, are likely to be misleading and should be avoided.

For risk assessment purposes, he suggests instead that spatial models be used to assess the population distribution of exposure and then apply established exposure–response relationships to predict the population distribution of disease risks. This does not obviate the need for spatial regression techniques, but then only in the dependent variable, since the analysis is conditional on the estimated exposure data. For grouped data, one might adopt a model of the form $\ln(\lambda_c) = \mathbf{Z}'_c\boldsymbol{\beta} + U_c + V_c$, as described earlier, incorporating both independent errors U_c and spatially correlated errors V_c. For individual data, one might model logit $\Pr(Y_i = 1) = \mathbf{Z}'_i\boldsymbol{\beta} + V_i$ with spatially correlated errors at the individual level in a similar manner.

Jerrett et al. (2005) used kriging to develop smooth exposure surfaces from 23 $PM_{2.5}$ and 43 O_3 monitors in the Los Angeles basin for assignment of individual exposures to the 22,905 participants from that region in the American Cancer Society cohort study of mortality. These were then used as individual-level covariates, along with personal and aggregate-level confounders, in a hierarchical Cox regression model that allowed for spatial correlations (Ma et al. 2003). This multi-level modeling approach for dealing with aggregate data will be discussed further in the following chapter. Briefly, the hazard rate model can be written as

$$\lambda_{csi}(t, \mathbf{Z}) = \lambda_{0s}(t)\exp(\mathbf{Z}'_{ci}\boldsymbol{\beta} + e_c)$$

where c indexes communities (based on zip codes), s strata (based on age, sex, race), and i individuals, e_c is a community-level random effect for spatially correlated residual mortality risk (with zero mean and variance and correlation between adjacent zip codes to be estimated), and $\mathbf{Z}$ is a vector of exposures and confounders at individual or aggregate levels. Elevated risks were found for all-cause, lung cancer, and cardiovascular disease mortality in relation to $PM_{2.5}$, and were not substantially affected by adjustment for O_3 or proximity to freeways. These associations were up to three times stronger at the intra-city level than previously reported results based on between-city comparisons (Burnett et al. 2001). Burnett et al.'s re-analysis of the full ACS cohort data also demonstrated strong spatial autocorrelation at the between-city level out to about 1000 km, but this was reduced by including a smooth surface in location. Adjustment for autocorrelation had little effect on the point estimates of the sulfate effect, although the confidence interval widened. Adjustment for the spatial smoother reduced the sulfate risk estimate more and more with increasing complexity of the surface, but it remained significant with a standard error that declined in proportion to the estimated coefficient.

Although not using formal geostatistical models *per se*, two analyses of the atomic bomb survivor data are revealing about the potential insights that spatial analysis can provide. Peterson et al. (1983) analyzed cancer mortality and Gilbert and Ohara (1984) analyzed acute radiation symptoms in relation to direction from the hypocenters, adjusting for distance and shielding in the former case, for radiation dose in the latter case. Details of their methods differ slightly, but both essentially entail a model of the form $\lambda(t, P, Z) = \lambda_{0Z}(t) \exp(\beta_P)$, where t denotes age, P location in eight directions from the hypocenter, and Z strata of distance/shielding or dose and other baseline factors. Both analyses reported significant excesses of the studied endpoints in the WSW sector in Nagasaki. In Hiroshima, there was a significant excess of mortality in the westerly direction (WNW and WSW) but of symptoms in a northerly direction (NNW and NNE).

Such excesses could be due either to underestimation of doses in those directions or to differences in other epidemiologic risk factors. The latter seems more likely for cancer, since the patterns were fairly similar inside and outside a circle 1600 m from the hypocenter. Indeed, the particularly large excess of lung cancer in WSW Nagaski could be due to employment in the shipyards (a source of asbestos exposure) located in the SW portion of the city, or to differences in smoking or socioeconomic status. This seems a less likely explanation for the acute symptoms, however, which are unlikely to be caused by anything other than radiation. These analyses were motivated in large part by concerns in the early 1980s about possible circular asymmetry in the Hiroshima doses due to the shape of the bomb and its trajectory at the moment of explosion, but subsequent dosimetry analyses have shown that any such asymmetry would have occurred only very close to the hypocenter where there were few survivors. The possibility of asymmetry due to subsequent fallout is more plausible, but largely unsupported by the available measurements. (Both bombs, being air bursts, generated relatively little fallout compared with the surface bursts at the Nevada Test Site and the Pacific Proving Grounds.)

These data would be worth revisiting using the modern geostatistical methods described above. Thus, one might posit a model of the form

$$\lambda(t, Z, p, \mathbf{W}) = \lambda_{0W}(t)s_1(\mathbf{p}) \left\{1 + \beta Z \exp[\gamma'\mathbf{W} + s_2(\mathbf{p})]\right\}$$

where $s_1(\mathbf{p})$ and $s_2(\mathbf{p})$ are spatially smooth functions (e.g., GAMs) in exact location (not just octants of direction from the hypocenter) describing residual spatial variation in baseline risk and excess risk due to radiation respectively. (As before, Z represents dose and $\mathbf{W}$ additional confounders and/or modifiers.) Plots of these smooth functions could then be examined for potential "hot spots" of risk factors or factors potentially contributing to dose errors. To avoid the appearance of data dredging, an alternative approach would be to use a 1 df spatial scan statistic to test for any pattern of departure from spatial homogeneity. Work along these lines is planned or currently underway by RERF statisticians.

Another approach to using spatial data suggested by Jarup is to use disease maps to characterize the baseline risk in a population and then in a prospective surveillance mode, observe changes in disease rates that could be related to subsequent changes in exposure.

Case clustering and surveillance

Tests of clustering behavior

Various authors have proposed *tests* of the clustering hypothesis. One of the earliest of these (Cuzick and Edwards 1990) is based on the proportion of the k nearest neighbors of each case who were also cases, tested against a null distribution obtained by randomly permuting the case-control indicators. When applied to data on childhood leukemia and lymphoma, they found significant clustering for k between 2 and 5, with the most significant (Monte Carlo $p < 0.003$) for $k = 3$. Tango (2007) subsequently provided a multiple-testing adjusted test for the minimum p-value across choices of k and for the same data obtained an overall significance level of only 0.085. See also Williams et al. (2001) for application of a closely related nearest-neighbor method to these data.

Diggle et al. (2003) (and earlier papers described therein) develop the distribution theory for the spatial correlation in inhomogeneous point processes. Letting $\lambda_1(\mathbf{p})$ denote the marginal hazard rate in some small neighborhood around $\mathbf{p}$ and $\lambda_2(\mathbf{p}, \mathbf{q})$ denote the joint hazard rate for a pair of events occurring in the neighborhoods of $\mathbf{p}$ and $\mathbf{q}$ respectively, then the spatial correlation $K(s)$ is defined as $\lambda_2(||\mathbf{p} - \mathbf{q}|| < s)/\lambda_1(\mathbf{p})\lambda_1(\mathbf{q})$. If the process were homogeneous, the expectation of $K(s)$ would be simply πs^2; values in excess of this would be indicative of spatial clustering (or conversely, values less than this of spatial repulsion, an unlikely situation in environmental epidemiology). The spatial correlation is estimated by

$$\widehat{K}(S) = \frac{1}{|A|} \sum_{i \neq j} \frac{I(D_{ij} < s)}{\lambda(\mathbf{p_i})\lambda(\mathbf{p_j})}$$

where $|A|$ denotes the area of the study region, the numerator is an indicator for whether the (i, j)-th pair is within some distance s of each other and the marginal spatial hazard function $\lambda(\mathbf{p})$ is estimated by a kernel density from the case-control locations (possibly adjusted for individual or location-specific covariates). For matched case-control data, Chetwynd et al. (2001) model these spatial correlations separately for cases and controls and then take their difference. The null distribution of this statistic is obtained by permuting the case-control indicators within matched sets. Diggle et al. (2007) extend this general approach to accommodate covariates, so as to address such questions as whether spatial clustering of cases can be explained by known risk factors, and to obtain estimates of exposure effects that take residual spatial dependence into account.

For grouped data, Moran's I is the correlation of the residuals ΔY_c between adjacent pairs $(c \sim d)$ of areas given by

$$I = \frac{n}{2A} \frac{\sum_{c \sim d} \Delta Y_c \Delta Y_d}{\sum_c (\Delta Y_c)^2}$$

where n is the number of areas and A is the number of adjacent pairs of areas.

For individual data, one can test this hypothesis by computing a statistic of the form

$$S = \sum_{i|Y_i=1} \sum_{j|Y_j=1} f(D_{ij})$$

for some function f of distance D_{ij} between all possible pairs of cases i and j, and compare this observed value of the test statistic against an empirical null distribution obtained by randomly permuting the case-control indicators. The proportion of random S values that exceeds the observed value is then the p-value. The choice of transformation $f(D)$ is arbitrary, but can have a big influence on the results. While $f(D) = D$ is an obvious choice, this gives heaviest weighted to the vast majority of pairs that are far apart and unlikely to share a common cause. Some measure of "closeness," such as $f(D) = 1/D$ is likely to yield a more powerful test. In particular, the function $f(D) = \exp(-\delta D)$ yields a test that is asymptotically equivalent to the score test for H_0: $\omega^2 = 0$ for any fixed value of δ, but there is no obvious way to implement a randomization test involving estimation of unknown parameters (here δ).

This permutation test is a special case of a more general class of statistics known as Hoeffding (1948) U-statistics, having the general form

$$S = \sum_{i \neq j} (Y_i - \mu_i) D_{ij} (Y_j - \mu_j)$$

Here $\mu_i = E(Y_i|Z_i)$, which could take the form of a linear or logistic regression, depending on the distribution of Y. This statistic is known to be asymptotically normally distributed with mean zero under the null hypothesis and variance that can be computed in closed form (Mantel 1967).

Unfortunately, this theoretical advantage is not all that useful in practice, since in even relatively large samples, the statistic can have a rather skewed distribution. Thus, reliance on asymptotic normally gives a poor indication of its true significance and a permutation test or calculation of the higher moments of the distribution is still needed (Siemiatycki 1978).

In using a case-control study in this way, controls are needed to provide a reference for the expected distribution of distances between pairs of cases under the null hypothesis of no clustering, and hence they should represent the geographic distribution of the population at risk. They should not, however, be geographically matched or there would be no differences to compare! If a matched design is used (matching on nongeographic risk factors like age, sex, race, or socio-economic status), then a matched test should be used. The test statistic is of the same form as given earlier—a sum over all cases of $f(D_{ij})$—but now the randomization is performed by randomly selecting one member of each matched pair to designate as the "case." Using this approach, Chetwynd et al. (2001) found no evidence of spatial clustering for childhood diabetes in a 1:2 matched case-control study.

Space-time clustering

Essentially, the procedures described in the previous section are testing whether case–case pairs tend to be closer to each other than random case or control–control pairs. Thus, the representativeness of the control group relative to the geographic distribution of the source population of cases is critical. A somewhat different approach applies the same type of test statistic only to cases, testing for an association between their spatial and temporal similarities. Suppose a disease were transmitted from person to person, say by an infectious agent. Then one would expect pairs of cases that shared a common cause to be close in both space and time. The first test of this hypothesis was described by Knox (1964) based on a simple 2×2 classification of pairs of cases as "close" in space or in time. Mantel (1967) generalized this approach to continuous functions of distances in space and time of the form using a statistic of the form

$$S = \sum_{i \neq j} f(D_{ij})g(T_{ij})$$

where D_{ij} is the spatial distance between case–case pair (i, j), $T_{ij} = |t_i - t_j|$ is the difference in their times of diagnosis, and f and g are some prespecified transformations to a scale of "closeness." He derived the asymptotic mean and variance of his test based on the theory of U-statistics (see the discussion of the asymptotic normality of such tests in the previous section). Pike and Smith (1974) showed that this test can have poor power when the disease has a long and variable latent period and proposed a more general formulation over sets of members of an acquaintanceship network. Such a network had been proposed for transmission of Hodgkin's disease among school children (Vianna and Polan 1973), but formal testing using such methods did not provide strong support for the hypothesis (Grufferman 1977; Isager and Larsen 1980; Scherr et al. 1984).

Kulldorff (2001) and Rogerson (2001) consider a somewhat different problem, namely, monitoring spatial clustering prospectively over time looking for changes in the pattern that might signal an emerging environmental hazard. Both of these involve variants of the original Knox test for space-time clustering, with consideration of where to set the threshold for multiple testing so as to have good power for detecting an change without excessive numbers of false alarms.

Evaluation of specific clusters

A first step in investigating a potential cluster is to decide whether any action is warranted. Considering the typically small number of cases, the vaguely defined source population, the large number of possible comparisons that could have led to such a cluster being discovered in the first place, the likelihood of clustering based on aggregation of well-known risk factors, and the *post-hoc* reasoning that led to suspicions about a possible environmental cause, an apparent cluster may represent nothing more than coincidence. For example, Neutra (1990) estimated that if one scanned the 5000 census tracts of California for "statistically significant" (at $p < 0.01$) excesses of each of 80 cancer sites, one might expect to find 2750 positive clusters a year! (See Box 9.2) It would clearly be infeasible to follow each of these up and unrewarding as the yield of true positives would be miniscule. Clearly, some more appropriate criterion is needed to guide the decision whether further investigation of a reported cluster is warranted.

Of course, in reality, the expected numbers would not be uniform across cells due to variation in the number of people at risk, their age distribution, and risk factors other than the suspected environmental cause, and there may be spatial correlations between nearby cells. Nevertheless, the basic observation is that the distribution of the maximum of a set of Poisson deviates is not easily interpreted, particularly when that observation is singled out for attention precisely *because* it is the largest or most significant excess. Furthermore, the narrower the circle that is drawn around the original cluster, the more "significant" it will appear as the denominator shrinks while the numerator remains about the same. (This is the origin of the term "Texas sharpshooter effect," in which a cowboy shoots randomly at the side of a barn, then draws a bull's-eye around his best cluster of shots!). Because one can often find an environmental source to blame in the neighborhood of any cluster, such *post-hoc* interpretations

9.2 How many clusters might one expect by chance?

Suppose in a given population, there are a total of Y cases distributed over C geographically defined units ("cells") and suppose for argument sake that the population at risk were uniformly distributed, so that the number of cases in each cell is independently Poisson distributed with mean $\mu = Y/C$. What is the probability of observing at least one cell with a "significant" excess? Let $\psi(y|\mu)$ denote the cumulative Poisson probability $\Pr[Y < y|\mu] = \Gamma(y, \mu)/\Gamma(y)$ where $\Gamma(y, \mu)$ denotes the incomplete Gamma function. Then the probability that at least one cluster has at least y cases is $1 - \psi(y|\mu)^C$. Now obviously we must do something about the multiple comparisons problem, since the more clusters we examine, the more opportunities there are to declare a significant excess in at least one of them. Suppose therefore that we use the normal approximation to the Poisson distribution with a simple Bonferroni correction, setting $y = \lceil \mu + Z_{\alpha/C}\sqrt{\mu} \rceil$ as the critical value for declaring significance, where $[\cdot]$ denotes the next larger integer. Does this solve the problem? Yes, if a sufficiently large continuity correction Δ is added to the critical value: we then obtain as the probability of at least one cell being called significant

$$1 - \left(\frac{\Gamma\left(\lceil \mu + Z_{\alpha/C}\sqrt{\mu} + \Delta \rceil, \mu\right)}{\Gamma\left(\lceil \mu + Z_{\alpha/C}\sqrt{\mu} + \Delta \rceil\right)} \right)^C$$

If $\Delta = 0$, this gradually increases in a sawtooth manner with increasing C, ultimately reaching 100% when the number of cells becomes much larger than the total number of cases. Adding $\Delta = 1$ to the critical value keeps this probability from diverging, but the accuracy of the test size becomes more variable with smaller and smaller cell sizes. Adding $\Delta = 2$ or using the exact Poisson critical values maintains the test size always below the nominal value α, but usually it is much smaller (i.e., highly conservative).

are very dangerous and little has been learned from investigation of such clusters (Rothman 1990). Indeed, Rothman and others have suggested that cluster observations should not be investigated at all, and that a public health agency might better devote its limited resources to replication of hypotheses generated by clusters in independent settings (different regions, different time periods).

Nevertheless, public, media, and political pressures to "do something" might force an agency to take some action (Wartenberg 2001; Elliott and Wartenberg 2004); indeed, notwithstanding the obvious difficulties of small numbers of cases and multiple comparisons, these authors have argued that cluster investigations can be scientifically rewarding and worthwhile under appropriate circumstances. As a start, one might begin by carefully verifying the existence of the cluster—confirming the diagnosis of cases, checking for possible double-counting, verifying the comparability of the boundaries used for numerators and denominators, computing standardized rate ratios against an appropriate comparison population, and so on. If confirmed as "real" and extending beyond the time period of the original observation, and if the numbers of cases are sufficient, and if there is a biologically plausible hypothesis, one might then follow up with a case-control study aimed at assessing the exposures of individuals to the alleged source of the problem, controlling carefully for possible confounding factors. Alternatively, a spatial correlation study like those described earlier in this chapter might be considered. Wartenberg (2001) provides a thoughtful discussion of criteria for pursuing cluster investigations, the rationale for doing them, options for further study, and approaches to routine surveillance of clusters. These issues were also discussed at a special meeting of the Royal Statistical Society on disease clustering and ecologic studies (see Wakefield et al. 2001; for the introduction to a special issue containing several papers on the subject).

Geographic information systems

The technological developments supporting the acquisition, integration, management, visualization, mapping, and analysis of geographic data are collectively known as GIS and have developed explosively in recent years. Particularly, well known are the tools offered by Environmental Systems Research Institute (ESRI, http://www.esri.com/) in Redlands, CA (the ArcGIS, ArcInfo and ArcView suite of programs). These have been widely used in many fields, such as geography, sociology, econometrics, urban planning, and so on. For authoritative treatments of the capabilities of GIS methods, the reader is referred to such textbooks as (Antenucci et al. 1991; Burrough and McDonnell 1998; DeMers 2000; Chrisman 2002; Maheswaran and Craglia 2004; Longley et al. 2005; Lawson 2006). The adoption of GIS methods by the environmental epidemiology community is relatively recent, with several high-profile conferences (Krzyzanowski et al. 1995; Lawson et al. 2000a; Jarup 2004; Pickle et al. 2006), numerous review articles (Vine et al. 1997; Beyea 1999; Jerrett et al. 2003; Krieger 2003; Willis et al. 2003b; Elliott and Wartenberg 2004; Nuckols et al. 2004; Briggs 2005), and several textbooks (Meade et al. 1988; Elliott et al. 1992; Pascutto et al. 2000) having been devoted to the subject. The National Cancer Institute also hosts a public-use website on GIS resources and their uses for cancer research, http://gis.cancer.gov/, with site-specific maps of cancer mortality available at http://www3.cancer.gov/atlasplus/type.html. International cancer incidence and mortality maps are available from the International Agency for Research on Cancer (Ferlay et al. 2001) (accessible at http://www-dep.iarc.fr/).

A key step in building a GIS is the acquisition of geographically referenced data. Fortunately, this is increasingly being facilitated by the availability of large-scale public databases, such as the U.S. Bureau of the Census TIGER/Line files of the boundaries of census units and other man-made features like roads, and environmental data like the Toxic Release Inventory maintained by the U.S. Environmental Protection Agency.

Complex objects can be stored as a set of line segments denoting their boundaries or as grid ("raster") of points. Superimposed on the geographic coordinates can be various kinds of "metadata" about the characteristics of the units, such as demographic or environmental data on each census block or dates, types, and intensities in pesticide application corridors. An important feature of GIS is its ability to overlay different sources of geographically coded data, perhaps containing a mixture of object types (points, lines, areas) recorded in differing coordinate systems, and its ability to interpolate between measured locations and to compute various kind of distance functions, buffers, polygon overlays, queries, and so on. The ease of automating such calculations is attractive for large scale epidemiologic applications; for example, a study of birth defects in relation to residence near landfills (Elliott and Wakefield 2001) used a simple distance metric to compute exposures for 1.5 million addresses in relation to 19,000 landfill sites over 16 years with two different assumptions about lags—a total of 10^{11} such computations!

Another key step is translating the narrative information like residence histories obtained from epidemiologic study subjects into geocodes that can be overlaid on geographic databases. Again, advances in software for address standardization and automated record linkage have greatly facilitated this process, although manual intervention to resolve unlinked addresses or ambiguous linkages can still be labor intensive. Computerized digitization of maps or aerial photographs (Thomas et al. 1992c; Brody et al. 2002; Rull and Ritz 2003) and Global Positioning Systems (GPSs) suitable for use in precisely locating objects in the field (Scott et al. 2001; Elgethun et al. 2003; Elgethun et al. 2007) have also revolutionized these tasks. Wearable monitoring devices are now available that may combine GPS devices with heart-rate monitors or other biophysical or environmental instruments with continuous downloadable recording devices for real-time time-activity studies. Needless to say, maintenance of the confidentiality of such highly personal identifying information needs to be carefully guarded and any publicly released data should have geocodes removed or corrupted in some fashion to preclude identifying individuals.

Ultimately, GIS is merely a (highly sophisticated) system for organizing spatial data and should be considered as a tool in support of the kinds of spatial analysis techniques discussed earlier in this chapter. A spatial analysis of GIS data can provide a prediction model that can be used to assign exposures (and uncertainties, as described in the Chapter 11), but some combination of predictions and actual measurements may be the best option. Different studies have balanced the use of sparse measurements for model building or validation quite differently (see Beyea 1999, for examples).

Illustrative of the complexity that is possible is an application by Gulliver and Briggs (2005) that used GIS to model traffic-related air pollution exposures of children as they traveled between home and school. Various proprietary traffic and pollution models were combined with their own models for regional and secondary pollutants, microenvironments, and time-activity patterns to estimate individual exposures over time (Figure 9.4). See Briggs et al. (2000); Bellander et al. (2001); Brauer et al. (2003) for other applications of GIS methods for modeling air pollution levels, combining available measurements with dispersion models. Van Atten et al. (2005) discuss the general principles involved and provide an extensive review of their applications.

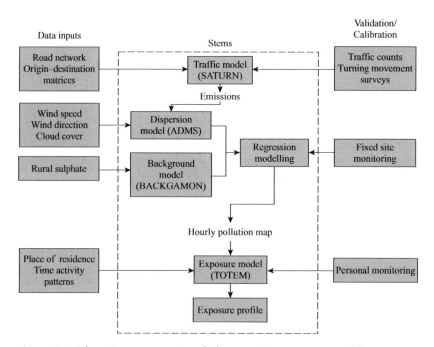

Figure 9.4. Schematic representation of the space-time exposure modeling systems (STEMS) of Gulliver and Briggs (2005). (Reproduced with permission from Briggs 2005.)

One particularly appealing use of GIS may be to target the selection of subjects in multi-phase case-control or cohort studies (Elliott and Wartenberg 2004). High-intensity exposures may be rare in the population as a whole—so that a standard case-control design would be inefficient— and a standard cohort design would be inefficient if the disease was also rare. A two-stage case-control design (Chapter 5) might use GIS methods to obtain a surrogate for exposure (say residence within some buffer surrounding an environmental exposure source) on all cases and controls and then select subsamples to oversample cases and controls within the highest surrogate-exposure strata. For example, Wartenberg et al. (1993) used GIS to identify and characterize populations living near high-voltage transmission lines for a subsequent epidemiologic study.

10 Ecologic inference

As discussed in the previous chapter, most exposures studied in environmental epidemiology are geographically distributed and hence spatially correlated. Although the ideal would be to evaluate exposure individually (exploiting or adjusting for the spatial structure of the data as needed) this is not always feasible. Frequently exposure can only be assigned at some group level—by place of residence, job title, or other groupings. In such situations, most of the information about exposure–disease associations may come from comparisons between groups rather than between individuals.

The simplest such design is known as an *ecologic correlation study*, in which disease rates (or means of some continuous outcome variable) are regressed on the mean exposure (or prevalence of a binary exposure variable) across the various groups under study, possibly adjusted for the means of other confounding variables. Unfortunately, this simple idea is potentially biased because associations across groups may not reflect associations across individuals; this phenomenon is known as the "ecologic fallacy." The term "ecologic correlation" is somewhat unfortunate, as it implies a focus on group-level variables (what we will call "contextual" variables below), whereas in most applications in environmental epidemiology the focus is really on trying to infer associations among individual-level variables using only aggregate data.

The classic example is Durkheim's (1951) analysis of suicide rates in the provinces of Germany in the nineteenth century (Figure 10.1). He found that suicide rates were highest in the counties with the highest proportion of Protestants. If extrapolated to a comparison of hypothetical provinces with 0% or 100% Protestants, the fitted regression line would predict a nearly eightfold difference in suicide rates between Protestants and non-Protestants. However, this form of comparison fails to inform us whether it was the Protestants or non-Protestants within these parishes who were committing suicide at higher rates, and indeed it was not implausible that it was religious minorities in the most heavily Protestant counties who might have the highest suicide rates, perhaps reflecting their feelings of alienation. Durkheim also compared the suicide rates in Protestants and others at the individual level and found the difference to be only twofold—still elevated, but clearly indicating some bias in the ecologic estimate.

This phenomenon appears to have been first recognized in a seminal paper in the sociology literature by Robinson (1950) and has been widely discussed in that field (Selvin 1958), and more recently in the epidemiologic

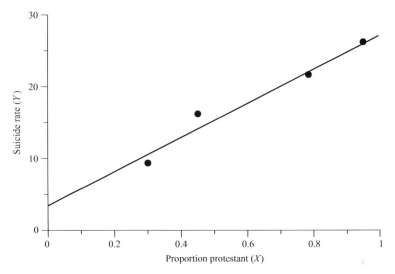

Figure 10.1. Relation between suicide rates and religion in four groups of Prussian parishes, 1883–1990. (Adapted from Durkheim 1951.)

field thanks to a critical review by Morgenstern (1982). Differences between individual- and group-level associations (sometimes called "cross-level" inference) can arise because of confounding or other biases within groups, confounding by group itself, or by effect modification by group (Greenland and Morgenstern 1989). Thus, although ecologic correlations can be a useful tool for hypothesis generation, epidemiologists generally feel that only studies of individual exposure–response relationships, appropriately controlled for both individual- and group-level confounders, can provide a secure basis for causal inference.

A major reason for the interest in ecologic inference in environmental epidemiology is that many environmental exposures (e.g., ambient air pollution) vary relatively little within geographic regions like cities, so that case-control or cohort studies conducted within a region or even multicenter studies adjusted for center will have little power. Furthermore, as we shall see in the following chapter, exposure measurement error is likely to bias individual exposure–response relationships towards the null, possibly quite severely. On the other hand, there may be a much larger range of variation between regions (leading to better power), average population exposure may be more readily estimated than for individuals, and disease rates can be based on very large numbers—entire populations rather than epidemiologic samples. In an important series of papers, these ideas were developed by Prentice and Sheppard (1989; 1990; 1991) in the context of evaluating the association between dietary fat and breast cancer, a situation where ecologic correlation studies suggested a strong positive association that has been difficult to confirm in case-control or cohort studies. They went on to propose a novel hybrid design, combining between-region comparisons with surveys of exposure and confounders in random samples from each region (Prentice and Sheppard 1995). We will return to their design at the end of this chapter. First, however, we consider the broad range of designs involving various combinations of measurements of exposure, disease, and confounders at the individual and group levels and how one might approach the analysis of clustered data involving both levels of comparison when complete data are available on all individuals.

Hybrid aggregate/individual designs

The fundamental problem with the ecologic correlation approach is that the joint distribution of exposure X, disease Y, and confounders Z is not observed. To remedy this problem, various forms of hybrid designs that combine individual- and group-level observations have been proposed, as mentioned at the beginning of Chapter 5. These can include single-phase studies in which some variables are measured at one level and some at another, or two-phase studies where aggregate data are obtained on a multiple population groups and individual-level data on subsamples of each group. These designs might also differ depending upon which variables— exposure, disease, confounders—are obtained at each level. A useful classification of the various possibilities is summarized in Table 10.1.

Consider first a single-phase study in which outcomes Y_{ci} and confounders Z_{ci} are obtained on all individuals $i = 1, \ldots, n_c$ within each of $c = 1, \ldots, C$ "clusters" (communities, groups, job titles, even time-periods or combinations of temporal, spatial, or other dimensions), but

Table 10.1. Conceptual classification of study designs involving individual and aggregate comparisons (Adapted from Kunzli and Tager 1997; Guthrie et al. 2002.)

Exposure assessment	Design and analysis	
	Group	Individual
Group	"Ecologic" $\left(\overline{Y}_c, \overline{X}_c, \overline{Z}_c\right)$	"Semi-individual" $\left(\overline{Y}_{ci}, \overline{X}_c, Z_{ci}\right)$
Individual	"Aggregate data" $\left(\overline{Y}_{ci}, X_{ci}, Z_{ci}\right)$	"Individual" (Y_{ci}, X_{ci}, Z_{ci})

exposure X_c is available only for the clusters (the "semi-individual" design in Table 10.1). Here, X_c might represent a measurement of exposure at some central site, the mean of a sample of area measurements, or an average exposure predicted by some dose reconstruction model. Suppose first that Y represents some continuous variable and consider the following hierarchical linear mixed model (Greenland 2002):

$$Y_{ci} = b_c + \mathbf{Z}_{ci}'\boldsymbol{\alpha} + e_{ci} \tag{10.1a}$$

$$b_c = \beta_0 + \beta_1 X_c + e_c \tag{10.1b}$$

or equivalently [substituting Eq. (10.1b) into (10.1a)],

$$Y_{ci} = \beta_0 + \beta_1 X_c + \mathbf{Z}_{ci}'\boldsymbol{\alpha} + e_{ci} + e_c \tag{10.2}$$

where e_{ci} and e_c are independent, normally distributed random errors. This model is similar in spirit to the GLMMs for longitudinal data discussed in Chapter 7, with clusters replacing individuals and individuals replacing time points. Indeed the same techniques for fitting the models and doing inference on their parameters can be applied. For handling longitudinal grouped data, the model can readily be expanded to a three-level model, as used in various analyses of the CHS data:

$$Y_{cik} = a_{ci} + b_{ci}t_k + \mathbf{Z}_{cik}'\boldsymbol{\alpha}_1 + e_{cik} \tag{10.3a}$$

$$b_{ci} = B_c + \mathbf{Z}_{ci}'\boldsymbol{\alpha}_2 + e_{ci} \tag{10.3b}$$

$$B_c = \beta_0 + \beta_1 X_c + e_c \tag{10.3c}$$

or the equivalent combined model with three random error terms. Here, time-varying confounders are adjusted for in Eq. (10.3a) and time-constant confounders in Eq. (10.3b), but exposure effects are evaluated only at the cluster level, Eq. (10.3c), as in the previous models, Eqs. (10.1) and (10.2).

Now suppose the outcome is binary or survival time data and suppose we wished to fit the corresponding logistic or hazard rate models,

$$\text{logit } \Pr(Y_{ci} = 1) = \beta_0 + \beta_1 X_c + \mathbf{Z}_{ci}'\boldsymbol{\alpha} + e_c$$

or

$$\lambda_c(t, \mathbf{Z}) = \lambda_0(t) \exp(\beta_1 X_c + \mathbf{Z}_{ci}'\boldsymbol{\alpha} + e_c)$$

These models would be standard logistic or Cox regression models, were it not for the additional cluster-level random effects e_c, requiring special fitting techniques. See, for example, Ma et al. (2003) for a discussion of random effects Cox models or Breslow (1984) for a discussion of random effects log-linear models. The outputs of such models include estimates of the variance of the regression coefficients α and β, which correctly reflect the additional group-level variability, as well as an estimate of the random effects variance itself.

Although this approach allows one to adjust the ecologic association between exposure and outcomes at the individual level, it does not get around the fundamental problem of the lack of knowledge of the joint distribution of exposure and confounders. Again, were it feasible to do so, one would have collected the X_{ci} data on the entire sample and the problem would be solved, but assuming this is not possible, one could adopt a two-phase sampling design like those discussed in Chapter 5. We defer further discussion of this idea until after the discussion of ecologic bias in the following section.

Ecologic bias

We now investigate more formally the validity of these procedures by comparing the parameters we would estimate from a study where complete data on the joint distribution of $(Y, X, \mathbf{Z})$ were available at the individual level with those we would estimate from a purely ecologic correlation study or one of the one- or two-phase hybrid designs. The basic technique is to compute what the model for the expected grouped data would look like from the corresponding individual-level model.

Linear models

In the classic "ecologic correlation" study, an investigator relates the rate of disease or some average health effect $\overline{Y}_c$ in a set of populations (typically geographically defined) to some measure of average exposure $\overline{X}_c$ (Figure 10.2), possibly adjusted for further covariates $\overline{\mathbf{Z}}_c$ also measured only at the group level. The so-called "ecologic fallacy" (Selvin 1958) or "cross-level bias" (Firebaugh 1978) concerns the difference between the estimated regression coefficient from such an analysis and that estimated from individual data, that is, a regression of Y_{ci} on X_{ci} and Z_{ci}. Greenland and Morgenstern (1989) and (Greenland 2002) describe three ways this difference can come about: (1) by within-group confounding that acts differentially across groups; (2) by confounding by group effects; and (3) by effect modification by group. Omitting covariates and writing a multi-level model as

$$Y_{ci} = \alpha + \beta_I(X_{ci} - \overline{X}_c) + \beta_E\overline{X}_c + e_{ci} + e_c$$

the absence of cross-level bias can be written as $\beta_I = \beta_E$; equivalently, rewriting the model as

$$Y_{ci} = \alpha + \beta_I X_{ci} + (\beta_I - \beta_E)\overline{X}_c + e_{ci} + e_c$$

we see that the absence of cross-level bias corresponds to no effect of $\overline{X}_c$ on Y_{ci} beyond its effect through X_{ci}. Such a group-level effect could arise, however, not as a direct causal effect of $\overline{X}_c$, but by confounding by some omitted group-level covariate $\overline{Z}_c$. Thus, cross-level bias actually represents the net effect of two separate phenomena: aggregation over individuals and misspecification of the form of the relationship at the group level induced by an individual level model. This understanding of ecologic bias appears first to have been expressed by Robinson (1950) and has been treated in numerous reviews (Selvin 1958; Morgenstern 1982; Richardson et al. 1987; Piantadosi et al. 1988; Greenland and Morgenstern 1989; Greenland and Robins 1994; Morgenstern 1995; Lasserre et al. 2000; Gelman et al. 2001; Greenland 2001; 2002; Wakefield and Salway 2001).

This phenomenon is illustrated in Figure 10.2. The individual data were simulated under the model $Y_{ci} = a_c + b_c X_{ci}$ where $X_{ci} \sim U[c, c+5], a_c \sim N(10 - 2c, 0.5^2)$ and $b_c \sim N(1, 0.1^2)$ for $c = 1, \ldots, 5$. Thus, even though the individual-level associations are positive in each group, the relationship among the group means is negative. Three circumstances must combine to produce this phenomenon: first, there must be some variation among

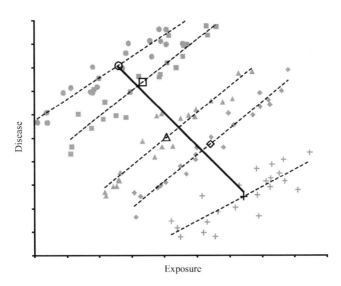

Figure 10.2. Simulated example of ecologic bias with five groups of individuals (solid gray symbols), each showing a positive exposure–response (dashed lines), while their group means (open symbols) show a negative exposure–response (solid line) due to confounding by group.

the group means $\overline{X}_c$ so that the distributions are separated horizontally; second, there must be some variation among the intercepts a_c, so that the distributions are separated vertically; third, and most importantly, there must be an association between $\overline{X}_c$ and a_c so that the associations go in different directions at the two levels. This example also illustrates a modest degree of effect modification by group (criterion 3 of Greenland and Morganstern), but this is not the cause of the ecologic bias in this example. Their criterion 1—within-group confounding acting differentially across groups—will generally lead to different slopes and is thus a special case of criterion 3.

Models for disease rates

Suppose we now assume a model for individual risk of disease of a linear relative risk form

$$\lambda(X_i, Z_i) = \lambda_0(Z_i)(1 + \beta X_i)$$

and consider the induced risk for a cluster of individuals c,

$$\overline{\lambda}_c = E_{i \in C}[\lambda(X_i, Z_i)] = E_{i \in C}[\lambda_0(Z_i)] + \beta E_{i \in C}[X_i \lambda_0(Z_i)]$$
$$\equiv \lambda_{0c}(1 + \beta \tilde{X}_c) \tag{10.4}$$

where

$$\tilde{X}_c = \frac{\Sigma_{i \in C} X_i \lambda_0(Z_i)}{\Sigma_{i \in C} \lambda_0(Z_i)}$$

is a *baseline-risk-weighted average* of the individual exposures, and λ_{0c} is the average of the individual risks in the group. The naïve regression

$$\overline{\lambda}_c = \lambda_0(\overline{Z}_c)(1 + \beta \overline{X}_c) \tag{10.5}$$

where $\overline{Z}_{Cc}$ and $\overline{X}_c$ are the unweighted means of Z_i and X_i, fails to capture the essence of Eq. (10.1) because $\lambda_0 \overline{Z}_c \neq \lambda_{0c}$ and $\overline{X}_c \neq \tilde{X}_c$.

If the relationship with exposure is nonlinear, say, a linear–quadratic form $\lambda(X_i, Z_i) = \lambda_0(Z_i)(1 + \beta_1 X_i + \beta_2 X_i^2)$, then the induced aggregate-level association will be a linear–quadratic function of the baseline-risk-weighted mean $\tilde{X}_c$ and variance $\tilde{V}_c$ of individual exposures,

$$\overline{\lambda}_c = \lambda_{0c} \left[1 + \beta_1 \tilde{X}_c + \beta_2 \left(\tilde{X}_c^2 + \tilde{V}_c \right) \right]$$

Now consider a linear model for a single confounder, $\lambda_0(Z_i) = \lambda_0(1 + \alpha Z_i)$. Substituting this expression into Eq. (10.2), this becomes

$$\overline{\lambda}_c = \lambda_0 \left(1 + \alpha \overline{Z}_c \right) \left(1 + \beta \overline{X}_c \right)$$

But the real model induced by Eq. (10.1) in this case is

$$\overline{\lambda}_c = \lambda_0 E_{i \in c} \left[(1 + \alpha Z_i)(1 + \beta X_i) \right]$$
$$= \lambda_0 \left[1 + \alpha \overline{Z}_c + \beta \overline{X}_c + \alpha\beta \left(\overline{Z}_c \overline{X}_c + C_c \right) \right] \qquad (10.6)$$

where $C_c = \text{cov}_{i \in c}(Z_i, X_i)$, the within-group covariance of the confounding and exposure variables. (Note that we can use simple averages of X and Z here because we are assuming the baseline risk λ_0 is constant conditional on Z.) It is precisely this last term that cannot be determined from aggregate data alone, and no amount of adding additional aggregate-level covariates in Z or X can overcome this difficulty (Lubin 2002). Of course, this difficulty arises because of the multiplicative form we have assumed in the linear relative risk model. A purely additive model would require neither the product term $\overline{Z}_c \overline{X}_c$ nor the within-county covariance term C_c, but multiplicative (i.e., relative risk) models are more widely used in epidemiology.

The within-group covariance could be estimated by separate population-based samples and combined with the aggregate-level data, as discussed later in this chapter. Alternatively, the groups could be stratified more finely, so that the covariance between exposure and confounders could be assumed to be zero within strata, but such finely stratified data are seldom available for most ecologic analyses without collecting individual-level data (which would defeat the purpose of doing an ecologic correlation study in the first place).

The appeal of the linear–additive model $\lambda_i = \lambda_0 + \boldsymbol{\alpha}'Z_i + \beta X_i$ is that it induces exactly the same form at the group level, $\overline{\lambda}_c = \lambda_0 + \boldsymbol{\alpha}'\overline{Z}_c + \beta \overline{X}_c$, but of course it should be remembered that such linear models are seldom used in epidemiology because of the strong nonlinear and multiplicative dependence generally seen with such factors as age. Hence, most of the analyses of individual data discussed elsewhere in this book have relied on logistic or loglinear models, so that the higher moments and covariance terms can become quite important in the induced models (Richardson et al. 1987; Dobson 1988; Greenland and Robins 1994; Prentice and Sheppard 1995; Lubin 1998). The Taylor series approximation of such models by a linear–additive one that has been used by some to justify ecologic analysis (e.g., Cohen 1990a) is valid only to the extent that the *within-group* variances of $\mathbf{W}_i$ and Z_i are small (Greenland 2001). Usually these will be much larger than the between-group variances of their means, which may be all that are available to support this claim.

In some instances, interest might focus on a "contextual" or "ecologic" variable—one that does not vary across individuals within groups, such as the ambient level of air pollution in a city. In this case, it would be tempting to argue that the association between λ_c and X_c from an ecologic analysis is unconfounded because the within-group correlation of Z_i and X_i is zero (since X_i does not vary with groups). This would be incorrect, however, as demonstrated by Greenland (2001; 2002), since both Y_i and Z_i are defined at the individual level, λ_i being simply a form of (person-time weighted) average of the individual Y_is. Thus, differences between groups in the average level of a risk factor could still confound the association between disease rates and a contextual exposure variable. Indeed, one of the key advantages of the semi-individual design discussed above is its ability to control for confounding at the individual level, so that the association with a contextual exposure variable is assessed using confounder-adjusted disease rates.

Before leaving this general discussion of ecologic bias, several other points are worth mentioning:

- Ecologic studies can be more sensitive to measurement error in exposure, although it may be less affected by measurement error in a confounder, as discussed in the following chapter.
- Frequently, the aggregate-level data available on exposure or confounders is not really the population average of the individual quantities but derived in some other way, such as inferring the average level of smoking from tobacco sales or the average exposure to air pollution from central-site ambient pollution measurements.
- Use of noncomparable restriction and/or standardization (e.g., age-adjusted disease rates and crude exposure variables) can lead to greater bias than using comparable measures (e.g., neither one being age adjusted) and allowing for confounders instead by including them as covariates (Rosenbaum and Rubin 1984).
- Unlike cohort and case-control studies, the means of exposure and disease (and possibly confounders) may be computed over overlapping periods of time, so that temporal sequence cannot be established.
- Multiple predictor variables may be more highly multi-collinear at the aggregate than at the individual level.
- Migration between groups can distort the group means being compared.

Examples

Domestic radon and lung cancer

Many of the insights about the pitfalls of ecologic inference in environmental epidemiology have derived from a particularly controversial series of papers on domestic radon. In 1987, Cohen (1987) proposed to test the linear–no-threshold hypothesis for low-dose radiation risks by means of

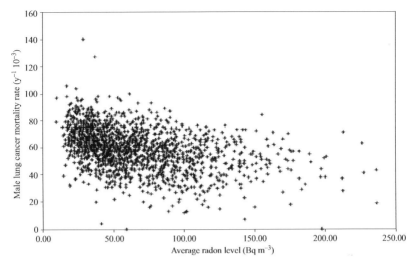

Figure 10.3. Male lung cancer rates for 1601 U.S. counties by average household radon levels. (Reproduced with permission from Richardson et al. 1995; Heath et al. 2004.)

an ecologic correlation study using lung cancer rates across U.S. counties and household radon levels from a database of measurements he had been assembling for many years (Cohen 1986; Cohen and Shah 1991; Cohen et al. 1994). The results (Cohen 1990b; 1995; 2001; Cohen and Colditz 1994) showed a strong negative correlation with radon (Figure 10.3), which could not be explained by the available county-level data on smoking or other confounders. The existence of this correlation is not disputed and has been replicated in several other countries (e.g., Haynes 1988 in the United Kingdom), but its interpretation has been questioned by many distinguished epidemiologists and statisticians (Greenland 1992; Stidley and Samet 1993; Greenland and Robins 1994; Piantadosi 1994; Stidley and Samet 1994; Doll 1998; Lubin 1998; Smith et al. 1998; Field et al. 1999; Goldsmith 1999; Gelman et al. 2001; Greenland 2001; Lubin 2002; Puskin 2003; Van Pelt 2003; Heath et al. 2004) and expert committees (NAS 1999; NCRP 2001), most of the full-length papers followed by further correspondence and rejoinders too numerous to list here. Joint analyses of case-control studies on domestic radon have clearly demonstrated a positive relationship that is generally consistent with the findings of the miner studies (Lubin and Boice 1997; NAS 1999; Darby et al. 2005; Krewski et al. 2005b; Krewski et al. 2006). In particular, two parallel analyses of ecologic and case-control data (Lagarde and Persha-gen 1999; Darby et al. 2001) demonstrated similar discrepancies, with the ecologic studies showing negative and the case-control studies positive associations. Similar issues have arisen with more recent reports

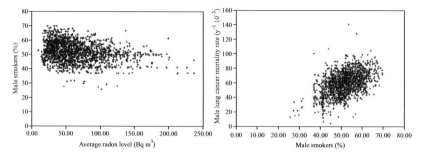

Figure 10.4. Prevalence of smoking by average household radon levels (left) and male lung cancer rates by prevalence of smoking (right) for 1601 U.S. counties. (Reproduced with permission from Heath et al. 2004.)

of ecologic associations of residential radon with childhood leukemia (Eatough and Henshaw 1993; Steinbuch et al. 1999; Kohli et al. 2000; Laurier et al. 2001; Evrard et al. 2005; 2006).

A central issue is the adequacy of control for confounders, notably for smoking in the lung cancer associations. Figure 10.4 displays the associations of smoking prevalence with radon (left) and with lung cancer (right) at the county level, demonstrating, as expected, a strong disease-smoking association, but only a weak (but negative) radon-smoking association. The question is whether adjustment for smoking prevalence alone at the group level would be adequate to control for confounding or whether more detailed individual-level data on the joint distribution of smoking and radon are needed. As mentioned above, no amount of adjustment for county-level smoking data or its correlates can be counted upon to remove confounding, even if the average values of all the relevant individual smoking variables were available.

Cohen's analyses are based on the simple approximation $\bar{\lambda}_{0c} = (1 - P_c)$ $\lambda_0 + P_c \lambda_1$, where P_c is the prevalence of smokers ($Z_i = 1$) and λ_0 and λ_1 are the baseline rates in nonsmokers and smokers respectively. (Here, prevalence was estimated from tobacco sales data by county.) Assuming a constant excess relative risk $\alpha = (\lambda_1 - \lambda_0)/\lambda_0$ for smoking and substituting this expression into Eq. (10.2) yields

$$\bar{\lambda}_c = \lambda_0 (1 + \alpha P_c)(1 + \beta \bar{X}_c)$$

But the real model induced by Eq. (10.1) includes the within-county covariance term in Eq. (10.3), which cannot be estimated by county-level aggregate data, so confounding by smoking cannot be completely eliminated no matter how many county-level smoking or socioeconomic correlates are included in the model.

The need to account for this within-community covariance arises because of the synergistic effect of radon and smoking and would not be needed in a purely additive model, but the available evidence strongly suggests a greater-than-additive (if not multiplicative) joint effect, and Cohen's analysis also assumes a multiplicative model. Indeed, Lubin (2002) provides a simple simulation showing how a truly linear relationship between $\bar{\lambda}_c$ and $\tilde{X}_c$ can be distorted to yield a spurious U-shaped relationship with $\overline{X}_c$, with a negative slope at low doses similar to that seen in Cohen's data.

Of course, these problems could in principle have been overcome by using data on the joint distribution of smoking and radon (and other confounders) from random samples of individuals within each county, but that would be a different study (see Multilevel Designs below). Or one could use age/sex/race/SES-adjusted disease rates and corresponding adjusted means of smoking and radon instead of crude values, but neither the smoking nor the radon data are available in this form.

In addition to these general issues with ecologic studies, various authors have noted other problems with Cohen's specific application of the method. For example,

- The quality of the data may be suspect: there is potential selection bias in the samples of homes that were measured; little detail is available on the means of other aspects of smoking (duration, intensity, filters, etc.); data on both smoking and radon generally followed the period covered by the cancer rates; and migration, changes in ventilation, and differences between concentration and exposure could produce systematic differences between the assigned county radon levels and the means of actual exposure.
- So far, we have focused on the lack of information on the distribution of individual-level data, particularly the joint distribution of exposure and confounders within counties, but "contextual" variables—those that vary between, but not within, counties can also be confounders. Three in particular have been suggested: rural-urban status (Darby et al. 2001), latitude (Lagarde and Pershagen 1999), and altitude (Van Pelt 2003). In particular, Darby et al. point out that such variables would have very little confounding effect in an appropriately matched case-control study, but a much larger effect on ecologic correlations.
- Cohen has repeatedly argued that the association remains despite adjustment for each of up to 500 socioeconomic variables and challenges the scientific community to come up with a plausible explanation for the finding. But such adjustment at the group level cannot be counted upon to eliminate confounding at the individual level.
- For some analyses, the data can be quite sparse (particularly analyses of rarer cancers), requiring a Poisson regression approach. By using an unweighted linear regression—or worse yet, restricting the analysis to an unrepresentative set of counties with higher rates—the results can be severely biased.

A particularly trenchant observation by Pushkin (2003) was that other smoking-related cancer sites showed a similar negative correlation with radon, whereas the nonsmoking-related sites did not, strongly suggesting residual confounding by smoking. These results differed from earlier analyses by Cohen, likely due to his inappropriate use of unweighted regression or exclusion of counties with zero or very low rates, which would have affected the rarer cancers disproportionately (as noted above).

In response to these various criticisms, Cohen has frequently claimed that his use of ecologic correlations is intended as a "*test* of the linear–no-threshold hypotheses," not as a means of *estimating* the low-dose slope. Under a linear model, he argues that the mean death rate is a function only of the mean radon concentration (but as noted above, this does not apply to nonlinear models). Hence, the predicted positive ecologic correlation remains unexplained, despite repeated attempts, rejection being a central feature of his use of the "Scientific Method." Nevertheless, the predicted positive correlation only holds if the within-community covariance of radon and confounders is zero and all between-community confounders have been adequately controlled in the analysis. Thus, the test is not really as sharp a test of the hypothesis as he claims. See Lubin (2002) for further discussion of the validity of Cohen's invocation of the scientific method.

Air pollution

In Chapter 11, some approaches to assessing inter-individual variation in personal air pollution exposures X_{ci} based on microenvironmental and spatial modeling are described. However, it is arguable that for public policy purposes, it is the "contextual" effect (Greenland 2001) of ambient pollution that is of greatest relevance, as it is more amenable to regulation than personal exposures are. Certain other confounding variables, such as altitude or weather, likewise have meaning only as contextual variables. Contextual variables could also interact with individual-level exposure or confounding variables. For example, it is possible that the effect of personal variation in exposure (due to time-activity patterns, indoor sources, or spatial variation in outdoor pollution within a community) has a relatively larger effect in low pollution communities than in high pollution communities. Such an effect has been particularly discussed in the context of income disparities (Pearce 2000). It is also possible that the effect of exposure measurement error could act differently at the different levels; see Brenner et al. (1992a); Wakefield and Elliott (1999) for a discussion of the effect of measurement error on ecologic regressions and (Greenland and Brenner 1993) for methods for correction; we will revisit this point in the following chapter. Clayton et al. (1993) discuss the use of spatial correlation analysis of residuals as a means of addressing ecologic bias.

With only 12 observations at the community level in the Children's Health Study, the prospects for including many such ecologic covariates are limited and there is some danger of "overadjustment"—controlling for variables which do not in fact have a causal effect on health outcomes, but are simply determinants of the pollution variables that are the real causal factors. Weather patterns, for example, are major determinants of pollution levels and thus one must think very carefully about whether they are indeed plausible risk factors for the health outcomes. There is abundant evidence that temperature and humidity are associated with mortality and hospitalization rates, independent of air pollution (Schwartz 1994c), so inclusion of such variables in the third level model might be justified. However, there is less evidence that wind is associated with health outcomes and since it is probably an even stronger determinant of pollution level than temperature, inclusion of wind in the model might constitute overadjustment. Furthermore, if temperature is measured with less error than air pollution levels, health endpoints may be more strongly associated with temperature than with air pollution, even if temperature is only a determinant of air pollution levels and has no direct impact on health. In an analysis of asthma incidence using a random effects Cox model, Jerrett et al. (2008) found that community-level humidity was a strong predictor of community-level incidence rates, reducing the residual community random effects variance substantially beyond that explained by personal risk factors. Including NO_2 or modeled traffic in the model further reduced the residual community effect virtually to zero. The estimated pollution effects become larger and their standard errors smaller after adjustment for humidity.

Another example is provided by personal income as a potential confounder of the association between air pollution and growth in MMEF (Berhane et al. 2004). Adjustment for income at the *individual* level [Eq. (7.2b)] did not alter the estimate of the NO_2 effect, even though it is significantly related to the outcome. Treating income as a *contextual* variable and adjusting the third-level model for the community-average income [Eq. (7.2c)] had virtually no effect on the point estimate of the NO_2 effect, but increased the standard error substantially. Adjustment at the school level (intermediate between individual and community) reduced the point estimate somewhat more and increased its standard error even more, even though the school-mean income was not itself a significant covariate. These discrepant results indicate that one must be careful in including community-level variables in an attempt to control for ecologic confounding.

Confounding by ecologic variables has been explored much more extensively in the reanalysis of the American Cancer Society cohort study (Pope et al. 2002) of mortality in 151 metropolitain areas across the United States. The main results of the study (Krewski et al. 2003) were described in the previous chapter, where we focused on spatial analyses. Here, we focus instead on the effect of ecologic variables as potential confounders or modifiers of the associations of sulfates with all-cause and cardio-vascular mortality. Willis et al. (2003b) described the basic strategy for selection of 22 contextual covariates (demographic, socioeconomic, residential segregation, access to health care, climate, physical environment, and co-pollutants), in addition to the 25 risk factors controlled at the individual level. Adjustment for these variables produced only relatively minor changes in the relative risk for all-cause mortality and sulfates (Jerrett et al. 2003), despite some highly significant associations of these variables with mortality and/or sulfates. The strongest adjustments from the model with no ecologic covariates (RR = 1.15) were from population change (RR = 1.06) and SO_2 (RR = 1.04). However, another analysis using the 513 counties making up the larger metropolitain areas was more robust to the effect of ecologic covariates (Willis et al. 2003a).

Latent variable models

If one were prepared to make additional modeling assumptions, it would in principle be possible to estimate the parameters of an individual model from purely aggregate data. Three such methods (along with one requiring some individual data) are reviewed by Cleave et al. (1995). To give a flavor of these approaches, we consider just one of them, the ecological logit model for binary X and Y data (Thomsen 1987), which introduces a latent variable W_i for each individual. The model assumes that the marginal probabilities $\Pr(X_i|W_i)$ and $\Pr(Y_i|W_i)$ have a logistic dependence on W_i and that the W_i are independently normally distributed with group-specific means μ_c and constant variance across groups. With these assumptions, it is then possible to estimate the joint probabilities π_{xyc} within each group c from the observed proportions $\overline{X}_c$ and $\overline{Y}_c$ and hence infer the relationship between X and Y at the individual level. Although identifiable in principle, the model relies on strong and untestable assumptions and appears to have led to somewhat biased estimates in an application to voter registration data. Somewhat weaker assumptions are required when subsample data are available on the joint distribution of X and Y for individuals within clusters, using a maximum entropy method (Johnston and Hay 1982). Their method is based on the iterative proportional fitting algorithm used for fitting loglinear models to categorical data, here by finding values of the unobserved π_{xyc} that are compatible with both the observed proportions $\overline{X}_c$ and $\overline{Y}_c$ in the aggregate data and with the subsample data on (X_{ci}, Y_{ci}) for each cluster. Cleave et al. provide an empirical comparison of these various methods on housing data.

Hierarchical models may offer a much more appealing approach to the problem. Wakefield (2004) provides a general framework in which the unobservable model parameters for the individual π_{xyc} are modeled in terms of a second-level model treating the πs as exchangeable across groups. For example, letting $p_{cx} = \Pr(Y_i = 1 | X_i = x, C_i = c)$, one can approximate the marginal likelihood, summing over the unobserved combinations $Y_{c0} + Y_{c1} = Y_c$, as

$$\Pr(Y_c | p_{c0}, p_{c1}) = \sum_{y_{c0}=0}^{Y_c} \text{Bin}(y_{c0} | N_{c0}, p_{c0}) \, \text{Bin}(Y_c - y_{c0} | N_{c1}, p_{c1}) \cong N[\mu_c, V_c]$$

where $\mu_c = N_{c0} p_{c0} + N_{c0} p_{c0}$ and $V_c = N_{c0} p_{c0}(1 - p_{c0}) + N_{c1} p_{c1}(1 - p_{c1})$. By itself, this likelihood is not identifiable, as it entails estimating two parameters (the p_{cx}) for each observation, so one must impose some structure through a prior distribution on the ps, such as a beta or logit-normal. Suppose, for argument sake, we take $\theta_{cx} = \text{logit}(p_{cx})$ as having normal distributions with a common mean θ_x and variance σ_x^2 across groups. Further credibility for the exchangeability assumption may be possible by regressing the θ_{cx} on additional ecologic covariates Z_c. Wakefield provides a thorough discussion of this and other choices of priors, including models involving spatial smoothing. Two other variants of this idea (Chambers and Steel 2001) involve local smoothing of the group-specific parameters across groups with similar X and Z values and a simple semiparametric approach not relying on any Z data.

Despite their adoption in other fields and their intuitive appeal, latent variable methods do not seem to have been used so far in epidemiologic applications.

Multilevel designs

To overcome the lack of information on the joint distribution of exposure and confounders within groups, one might use a two-phase sampling scheme. Thus, suppose in the main study one observed only $\overline{Y}_c$ and in a substudy one observed (X_{ci}, Z_{ci}) for simple random samples of individuals from each group (the "aggregate data" design in Table 10.1). In this case, one could use the subsample to build an individual-level regression model for $E(X_{ci} | Z_{ci}) = a_c + Z'_{ci} d_c$, and then fit the ecologic data to

$$\overline{Y}_c = \beta_0 + \overline{Z}'_c \alpha + \beta E\left(\overline{X}_c | \overline{Z}_c\right) + e_c$$

(Similar models are possible for binomial, Poisson, or survival time data.) Thus, the estimates of the exposure effect are adjusted for confounders at the cluster level, yet exploit the relationship between exposure and confounders as assessed at the individual level, without requiring that relationship to be constant across clusters.

This idea was first developed by Prentice and Sheppard (1995) in the context of a loglinear model for $\mu_i(\beta) = \Pr(Y_i = 1 | X_i) = p_0 \exp(\beta X_i)$, showing that the induced model $\overline{\mu}_c(\beta) = E(Y_c) = E_{i \in c} [\mu_i(\beta)]$ could be fitted by replacing $\overline{\mu}_c(\beta)$ by sample averages $\hat{\mu}_c(\beta) = p_0 \Sigma_{i \in c} e^{\beta X_i} / n_c$. They provided a generalized estimating equation for this purpose,

$$\sum_c \left(Y_c - \hat{\mu}_c(\beta)\right) \hat{V}_c^{-1}(\beta) \hat{D}_c(\beta) = 0$$

where

$$\hat{D}_c(\beta) = \begin{pmatrix} \hat{\mu}_c(\beta) \\ p_0 \Sigma_{i \in c} X_i e^{\beta X_i} / n_c \end{pmatrix}$$

is a vector of derivatives of $\hat{\mu}_c(\beta)$ and

$$\hat{V}_c(\beta) = \text{var}\left(Y_i | \{X_i\}_{i \in c}\right) = \hat{\mu}_c(\beta) - p_0^2 \Sigma_{i \in c} e^{2\beta X_i} / n_c$$

They also provide a correction for bias due to small subsample sizes. In the presence of classical measurement error, where the true covariate X is measured with error as $Z \sim N(X, \sigma^2)$, estimates of β from individual-based cohort or case-control studies would be biased towards the null by a factor that depends upon the measurement error variance σ^2, as discussed in the following chapter. However, Prentice and Sheppard showed that there would be virtually no bias from using Z in place of X in their aggregate data model if Z were available for the entire population. Even if it was measured only on modest-sized subsamples of each group, the bias towards the null would be much smaller than in an individual-based study. See Guthrie and Sheppard (2001) for simulation studies describing the performance of the Prentice and Sheppard method in the situations of confounding, nonlinearity, nonadditivity, and measurement error, and Guthrie et al. (2002) for extensions incorporating spatial autocorrelation. Wakefield and Salway (2001) provide a general parametric framework, allowing for measurement error, confounding, and spatial dependency.

Plummer and Clayton (1996) consider the design of such studies in terms of the trade-off between the sample sizes needed for adequately precise estimation of the group means and the number of groups needed for fitting the aggregate data model. They conclude that the substudy sample sizes in each group should be proportional to the corresponding expected number of cases in the main study, with an optimal ratio that increases with the size of the exposure relative risk from about 0.2 for RR = 1.5 to 2 for RR = 3.5. They also argue that the sample should be stratified by important risk factors like age and sex. Sheppard et al. (1996) also considered design issues and showed that increasing the number of groups from 20 to 30 or 40 with subsamples of 100 in each yielded greater improvements in power than corresponding increases in subsample sizes keeping the number of groups fixed.

This need for relatively large samples in each group may be reduced by using Bayesian methods to estimate the group means. Salway and Wakefield (2008) describe an approach using with a Dirichlet process prior to stabilize the estimated group means from small samples using a parametric distribution for the within-group values. Using data on radon and smoking status from 77 counties in Minnesota with at least one household radon measurement (generally quite sparse, ranging from 1 to 122 measurements per county), they demonstrated that the weak negative correlation seen with the standard ecologic analysis became significantly positive using their hybrid model.

The Prentice and Sheppard approach extends naturally to multiple risk factors, although the main focus of their simulations (as well as those of Plummer and Clayton and of Salway and Wakefield) is on a single exposure variable, not on control of confounding. But a key advantage of the hierarchical design is its ability to provide information on the joint distribution of exposure and confounders that is lacking in a standard ecological correlation study. If subsamples are not available at all, Lasserre et al. (2000) showed by simulation that standard ecologic regressions can be improved by including products of the marginal means for exposure and confounders in addition to their main effects, at least for binary risk factors.

Jackson et al. (2006; 2008) consider a different form a joint individual and aggregate data where information on outcomes as well as predictors are available from the subsamples (e.g., from a multicenter cohort or case-control study). Their hybrid analysis combines a logistic model for the individual data with a binomial model for aggregate data using the outcome probabilities induced by the individual model. They provide expressions for these probabilities in relation to the proportions of a binary covariate or the means of a continuous covariate, also incorporating information on their variances and correlation between covariates if either are available. In an application to the data on birth weight and water chlorination by-products described in the previous chapter, Molitor et al. (2008) used the aggregate data to control the potential confounding by maternal smoking and ethnicity—variables that were not available on the individual-level dataset.

Although the focus of most of the literature on ecologic inference has been on geographic comparisons, the groups can be formed on any basis, including temporal dimensions or combinations of dimensions. See (Sheppard 2005; Stram 2005) for a discussion of the relevance of these hybrid individual/aggregate designs to time-series studies of air pollution and the influence of exposure measurement error, for example.

11 Measurement error and exposure models

Exposure assessment is often one of the biggest challenges in environmental epidemiology. The topic is complex and very specific to the particular factors under study; entire books have been written on the general principles (Checkoway et al. 1989; Armstrong et al. 1992; Steenland and Savitz 1997; Nieuwenhuijsen 2003) as well as specialized texts in specific fields (e.g., Till and Meyer 1983 for exposures to radiation). Rather than discuss exposure assessment methodology in detail here, we briefly review the various general types of methods that are used, so as to motivate a treatment of the problem from a statistical point of view. Thus, this chapter will discuss approaches to exposure modeling, the influence of uncertainties in exposure estimates on exposure–response relations, and methods of correction for them.

The terms "measurement error" and "misclassification" are often used interchangeably, although generally we favor the former when considering continuous variables and the latter for categorical variables. Despite this similarity in concepts, however, the implications of errors and the available methods of correction are quite different, so subsequent sections of this chapter will treat the two situations separately.

Broadly speaking, exposure assessment methodologies fall into two classes: direct measurement; or prediction models. Examples of direct measurement are personal radiation dosimeters and questionnaires. Dose reconstruction systems, based on records of environmental releases and models for their dispersion, deposition, and uptake by humans, would be examples of prediction models. Often, an exposure assessment protocol may entail elements of both types. For example, one might need to combine area measurements of exposure concentrations in various places where people could have been exposed with questionnaire information about time-activity patterns to build a model for personal exposures. Measurements may be spatially and temporally distributed, but so sparse in either dimension that assignment of exposures to individuals at times and places not directly measured may require spatial/temporal interpolation models, using techniques like those discussed in Chapter 9. Determinants of exposures, such as traffic counts (for air pollutants), power line configurations (for electromagnetic fields), or local geology and housing characteristics (for domestic radon) may be used to build exposure prediction models, calibrated against actual measurements. Measurements of excreted metabolites may require extensive physiological modeling to translate them to estimates of intake or body burdens (see Chapter 13).

In addition to the distinction between measurement and prediction models, three other characteristics need to be borne in mind when evaluating the implications of measurement error: whether the errors are systematic or random; whether differential or not; and whether shared or not.

Systematic errors are those where all subjects' exposures are biased by a similar amount. If the nature of such errors is known, they are of less concern because they can be corrected for systematically and may have no influence on the association with disease anyway. For example, if it amounts to a consistent shift in the exposure distribution by a constant additive amount, then the slope of a exposure–response relationship will be unchanged. (Of course, a systematic doubling of all exposures would yield a halving of the corresponding slope coefficient.) In this chapter, we are thus more concerned with random errors.

Loosely speaking, we say errors in exposure are differential if they depend upon the outcome under study (a more formal definition is provided below). For example, recall bias in a case-control study arises when cases and controls tend to recall their past exposures differently. Techniques like blinding the interviewer to whether the subject is a case or a control can help minimize this form of differential measurement error. Of course, disease outcomes are also subject to error; likewise, we would call disease errors differential if they were related to exposure (or other covariates).

In settings where measurements relate to individuals, it may be reasonable to assume that the errors are independent across subjects—one person's dosimeter is not influenced by errors in another's. In many circumstances, however, such as when area measurements are applied to all the people who are exposed to that area (as in a job–exposure matrix) or when prediction models are used that have elements in common across individuals (as in the magnitudes of releases from point sources in a dispersion model), measurement errors can be correlated between individuals. In addition to the bias in effect estimates that can result from measurement error, these correlations can lead to underestimation of standard errors and exaggerated significance tests, unless properly allowed for in the analysis.

These various exposure scenarios can have very different statistical properties and require different approaches for allowing for uncertainties in their exposure assignments, as discussed in the next section. Having established this general statistical framework, we will then revisit the various dosimetry systems and describe some of them in greater detail. For a more thorough treatment of methods of correction, see such textbooks as (Fuller 1987; Carroll et al. 1995) and review articles (Thomas et al. 1993; Thompson and Carter 2007).

A general framework for exposure measurement error

No system of exposure assessment is ever perfect, so it is useful to distinguish between the "true" exposure X and the estimated exposure Z assigned by the dosimetry system, whether based on direct measurements, modeling, or some combination of the two. For the present purposes, let us ignore the temporal element and assume that X_i and Z_i are single quantities for each individual i. In addition, let us assume we have some outcome variable Y_i and, in some contexts, a vector of determinants $\mathbf{W}_i$ for each person's true exposure. Finally, since in many contexts, exposure may be assigned not to individuals but to groups, let us follow the notation of the previous chapter and denote such clusters by the subscript c.

A very general framework for thinking about exposure measurement issues (Clayton 1991) is displayed in Figure 11.1. In this graphical framework, measured quantities are represented by boxes and unobserved random variables and model parameters by circles. The full model is thus composed of three submodels:

$$\text{Exposure model:} \quad \Pr(X|\mathbf{W};\alpha,\tau^2)$$

$$\text{Measurement model:} \quad \Pr(Z|X;\sigma^2) \quad\quad (11.1)$$

$$\text{Disease model:} \quad \Pr(Y|X;\beta)$$

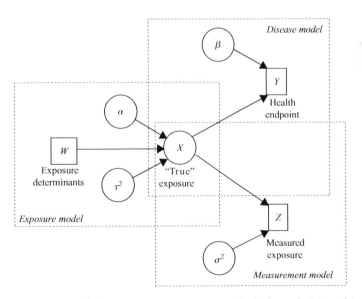

Figure 11.1. Framework for exposure measurement error in the form of a Directed Acyclic Graph.

The specifics of each submodel will depend on the context. Analogous to the distinction between exposure assessments based on measurements or predictions, statisticians distinguish between two specific models for measurement error:

- The *classical error* model does not involve any exposure predictors $\mathbf{W}$ and in particular assumes that $E(Z|X) = X$, in other words, the measurement instrument is calibrated in an unbiased fashion so that the measured values are distributed around the true exposure with expectation equal to the true exposure and errors $e = Z - X$ that are independent of X. An example would be $Z \sim N(X, \sigma^2)$, but normality is not the central feature of this model; the true doses and measurements could be lognormally distributed, for example.
- The *Berkson error* model is based instead on an "instrumental variable" W which influences an individual's exposure, although individuals with the same W could differ in their true exposures due to unmeasured personal characteristics. In particular, the Berkson model assumes $E(X|W) = W$, now with errors that are independent of W. Here, W might represent the average exposure of a group with similar characteristics; for example, taking place of residence c as a grouping characteristic, one might use the measured air pollution level Z_c at a central site or the average $\overline{X}_c$ of a sample of personal measurements in that community. More generally, we will view the result of a model that gives a prediction $\hat{X}(\mathbf{W}) \equiv E(X|\mathbf{W}; \boldsymbol{\alpha})$ as a form of Berkson error model, where $\mathbf{W}$ could include several variables describing sources (e.g., traffic density), and personal modifiers (distance, household ventilation, activities, etc.). For example, $\hat{X}(\mathbf{W})$ could be the result of a linear regression of a set of measurements Z_j on a sample of individuals or locations j of the form $Z_j = \mathbf{W}'_j \hat{\alpha} + e_j$.

Just as the exposure assessment protocol can involve a combination of measurements and predictions, so the statistical measurement error model can involve a combination of classical and Berkson error components. In general, we will call a "complex dosimetry system" unbiased if $E(Z|X) = X$ and $E(X|\mathbf{W}) = \hat{X}(\mathbf{W})$.

Classical and Berkson models have different implications for the estimation and testing of the parameters β of the exposure–response relationship $Pr(Y|X; \beta)$ that is of primary interest. The simplest case is when all the relationships are linear and all the variables are normally distributed, that is,

$$Y \sim N(\beta_0 + \beta_1 X, \omega^2)$$

$$Z \sim N(X, \sigma^2)$$

$$X \sim N(\mathbf{W}'\boldsymbol{\alpha}, \tau^2)$$

Of course, both the distribution of $Y|X$ and of $Z|X$ could depend upon additional covariates; for example, the mean or variance of $Z|X$ could depend upon personal characteristics like age or measurement characteristics like the time of day the measurement is performed. For simplicity, we ignore such dependencies in the notation that follows.

An important distinction is between the so-called "structural" and "functional" approaches to the problem (Pierce et al. 1992; Schafer 2001). In Figure 11.1 and Eq. (11.1), we have described a structural model, in which X is viewed as a random variable having some distribution (possibly depending upon covariates $\mathbf{W}$). In functional models, on the other hand, X is treated as a nuisance parameter—a fixed quantity—for each subject. A difficulty with the functional approach is the large number of parameters, leading to a failure of asymptotic consistency of the estimator of the exposure–response parameters, as first noted by Neyman and Scott (1948). This problem can be overcome by treating the Xs as random variables with a distribution that depends upon only a small number of parameters. As we shall see below, however, progress has been made using "semiparametric" methods that treat the distribution of X as completely unknown, to be estimated along with the other parameters. This idea dates back to a theoretical proof of asymptotic consistency provided by Kiefer and Wolfowitz (1956), but no practical implementation was available until decades later. Key to the success of these approaches is that X is still treated as a random variable rather than a parameter. Whether X is treated as a parameter or a random variable is a separate issue from whether the model is specified in terms of $\Pr(X|Z)$ directly or its separate components $\Pr(Z|X)$ and $\Pr(X)$. Different methods discussed below use one or the other of these two approaches.

Consider first the classical error model, with $\mathbf{W}'\boldsymbol{\alpha} \equiv \mu$ for all subjects. This would be appropriate for a pure measurement system, such as one where each individual's exposure assignment is based on some kind of a personal dosimeter. Now, of course, we do not observe X, only Z, so we consider the "induced" relationship between the observable variables Y and Z:

$$\Pr(Y|Z) = \int \Pr(Y|X = x)\,\Pr(X = x|Z)\,\mathrm{d}x$$

$$= \int \Pr(Y|X = x)\,\frac{\Pr(Z|X = x)\,\Pr(X = x)}{\Pr(Z)}\,\mathrm{d}x$$

$$= \frac{1}{\varphi\left(\frac{Z-\mu}{\sigma^2+\tau^2}\right)}\int \varphi\left(\frac{Y - \beta_0 - \beta_1 x}{\omega^2}\right)\varphi\left(\frac{Z - x}{\sigma^2}\right)\varphi\left(\frac{x - \mu}{\tau^2}\right)\mathrm{d}x$$

Straight-forward calculus shows that this integral is simply another normal density with mean and variance given by:

$$E(Y|Z) = \beta_0 + \beta_1 E(X|Z) = \beta_0 + \beta_1[\mu + C(Z - \mu)] = b_0 + b_1 Z \quad (11.2)$$

where

$$b_0 = C\beta_0 + (1 - C)\beta_1\mu$$
$$b_1 = C\beta_1$$

with $C = \tau^2/(\sigma^2 + \tau^2)$, and

$$\text{var}(Y|Z) = \sigma^2 + \beta_1^2\text{var}(X|Z) = \sigma^2 + \beta_1^2(1 - C)\tau^2$$

Thus, the regression coefficient of interest β_1 is attenuated by the multiplicative factor C, which depends upon the ratio of the variance of X to the variance of Z. The intuition behind this result is that the measurements are "overdispersed", with random measurement error superimposed on top of the variability in the true exposures: $\text{var}(Z) = \text{var}(X) + \text{var}(Z|X) = \sigma^2 + \tau^2$. Thus, one can think of the X-axis as being "stretched" by the factor $1/C$, while the Y-axis is left intact, leading to a flattening of the slope of the exposure–response relationship (Figure 11.2, left). Note also that the variance of $Y|Z$ is also inflated by an amount that depends upon β_1.

Now consider the Berkson error model (Figure 11.2, right). The derivation is similar to that given above, except that there is now no need to

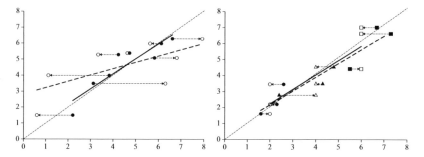

Figure 11.2. Simulated examples of classical and Berkson error. Left: classical error with $X \sim N(4,2)$, $Z \sim N(X,2)$, $Y \sim N(X,1)$, so the population regression (dotted line) has slope $\beta = 1$; solid circles and solid line are the regression of Y on X with expected slope equal to the population slope, $\beta = 1$; open circles and dashed line are the regression of Y on Z with expected slope $= \beta \times 2/(2 + 2) = 0.5$; this results from the overdispersion of the predictor (arrows connecting selected X values to corresponding Z values). Right: Berkson error with a categorical instrumental variable W taking values 3, 4, 5, 6, $X \sim N(W,1)$, $Z = E(X_w)$, $Y \sim N(X,1)$; solid markers are X, the shapes distinguishing the categories of W; open circles are the corresponding values of Z_w; solid line is the regression of Y on X, dashed line is the regression of Y on Z_w.

apply Bayes theorem to express $\Pr(X|Z)$ in terms of $\Pr(Z|X)$, since we have a model for X directly, namely $\Pr(X|W)$. As a result, it is easy to show that

$$E(Y|\mathbf{W}) = \beta_0 + \beta_1 E(X|\mathbf{W}) = \beta_0 + \beta_1 \hat{X}(\mathbf{W})$$

and

$$\text{var}(Y|\mathbf{W}) = \sigma^2 + \beta_1^2 \text{var}(X|\mathbf{W}) = \sigma^2 + \beta_1^2 \tau^2$$

Thus, the regression coefficient of interest β_1 is not attenuated at all, although the residual variance of $Y|\mathbf{W}$ is inflated, as in the classical error case, and again depends upon β.

Similar results can be derived for many other models that are widely used in epidemiology, such as the logistic, Poisson, and Cox regression models (Carroll et al. 2006). However, it must be appreciated that the widely quoted result that, in classical error models, regression coefficients are biased toward the null, whereas under Berkson error they are unbiased, depends upon a number of assumptions including normality and linearity. In addition, an implicit assumption in the whole framework is that errors are "nondifferential." In the epidemiological literature, this assumption is often expressed as "the measurement error distribution does not depend upon the outcome variable." For example, in a case-control study, recall bias—where the reported exposures would have a different relationship to the truth in cases and in controls—would violate this assumption. What this expression fails to adequately capture is that the "measurement error distribution" $\Pr(X - Z)$ is not the same thing as the conditional distributions of $X|Z$ or $Z|X$ on which the framework is based. A more precise way of expressing the concept of nondifferential measurement error is that "the outcome is conditionally independent of the measurements given the true exposures" or equivalently $\Pr(Y|X, Z) = \Pr(Y|X)$ (in mathematical shorthand: $Y \perp Z|X$). In other words, if we knew the true exposures, the measurements themselves would not provide any additional information about the outcomes. A corollary of this definition is that $\Pr(Z|X, Y) = \Pr(Z|X)$, which is the real meaning of the statement that "measurement errors do not depend on the outcome." However, the converse, $\Pr(X|Y, Z) = \Pr(X|Z)$, is not true, since Y really is related to X, not Z.

Another important special case is when all variables are binary. In the epidemiological literature, this is generally called the problem of exposure misclassification rather than exposure measurement error. For binary variables, neither the classical nor the Berkson error models can apply, since $E(X|Z) = \Pr(X = 1|Z)$ which cannot equal Z except in the complete absence of error (and similarly for $E(Z|X)$). Nevertheless, some properties analogous to those discussed earlier are possible. Here it is more

Table 11.1. Expected $2 \times 2 \times 2$ table under exposure misclassification, where $S = 1 + p(\psi - 1)$ and $p = \Pr(X = 1|Y = 0)$

X	Z	Y = 0	Y = 1
0	0	$(1 - p)\alpha_0$	$(1 - p)\alpha_0/S$
	1	$(1 - \alpha_0)(1 - p)$	$(1 - p)(1 - \alpha_0)/S$
	Total	$1 - p$	$(1 - p)/S$
1	0	$p(1 - \alpha_1)$	$p(1 - \alpha_1)\psi/S$
	1	$p\alpha_1$	$p\alpha_1\psi/S$
	Total	p	$p\psi/S$
Total	0	$(1 - p)\alpha_0 + p(1 - \alpha_1)$	$[(1 - p)\alpha_0 + p(1 - \alpha_1)\psi]/S$
	1	$(1 - p)(1 - \alpha_0) + p\alpha_1$	$[(1 - p)(1 - \alpha_0) + p\alpha_1\psi]/S$

convenient to specify the relationship between X and Z in terms of sensitivity $\sigma_1 = \Pr(Z = 1|X = 1)$ and specificity $\sigma_0 = \Pr(Z = 0|X = 0)$. Thus, we would call the misclassification nondifferential if σ_1 and σ_0 do not depend on Y. Then, we express the relationship between X and Y in terms of the true odds ratio (OR)

$$\psi = \frac{\Pr(Y = 1|X = 1)/\Pr(Y = 0|X = 1)}{\Pr(Y = 1|X = 0)/\Pr(Y = 0|X = 0)}$$

and the induced (or misclassified) OR ψ^* for the relationship between Y and Z. Table 11.1 illustrates the relationships amongst the three variables. From the bottom two rows of the table, one can easily compute the misclassified OR as

$$\psi^* = \frac{[(1 - p)(1 - \sigma_0) + p\sigma_1\psi][(1 - p)\sigma_0 + p(1 - \sigma_1))]}{[(1 - p)\sigma_0 + p(1 - \sigma_1)\psi][(1 - p)(1 - \sigma_0) + p\sigma_1]}$$

which does not have any simple expression in terms of ψ like the one for continuous variables given earlier. Nevertheless, ψ^* is always biased toward the null, that is, $\psi > \psi^* > 1$ if $\psi > 1$ or conversely (Figure 11.3).

Some other effects of measurement error

So far, we have focused mainly on the effects of nondifferential measurement error on slopes of exposure–response relationships (or relative risks for a binary exposure variable). To recap:

- Classical error for a continuous variable generally induces a bias toward the null in a exposure–response relationship because $\mathrm{var}(X) < \mathrm{var}(Z)$. Confidence limits will tend to shrink roughly in proportion to the estimate, so that the significance test will not generally be invalidated, although may lose power.

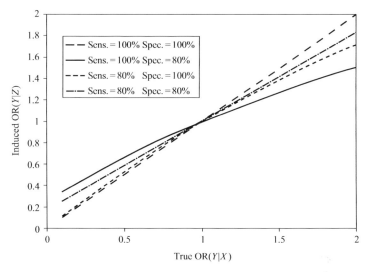

Figure 11.3. Induced ORs for Y|Z as a function of the true OR for Y|X for various levels of sensitivity and specificity, Pr(X|Z).

- Berkson error for a continuous variable may not bias the slope of a exposure–response relationship, but only under certain restrictive conditions (additive errors with constant variance, a linear exposure–response, etc.). However, because var(Z) < var(X) in this model, power will be reduced and confidence intervals widened, relative to what would have been found in the absence of error.
- Nondifferential misclassification of a binary exposure variable will always produce a bias in the relative risk toward the null and reduce the power of the test.

In addition, however, there can be a number of other effects, some less obvious:

- The shape of a exposure–response relationship can be affected: for example, multiplicative errors will tend to reduce the magnitude of a quadratic component of a linear–quadratic dose–response curve (Schafer and Gilbert 2006).
- Measurement errors in extended exposure histories can also distort the influence of temporal factors like latency (Thomas 1987) and dose–rate effects (Stram et al. 1999; Stram et al. 2000), particularly when the magnitude of the errors varies over time.
- Nondifferential misclassification of a categorical covariate can produce a bias *away* from the null for comparisons between particular pairs of categories, and can also bias an estimate of trend away from the null (Dosemeci et al. 1990; Flegal et al. 1991).

- In multivariate models, nondifferential error can cause some loss of control of confounding or distort a measure of interaction (Greenland 1980; Fung and Howe 1984; Armstrong 1998); for example, error in one covariate can lead to transfer of some or all of a causal effect of that variable to another better-measured one with which it is correlated (Zeger et al. 2000).
- In ecological correlation studies, exposure associations will not necessarily be biased by nondifferential misclassification of a confounder in the way they are in individual studies (Brenner et al. 1992a; Carroll 1997), but they can be severely biased away from the null when the exposure variable itself is misclassified (Brenner et al. 1992b).
- Errors that are correlated between individuals, such as those in a exposure prediction model (e.g., a job–exposure matrix) that are shared by some individuals, will tend to produce confidence limits that are too narrow (Stram and Kopecky 2003).
- Blind assignment of exposure may not prevent differential misclassification when categories with different effects are combined (Wacholder et al. 1991).
- Relative risks (Brenner et al. 1993) and attributable risks (Vogel et al. 2005) can be overestimated by correlated misclassification of exposure and disease, even if both errors are nondifferential.
- Methods of correction can lead to bias in the opposite direction from the uncorrected estimate (i.e., over-correction) when the validation data used for this purpose is itself less than perfectly measured (Flegal et al. 1991; Wacholder et al. 1993).
- Likewise, a correction for nondifferential error that assumes the errors are uncorrelated with the true value can produce unpredictable results when this assumption is violated, such as measures based on self-report; in some instances, this can lead to a positive association when there is no association with true exposure (Wacholder 1995; Armstrong 1998).

Epidemiologists have sometimes argued that if they have observed an association between an outcome and some measured exposure that may be subject to error or misclassification, then the true exposure–response relationship must be even stronger than that observed. Such a claim is not warranted by the theoretical developments described above. The theory describes the *expectation* of various random variables, including the estimates of the induced parameters b or ψ^* in relation to the true parameters β and ψ, over hypothetical replications of the study. In any single finite dataset, the observed values of these estimates could differ upwards or downwards, so that it is possible that the estimates of the induced parameters could be more extreme than for the population parameters for the true exposures. For example, simulation studies (Thomas

1995) have shown that this could happen in a substantial proportion of
replicate datasets, depending upon their sample size and other factors.

Methods of correction for measurement error

Continuous exposure variables

There is a large literature on methods of correction for measurement error,
which can be reviewed only briefly here. The appropriate method will
depend in part on the type of information available about the distribution
of measurement errors and the form of the assumed model. In general,
however, most such methods entail some form of replacement of the available data on W and/or Z by an estimate of the corresponding X and using
that estimate as if it were the truth in an analysis of its relationship to Y,
or integration over the distribution of X given W and Z.

Parametric regression calibration

To illustrate this idea, we begin with one of the simplest approaches,
known as the "regression calibration" method (Rosner et al. 1989; 1990;
1992). This approach is appropriate for the classical error model where
information on the error distribution comes from a separate "validation"
study. Given observations of (X, Z) and (Y, Z) from separate datasets, one
first performs two regressions, $E(X|Z) = \alpha_0 + \alpha_1 Z$ and $E(Y|Z) = \gamma_0 + \gamma_1 Z$,
and then, based on Eq. (11.2), estimates the corrected coefficient of the
regression of Y on X as $\hat{\beta}_1 = \hat{\gamma}_1 / \hat{\alpha}_1$, with variance

$$\text{var}\left(\hat{\beta}_1\right) = \frac{1}{\hat{\alpha}_1^2}\text{var}\left(\hat{\gamma}_1\right) + \frac{\hat{\gamma}_1^2}{\hat{\alpha}_1^4}\text{var}\left(\hat{\alpha}_1\right)$$

When the validation data is a subset of the main study, then the covariance
between $\hat{\gamma}_1$ and $\hat{\alpha}_1$ must be taken into account (Spiegelman et al. 2001).
Similar approaches can be applied in logistic or other kinds of exposure–
response models; indeed, the original regression calibration method was
introduced in the context of logistic regression for unmatched studies, but
in this case requires an assumption that the measurement error variance σ^2
is small compared with the variance of true exposures τ^2. This two-stage
approach has the appeal of being relatively simple and does not require
data on all three variables for everybody (if it were, there'd be no need
for the Z data anyway, as one could simply model the $Y|X$ relationship
directly!). The first step can also be readily extended to incorporate other
variables, nonlinearities, transformations to normality and additivity,
and so on.

Regression substitution

A variant of this idea uses a single imputation approach, similar to that described in Chapters 4 and 7 for dealing with missing data, but using the estimated α coefficients to compute $\hat{X}_i = E(X|Z_i; \hat{\alpha})$ for each main study subject and then regress the Y_i on these $\hat{X}_i$ estimates rather than on their Z_i. Honest variance estimation requires that the uncertainties in the $\hat{X}_i$s be taken into account, however, and in nonlinear models, this approach cannot be counted upon to completely eliminate the bias due to measurement error unless the error variance is relatively small. In general, what is needed is $E_{X|Z}(Y|X)$. For example, suppose one wanted to fit a linear–quadratic model $E(Y|X) = \beta_0 + \beta_1 X + \beta_2 X^2$; then one would require both $E(X|Z)$ and $E(X^2|Z)$, which could be estimated by two separate regressions of the substudy data or by computation of $E(X^2|Z) = E^2(X|Z) + \mathrm{var}(X|Z)$ from a fitted model for $\Pr(X|Z)$. These problems can be avoided using multiple imputation, generating several data sets with X_i data randomly sampled from the predicted distribution using the estimated $\mathrm{var}(X|Z)$, and then combining the estimates of $\hat{\beta}$ and $\mathrm{var}(\hat{\beta})$ from each dataset, as described in Chapter 4 for the multiple imputation method of dealing with missing data.

SIMEX method

A remarkably simple method introduced by Cook and Stefanski (1994) and described in the textbook by Carroll et al. (2006) is based on the tendency for nondifferential classical error to attenuate a exposure–response relationship. Essentially, one takes the observed data on (Y, Z) and adds additional random errors to produce $Z^* = Z + e$ with $\mathrm{var}(e) = \sigma^2 - \sigma_0^2$ (where σ_0^2 is the actual measurement error variance) across a range of values of σ^2. For each choice of σ^2, one performs the naïve regression of Y on Z^* to estimate β_σ and then plots β_σ against σ. Extrapolating the resulting curve $\beta_\sigma = \hat{\beta}_0 \sigma^2/(\sigma^2 + \hat{\tau}^2)$ back to $\sigma = 0$ (Figure 11.4) yields an estimate of the measurement-error corrected slope!

Parametric likelihood and quasi-likelihood methods

Now suppose one does not have individual data on (Z, X), only literature-based estimates of $\sigma^2 = \mathrm{var}(X)$ and $\tau^2 = \mathrm{var}(Z|X)$. The obvious estimator of the corrected slope coefficient for a linear relationship, based on Eq. (11.2), would then be simply

$$\hat{\beta} = \hat{b}_{Y|Z}/\hat{C} = \hat{b}_{Y|Z}\left(1 + \frac{\hat{\tau}^2}{\hat{\sigma}^2}\right)$$

In both this approach and the regression calibration/substitution ones, estimation of the measurement error parameters is separate from estimation

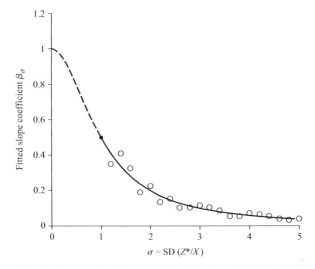

$\sigma = \mathrm{SD}\,(Z^*/X)$

Figure 11.4. Simulated example of the use of the SIMEX method. Solid circle represents the fitted slope coefficient β at the known measurement error standard deviation σ; open circles represent the fitted slopes after addition of further measurement error; the dashed line gives the extrapolation back to no measurement error.

of the exposure–response parameters. One would expect that joint estimation in a single stage would yield more efficient estimates and variance estimates for β that properly accounted for the uncertainties in α. Suppose one has measurements of (X, Z) on a subset of subjects S, and only (Y, Z) on the remainder $N \backslash S$. The full likelihood of the combined data is then

$$L(\alpha, \beta, \sigma) = \prod_{i \in S} p_{\alpha,\sigma}\,(X_i | Z_i, W_i) \times \prod_{j \in N \backslash S} \int p_\beta\,(Y_j | x)\, p_{\alpha,\sigma}\,(x | Z_j, W_j)\, \mathrm{d}x$$

$$(11.3)$$

where

$$p_{\alpha,\sigma}(X | Z, W) = \frac{p_\alpha(X | W)\, p_\sigma(Z | X)}{\int p_\alpha(x | W)\, p_\sigma(Z | x)\, \mathrm{d}x}$$

If all these component probabilities are normal distributions, it is relatively straightforward to express this likelihood in closed form, and this turns out to be equivalent to either the regression calibration or regressions substitution methods described above. See Fearn et al. (2008) for an application to domestic radon and lung cancer, showing that the regression calibration approach provides a very good approximation to this full likelihood.

One other special case is also worth mentioning: if Y is binary, then there is no closed form solution if $\Pr(Y | X)$ is assumed to be logistic, but a closed form solution can be derived if a probit model is used instead. The probit model can be thought of as involving a latent variable Y^* which is normally distributed with mean $\beta_0 + \beta_1 X$ and unit variance, with $Y = 1$ if $Y^* > 0$, zero otherwise. Thus, $\Pr(Y = 1|X) = \Pr(Y^* > 0|X) = \Phi(\beta_0 + \beta_1 X)$ where $\Phi(\cdot)$ is the standard cumulative normal distribution function. This model is very similar in shape to the logistic function (Figure 4.1), and with a rescaling of the coefficient $\beta_1 = 0.607\alpha_1$ nearly identical to a logistic with slope coefficient α_1. The appeal of the probit model is that it extends naturally to the situation where X is also a latent variable, since if $X \sim N(\mathbf{W}'\alpha, \sigma^2)$, then $Y^*|\mathbf{W} \sim N(\beta_0 + \beta_1 \mathbf{W}'\alpha, \sigma^2 + 1)$ and

$$\Pr(Y = 1|\mathbf{W}) = \Phi\left(\frac{\beta_0 + \beta_1 \mathbf{W}'\alpha}{\sqrt{\sigma^2 + 1}}\right)$$

and analytic expressions for the joint distribution of $(Y, Z|\mathbf{W})$ are also relatively straight-forward to derive. With this, maximum likelihood estimation of β jointly with the measurement error parameters (α, σ^2, τ^2) is straight-forward.

A relatively simple approach to combined Berkson and classical error was provided by Reeves et al. (1998). Their approach is parametric, assuming both X and $Z|X$ are normally distributed (on a natural or log scale), allowing exact expressions for a normally distributed outcome or a good approximation using the probit model for a binary outcome. In an application to a case-control study of lung cancer in relation to domestic radon, exposure was defined as the average concentration over a 30-year period, during which subjects may have lived in several homes, not all of which could be measured directly. This requires forming a time-weighted average of $E(X|Z)$ for directly measured homes or of $\text{Avg}(X|P)$ over homes in the neighborhood at location P for those homes where direct measurements were not available. The former has a classical error structure, the latter a Berkson error structure. Estimates of the measurement error variance needed for these calculations derived from a subset of homes with replicate measurements separated by up to 10 years. They found that measurement error correction increased the slope estimate from 0.09 ± 0.16 to 0.15 ± 0.25 per 100 Bq/m^3.

Approximate solutions can be used when the measurement error variance is sufficiently small (Whittemore and Keller 1988; Carroll and Stefanski 1990). These generally entail a first-order Taylor series expression for the score equation as a function of $E_X(Y|Z)$ and $\text{var}_X(Y|Z)$ and solving this expected estimating equation using GEE or quasi-likelihood methods. In fact, the regression calibration methods discussed earlier can be seen as a special case of such approximations to the full likelihood, replacing the nuisance parameters by estimates obtained from the validation subsample. A "pseudolikelihood" (Carroll et al. 1984; Schafer 1987) results from estimating the nuisance parameters (α, σ) in the full likelihood [Eq.(11.3)] using only the validation data and substituting them in the part of the likelihood for the main study data, treating these nuisance parameters as if they were the true values.

Semiparametric efficient estimator method

The regression calibration and likelihood-based methods all involve parametric assumptions about the distributions of $Z|X$ and X (or equivalently, about $X|Z$), such as normality and linearity of the relationship between the two variables. In addition, the regression calibration method is only approximate, requiring the additional assumption that the measurement error variance is relatively small. In an effort to weaken these requirements, there has been great interest in developing various methods generally known as "semiparametric." These methods take a nonparametric approach to modeling the joint distribution of (X, Z), while continuing treat the $(Y|X)$ relationship parametrically.

Pepe and Flemming (1991) introduced the simplest of these approaches for the case where Z is discrete. In this case, one can write the likelihood contribution for an individual j in the main study with unobserved X as

$$\Pr(Y_j|Z_j) = E[\Pr(Y_j|X)|Z_j] = \sum_{i \in S} \frac{I(Z_j = Z_i)}{n_S(Z_j)} \Pr(Y_j|X_i)$$

where $n_S(Z_j)$ is the number of substudy subjects with $Z = Z_j$. The overall likelihood is thus formed as a sum of main and substudy contributions

$$L(\beta) = \prod_{i \in S} \Pr(Y_i|X_i) \times \prod_{j \in N \setminus S} \Pr(Y_j|Z_j)$$

and they provide a robust sandwich estimator to allow for the additional variability due to the estimation of $\Pr(X|Z)$ from a finite subsample for use in $\Pr(Y|Z)$.

Carroll and Wand (1991) extended this approach to continuous Z variables approach using a kernel density estimator for the distribution of $[X|Z]$. The model can be described as

$$\Pr(Y|Z) = E[\Pr(Y|X)|Z] = \frac{1}{n_v} \sum_{i \in S} K\left(\frac{Z - Z_i}{\sigma}\right) \Pr(Y|X_i)$$

where $K(\cdot)$ denotes any symmetric density function like the normal with standard deviation σ. The only nuisance parameter here is σ in the smoothing kernel, for which they describe an optimal estimator under suitable parametric assumptions. The model can be extended to include covariates W in a model for $\Pr(X|W)$. See also Roeder et al. (1996) for a similar approach and extensions to case-control studies with additional covariates.

Both these methods entail additional assumptions, such as that membership in the substudy is at random and unrelated to Y, that X is Missing at Random, and that the dimensionality of W is small so that nonparametric estimation of $\Pr(X|W)$ is feasible. This basic approach was generalized by Robins et al. (1995a, b) to a "semiparametric efficient estimator," which includes the Pepe–Flemming and Carroll–Wand ones as special cases but relaxes these assumptions and yields the most efficient estimator in the class of all possible semiparametric estimators. The basic idea entails using an estimating equation combining contributions from main and substudy contributions of the form

$$U(\beta) = \sum_{i \in S} u_\beta(Y_i | X_i, W_i) + \sum_{j \subset N \setminus S} \tilde{u}_\beta(Y_j | X_j, W_j)$$
$$- \sum_{i \in S} \frac{1 - \pi(Y_i, Z_i, W_i)}{\pi(Y_i, Z_i, W_i)} E_{y|X_i} \left[\tilde{u}_\beta(y | Z_i, W_i) \right]$$

where $u_\beta(Y|X, W)$ are score contributions derived from the true $\Pr(Y|X, W)$ for the validation study subjects, for whom X is observed, $\tilde{u}_\beta(Y|Z, W)$ is any convenient nonparametric estimator of $u(Y|Z, W)$, and $\pi(Y, Z, W)$ is the probability of selection into the subsample. Subtraction of the last term ensures that the resulting estimating equation will yield a consistent estimator of β for any choice of $\tilde{u}_\beta(Y|Z, W)$. The closer the latter is to the true model, the more efficient it will be. Robins et al. also provide an estimator that is locally fully efficient, provided the dimensionality of W is not too large. See Spiegelman and Casella (1997); Chatterjee and Wacholder (2002); Sturmer et al. (2002) for simpler descriptions of the method and comparisons of the performance of this and alternative approaches.

Mallick et al. (2002) discuss a quite different kind of semiparametric model in the context of a mixture of Berkson and classical errors. Like the parametric approach of Reeves et al. discussed above, they introduce a latent variable L such that $\Pr(X|L)$, $\Pr(Z|L)$, and $\Pr(L)$ are all parametric lognormal distributions (at least conditional on additional predictors W), but treat $\Pr(Y|X, W)$ nonparametrically as a monotonic spline function. They also consider nonparametric Bayesian estimation of $\Pr(L)$. Applications of this approach to thyroid tumors in residents downwind of the Nevada Test Site are described below.

Schafer (2001) and Pierce and Kellerer (2004) take yet another approach, based on the factorization $\Pr(X, Z) = \Pr(Z|X)\Pr(X)$ and estimating the second factor nonparametrically, assuming a parametric form for the first. Their approaches are applicable in settings where there is no external validation study giving direct measurements of the joint distribution of (X, Z), but only the marginal density $\Pr(Z)$ is observed and the measurement error distribution $\Pr(Z|X)$ can be treated as known. The problem then reduces to deconvoluting the integral

$$\Pr(Z) = \int \Pr(Z|X)\,\Pr(X)\,\mathrm{d}X$$

to estimate $\Pr(X)$. Schafer used maximum likelihood with nonparametric density estimation for $\Pr(X)$, an approach first suggested by Laird (1978) and generally implemented with the E–M algorithm or variant thereof, computationally a rather intensive process. Pierce and Kellerer instead take a more general regression substitution approach using the first few momemts $E(X^k|Z)$ to substitute into a better approximation to $E[\Pr(Y|X)|Z]$, assuming only that the distribution of X is "smooth." Their solution to the computational problem is somewhat complex, but basically entails the use of Laplace approximations for log $\Pr(X)$ and its relationship with log $\Pr(Z)$. In their application to the atomic bomb survivor study, they assume $Z|X$ is lognormally distributed and perform a sensitivity analysis with different assumptions about $\mathrm{var}(Z|X)$. The results of their analysis are discussed in the applications section below.

Bayesian methods

The Bayesian analysis of the classical measurement error model was introduced by Clayton (1988) and has been greatly facilitated by the advent of Markov chain Monte Carlo (MCMC) methods for fitting models with many latent variables or nuisance parameters. The basic form of the model is the same as described earlier in Figure 11.1 and Eqs. (11.1), but is supplemented with prior distributions on the parameters $\theta = (\alpha,\ \beta,\ \tau^2,\ \sigma^2)$. Inference on the slope parameter of interest β is obtained by integrating not just over the conditional distribution of the unobserved Xs (as in likelihood-based methods), but also over the other parameters. This is readily accomplished by the following two-step iterative procedure: at iteration r,

- For each subject, sample a random value of X_i from its full conditional distribution

$$\Pr(X_i^{(r)}|\mathbf{W}_i,\ Z_i,\ Y_i,\ \boldsymbol{\theta}^{(r)}) \propto \Pr(Y_i|X_i;\ \beta^{(r)})$$
$$\times \Pr(X_i|\mathbf{W}_i,\ \boldsymbol{\alpha}^{(r)},\ \sigma^{(r)})\,\Pr(Z_i|X_i;\ \tau^{(r)})$$

- Randomly sample the model parameters from their respective full conditional distributions, treating the current assignments of true exposures

$X^{(r)}$ as if they were real data:

$$\Pr(\beta^{(r+1)}|\mathbf{Y}, \mathbf{X}^{(r)}) \propto \left\{ \prod_i \Pr(Y_i|X_i^{(r)}; \beta) \right\} \Pr(\beta)$$

$$\Pr(\alpha^{(r+1)}, \sigma^{(r+1)}|\mathbf{X}^{(r)}, \mathbf{W}) \propto \left\{ \prod_i \Pr(X_i^{(r)}|\mathbf{W}_i; \alpha, \sigma) \right\} \Pr(\alpha, \sigma)$$

$$\Pr(\tau^{(r+1)}|\mathbf{X}^{(r)}, \mathbf{Z}) \propto \left\{ \prod_i \Pr(Z_i|X_i^{(r)}; \tau) \right\} \Pr(\tau)$$

The algorithm continues for many iterations, and one tabulates the distributions of the quantities of interest (e.g., the parameters θ), following a suitable "burn-in" period to allow for convergence in probability. If the various component distributions are conjugate, then these conditional distributions may be easy to sample from, but methods such as adaptive rejection sampling (Gilks and Wilde 1992) or the Metropolis–Hastings algorithm (Chib and Greenberg 1995) may be used to sample from arbitrary combinations of distributions (see Chapter 4). It is thus a very flexible approach to fitting models of great complexity for which exact solutions may not exist and the numerical integrations that would otherwise be needed for maximum likelihood methods. For a full discussion of MCMC methods, see the book (Gilks et al. 1996).

Richardson and Gilks (1993a,b) describe applications to complex measurement error settings involving job–exposure matrix approaches to occupational exposure assessment. Their scenario entails exposure assessment for job titles based on area measurements that have a classical error structure, followed by assignment of exposures to individuals given their job titles with a Berkson error structure.

An even more complex application is provided by an analysis of the atomic bomb survivor data taking account of the grouped nature of the available data (Deltour et al. 1999; Little et al. 2000; Bennett et al. 2004). Here, for each group c, defined by a range of values of estimated dose $Z_i \in [\zeta_c, \zeta_{c+1})$ and strata s of other factors (age, gender, city, latency, etc.), the challenge is to sample possible values of $\overline{X}_{cs} = E_{i \in (c,s)}(X_i|\overline{Z}_{c,s})$, where $\overline{Z}_{c,s}$ is the weighted average dose provided in the dataset for that stratum. Bennett et al.'s analysis includes such additional complexities as a flexible dose–response relationship combining linear splines at low doses with a linear–quadratic–exponential relationship at high doses (Figure 11.5), combined with a lognormal measurement error model and a Weibull distribution of true doses, a combination for which no closed-form solution would be feasible.

See Muller and Roeder (1997) and Schmid and Rosner (1993) for other examples of Bayesian measurement error approaches; the former is notable for adopting a semiparametric approach using mixture models to specify the distribution of true doses.

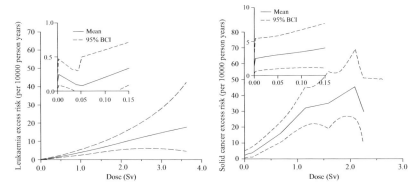

Figure 11.5. Posterior estimates of the dose–response for leukemia (left) and solid cancer (right) mortality among the atomic bomb survivors; insets expand the low-dose region modeled by linear splines. (Reproduced with permission from Bennett et al. 2004.)

MCMC methods can also be applied in a non-Bayesian manner as simply a way to approximate the likelihood. For example, one could generate a random sample of $\{X_{js}\}_{s=1,\dots,s}$ values for each subject j given Z_j and then approximate the integral in the likelihood by $\Pr(Y_j|Z_j) = \sum_s p_\beta(Y|Z_{js})/S$ as a function of β, and then maximize this likelihood using standard methods. An example of an application of this approach to a complex mixture of classical and Berkson errors in the Hanford study is described below.

Differential measurement error

Up till now, we have restricted attention to the case of nondifferential measurement error, that is, $\Pr(Z|Y, X) = \Pr(Z|X)$ or equivalently $\Pr(Y|X, Z) = \Pr(Y|X)$. Now suppose this assumption does not hold, as might easily occur, say, in a case-control study with recall bias. Given the potential importance of differential measurement error, particularly in case-control studies, it is surprising that this problem has received relatively little attention in the statistical literature until recently (beyond various studies of the impact of violation of the assumption of nondifferential error in the standard methods). This lack of attention probably derives in part from a fundamental problem of identifiability in the differential error case, arising from the confounding of the causal effect of X on Y with the bias in $Z|X$ due to Y. The full retrospective likelihood for case-control data can be decomposed as

$$\Pr(Z|Y) = \frac{\Pr(Y, Z)}{\Pr(Y)} = \frac{\int p_\beta(Y|x)p_{\delta,\tau^2}(Z|x, Y)p_{\alpha,\sigma^2}(x)\,\mathrm{d}x}{\int p_\beta(Y|x)p_{\alpha,\sigma^2}(x)\,\mathrm{d}x}$$

where $\Pr(Z|X, Y)$ now depends upon differential error parameters δ in addition to the usual error variance τ^2. A fully parametric model is in principle straight-forward, although computationally burdensome and relies on assumptions about $\Pr(X)$ that one would rather avoid. Most treatments of this problem therefore rely on a pseudolikelihood approximation to the prospective likelihood $\Pr(Y|Z) = \int p_\beta(Y|x)\tilde{p}_{\sigma,\tau^2}(x|Z, Y)\,dx$, where $\tilde{p}_{\delta,\tau}(x|Z, Y)$ is estimated from the validation substudy data on cases and controls and treated as known. One might, for example, assume linear model of the form $Z \sim N(X + \delta_Y, \tau^2)$. However, naively regressing X on Z and Y—as in the standard regression calibration approach—is not correct in the presence of differential error, because it effectively uses the $p_\beta(Y|X)$ factor twice—first to estimate $X|Z,Y$ and then to fit $Y|X$—thereby biasing β away from the null. Instead one must reverse the regression calibration to fit $\Pr(Z|X, Y)$, develop a model for $\Pr(X)$, and then combine the two submodels to estimate

$$\tilde{p}(x|Z, Y) = \frac{p_{\hat{\delta},\hat{\tau}^2}(Z|X, Y)\hat{p}(X)}{\int p_{\hat{\delta},\hat{\tau}^2}(Z|x, Y)\hat{p}(X)\,dx}$$

Here $\hat{p}(X)$ could in principle be estimated either parametrically or nonparametrically, as described above in the section on the semiparametric efficient estimator method. Since one typically knows very little about the form of $\Pr(X)$ and has no particular interest in this part of the model, nonparametric estimation is attractive so as to achieve robustness. However, if there are more than about two covariates W in $\Pr(X|W)$, this can be difficult and a parametric model might be preferred. See Spiegelman and Casella (1997) for examples of both approaches. Closed form expressions are possible in the case of a probit model for binary outcomes with normally distributed X and Z, but the estimates are somewhat biased (and variances underestimated) unless the uncertainties in the nuisance parameters are appropriately accounted for. Although the problem of differential error is addressed in the Carroll and Wand (1991) and Pepe and Flemming (1991) papers, their approaches both assume random, not case-control sampling, for the validation substudy, whereas the Robins et al. (1995a; 1995b) semiparametric method is fully general, and would even allow two-phase stratified sampling conditional on any combination of Y, Z, and W.

In simple cases, the regression substitution and regression calibration approaches provide attractive alternatives. In the case where $[Y|X]$, $[Z|X, Y]$, and $[X]$ are all normally distributed, the attenuated slope β^* becomes

$$\beta^* = \frac{\sigma^2\beta + \omega^2\delta + \delta\sigma^2\beta^2}{2\delta\sigma^2\beta + \omega^2\delta^2 + \sigma^2\delta^2\beta^2 + \tau^2 + \sigma^2}$$

Thus, given estimates of the parameters $(\delta, \sigma^2, \omega^2, \tau^2)$ from a validation sample and β^* from the naïve regression of Y on Z in the main study, one could solve the resulting quadratic equation for the corrected slope coefficient β and compute a variance estimator using the delta method. To use the regression substitution approach, one must reverse the usual regression to estimate the parameters (δ, τ^2) in $[Z|X, Y]$ along with a model for $\hat{p}(X)$, as described above, and compute an "adjusted" expectation from $\int x \tilde{p}(x|z, Y) \, dx$. This expectation is then used in place of Z in the analysis of the main study data. Carroll et al. (1993) formalize this idea in a pseudolikelihood method for differential error in case-control data. See Sturmer et al. (2002) for simulation studies of the performance of regression calibration and semiparametric methods in the presence of differential and nondifferential error.

Multivariate models

So far, we have focused on measurement error in a single variable, although other covariates measured without error may have been included in one or more components of the model. As noted earlier, when multiple factors are subject to measurement error, the usual tendency for classical error or nondifferential misclassification to bias associations toward the null may no longer hold. In particular, some of the causal effect of a poorly measured variable can be transferred to a better measured variable it is correlated with. To understand this phenomenon, consider the case where $[X]$ and $[Z|X]$ are multivariate normally distributed and $[Y|X]$ is given by a normal linear regression with vector of slope coefficients β. Then the multivariate generalization of the slope attenuation factor C in Eq. (11.2) becomes

$$E(Y|Z) = \beta \Omega (\Omega + T)^{-1} Z$$

where $\Omega = \text{cov}(X)$ and $T = \text{cov}(Z|X)$ (Armstrong et al. 1989). Zeger et al. (2000) provide tables of the expected slopes in a bivariate model under a variety of scenarios involving causal effects of one or both variables with a range of correlations in Ω and T. Many of the techniques for measurement error correction described earlier in the univariate case have straight-forward multivariate generalizations, although some (like the semiparametric methods) can become unwieldy. For example, the regression substitution method can be applied by using the subsample data to estimate $E(X|Z)$ and then replacing Z by this quantity in the analysis of the main study (Kuha 1994; Fraser and Stram 2001).

An interesting application of this idea is as an adjustment for uncontrolled confounding. Sturmer et al. (2005) treat this problem as one of measurement error in a propensity score framework. Specifically, suppose one has a vector $\mathbf{W}$ of potential confounders that are measured on the main study of the relationship between Y and some exposure of interest X, but some larger vector of confounders $\mathbf{V}$ is available only on a substudy. They first regress X on $\mathbf{W}$ and $\mathbf{V}$ in the substudy to obtain a propensity scores $p_{\mathbf{W}}$ and $p_{\mathbf{V}}$ and the covariance between them. The estimated propensity scores $p_{\mathbf{W}}$ are then included in the regression of Y on X in the main study and the estimated slope β_X is corrected for the measurement error in treating $p_{\mathbf{W}}$ as a flawed surrogate for $p_{\mathbf{V}}$.

Binary exposure variables

Single imputation of X by its expectation does not work for binary variables, since a fractional value is not a valid possibility. However, one can perform the analogous calculation on the matrix of cell counts in the 2×2 table of observed counts N_{yz} to obtain the table of expected counts n_{yx} and then analyze these as if they were the real data. As in the continuous case, however, some adjustment to the variance is needed to account for the uncertainty in the imputation.

The basic approach was described by Greenland and Kleinbaum (1983). Let $\mathbf{N} = (N_{00}, N_{01}, N_{10}, N_{11})$ denote the vector of observed cell counts and $\mathbf{n}$ the corresponding vector of true cell counts for the X-Y relationship. Furthermore, let $\mathbf{M} = (M_{xz})$ where $M_{xz} = \Pr(X = x | Z = z)$ is the matrix of misclassification probabilities (sensitivity and specificity), which can be estimated from a validation subsample. Then it is easy to see that $E(\mathbf{N}) = \mathbf{Mn}$, which suggests a natural estimator of $\hat{\mathbf{n}} = E(\mathbf{n}|\mathbf{N}) = \hat{\mathbf{M}}^{-1}\mathbf{N}$, from which the OR can be computed directly. Alternatively, one could parameterize the problem in terms of positive and negative predictive values $\mathbf{m} = (m_{zx})$ where $m_{zx} = \Pr(Z = z | X = x)$, leading to a direct estimator $\hat{\mathbf{n}} = E(\mathbf{n}|\mathbf{N}) = \hat{\mathbf{m}}\mathbf{N}$, which turns out to be more efficient (Marshall 1990). Calculation of the asymptotic variance of the OR is based on straight-forward application of the delta method to the sampling variances of $\mathbf{M}$ or $\mathbf{m}$ and $\mathbf{N}$ (Greenland 1988b). See Chen (1989); Morrissey and Spiegelman (1999) for further discussion. Of course, multiple imputation can also be used, as described above, where random samples of $X|Z$ are treated as observed data and the log ORs from several such samples then averaged and the within- and between-sample variances used to compute the variance of the log OR.

Maximum likelihood provides a direct approach to estimating both the OR parameter and the misclassification probabilities. Suppose in addition to the observed (Y, Z) data $\mathbf{n}$, we also had an independent "validation" sample of (X, Z) data $\mathbf{M}$. Letting $\pi_{yx} = \Pr(Y = y | X = x)$ and $\mu_{xz} = \Pr(X = x | Z = z)$ denote the underlying probabilities in the tabulations $\mathbf{n}$ and $\mathbf{M}$ described above, the likelihood can be expressed as

$$L(\pi,\mu) = \prod_{y=0}^{1}\prod_{z=0}^{1}\left(\sum_{x=0}^{1}\pi_{yx}\mu_{xz}\right)^{N_{yz}} \times \prod_{x=0}^{1}\prod_{z=0}^{1}(\mu_{xz})^{M_{yz}}$$

where the πs could easily be re-expressed in terms of more natural parameters, such as the logit of the baseline risk and the log OR.

In some circumstances, no "gold standard" measurement of X may be available, even on a substudy, but the measurement error parameters in Z can be estimated indirectly using multiple measurements Z_k if they can be assumed to be independent conditional on X (Elton and Duffy 1983; Marshall 1989; Spiegelman et al. 2001). For example, if X, Z, and Y are all binary with two measurements of Z, the complete data can be viewed as the cell counts in a $2 \times 2 \times 2 \times 2$ contingency table of (X, Z_1, Z_2, Y), for which only the $2 \times 2 \times 2$ table collapsing over X are observed. Maximum likelihood fitting of the full model involving the OR for the X-Y association, along with nuisance parameters for $\Pr(X)$ and $\Pr(Z|X)$, assuming independence of $[Z_1, Z_2|X]$ is straight-forward using the E–M algorithm, as described in the section on missing data in Chapter 4.

Aggregate, time series, and spatial studies

As noted earlier, measurement errors can have quite different effects on analyses of aggregate data, such as time series and ecologic correlation studies. For example, Brenner et al. (1992b) showed that nondifferential misclassification of a binary individual exposure variable—which would be expected to produce a bias *toward* the null in individual-level analyses—actually biases an ecologic association *away* from the null. This essentially results from the tendency for grouping on a variable unrelated to exposure to shrink the misclassified group prevalences toward the center of the distribution (as in Figure 11.2, right), while the groups' disease rates remain unchanged, thereby increasing the slope. Using external information on the individual-level true and false positive rates α_1 and $1 - \alpha_0$, respectively, Greenland and Brenner (1993) provided a simple correction to the ecologic estimate of the relative risk by regressing Y_g on $E(\overline{X}_g|\overline{Z}_g) = (\overline{Z}_g - (1 - \alpha_0))/(\alpha_1 - (1 - \alpha_0))$, where $\overline{Z}_g$ and $\overline{X}_g$ are the true and misclassified exposure prevalence in group g, and derive appropriate variance estimates using the delta method. This situation is somewhat unique to binary exposure variables, however, as grouping a continuous variable would not tend to produce this shrinkage. In a simulation study of domestic radon and lung cancer rates, Stidley and Samet (1994) showed that the slope of the ecologic regression and its standard error were actually reduced by nondifferential measurement error.

Similar issues can arise in time-series analyses where daily mortality rates are regressed on daily ambient pollution levels. Zeger et al. (2000) discuss the effect of measurement error in time-series studies by computing the induced population association from a model for individual risk. Letting $\lambda[t, X_i(t)] = \lambda_{0i}(t) \exp(\beta X_i(t))$ denote the hazard rate for individual i in relation to personal exposure $X_i(t)$, the observable population rate can be decomposed as

$$\lambda[t|Z(t)] = E\left(\left\{\left[\lambda_{0i}(t) - \overline{\lambda}_0(t)\right] + \overline{\lambda}_0(t)\right\}\right.$$
$$\left. \times \exp\left\{\beta\left[X_i(t) - \overline{X}(t)\right] + \left[\overline{X}(t) + Z(t)\right] + Z(t)\right\}\right)$$
$$\cong \overline{\lambda}_0(t) \exp[\beta Z(t)] + \text{additional terms}$$

where $\overline{\lambda}_0(t)$ is the population average baseline risk, $\overline{X}(t)$ is the population average personal exposure, and $Z(t)$ is the ambient level (they also include terms for deviations between the true and measured ambient levels, the effects of measured confounders, and a smooth function in time, which are omitted here to simplify the notation). The various additional terms involve the covariance of individual baseline risks and individual exposures and the variance of personal exposures, as well as any systematic deviations between the population average personal exposure and the ambient concentration. They discuss the contributions of these various terms and show that a regression on a risk-weighted average of personal exposures

$$X^*(t) = \frac{\sum_i \lambda_{0i}(t)\, X_i(t)}{\sum_i \lambda_{0i}(t)}$$

would in principle be the desired variable to use in time-series analyses of aggregate data. This quantity could differ appreciably from $Z(t)$, however, because high risk individuals could have different exposures from the general population and because of the contributions of indoor sources. Using data from an intensive study of personal exposures in 178 nonsmoking California residents (Ozkaynak et al. 1996), they estimated the potential magnitude of these various deviations and derived a simple adjustment for measurement error, increasing the slope of the regression on $Z(t)$ from 0.84 (95% CI −0.06, 1.76) to 1.42 (−0.11, 2.95) for the regression on $X^*(t)$. In a similar vein, Brauer et al. (2002) demonstrated that aggregation across subjects in a time-series analysis could weaken the evidence for a threshold effect and lead to underestimation of the location of a threshold, even if personal thresholds were the same across subjects (this phenomenon is discussed further in Chapters 6 and 15). For further discussion of the effects of measurement error in time-series, panel, and ecological studies, see (Sheppard and Damian 2000; Guthrie et al. 2002; Sheppard 2005; Stram 2005) and the description in the previous chapter of sampling designs for augmenting an ecologic study with individual samples from each of the areas compared.

In Chapter 8, we introduced a "meta-regression" approach to hierarchical modeling of multi-city time-series data (Schwartz and Coull 2003), which we now describe more formally. Let i subscript the cities and consider a linear regression model $Y_{it} = b_{0i} + b_{1i}X_{1it} + b_{2i}X_{2it} + e_{it}$ where $\mathbf{b}_i = (b_{0i}, b_{1i}, b_{2i}) \sim N_3(\boldsymbol{\beta}, \boldsymbol{\Sigma})$ and further suppose a classical error model $\mathbf{Z}_{ct} \sim N_2(\mathbf{X}_{ct}, \boldsymbol{\Omega})$. Instead of fitting the bivariate hierarchical model directly, they perform two regressions for each city,

$$Y_{it} = a_{0i} + a_{1i}Z_{1it} + e'_{it} \quad \text{and} \quad Z_{2ct} = c_{0i} + c_{1i}Z_{1it} + e''_{it}.$$

In the second stage, they then regress

$$\hat{a}_{1i} = \beta'_1 + \beta'_2\hat{c}_{1i} + d'_i$$

They show that $\hat{\beta}'_2$ is a consistent estimator of β_2, even though $\hat{\beta}'_1$ would be attenuated by measurement error. Reversing the roles of the two variables would lead to a consistent estimator of β_1.

Efficient design of validation studies

For relatively simple measurement error models, it is possible to derive the optimal sampling fraction that would minimize the variance of the corrected exposure–response parameter β, subject to a constraint on the total cost of the study. Greenland (1988a) described the basic idea for binary exposure and disease variables and demonstrated that under some circumstances, the most efficient design might entail simply collecting the more expensive measure X on a smaller study without using a two-phase (main study/validation substudy) design at all. Spiegelman and Gray (1991) describe similar calculations for a logistic disease model with X and Z normally distributed and provide extensive results on the dependence of the optimal design on cost and variance ratios. See also Thomas (2007a) for situations where closed-form optimization results are possible for various exposure– and/or disease-based sampling schemes (all variables binary, all continuous, binary outcomes with continuous exposures) and simulation results for spatially correlated exposure data.

These results can be derived in a number of essentially equivalent ways. The likelihood approach (Thomas 2007a) uses the $(\beta\beta)$ element of the inverse of the full Fisher information derived from the likelihood

[Eq. (11.3)] to derive an expression for the asymptotic unit variance

$$V(\mathbf{s},\theta) \equiv var\left(\hat{\beta}_{(1,s)}\right)\Big|_\theta = \lim_{N\to\infty} N var\left(\hat{\beta}_{(N,s)}\right)\Big|_\theta$$

per main study subject as a function of the model parameters θ and the sampling fractions $\mathbf{s} = (s_{yw})$ for the substudy subjects in each stratum of Y and W on whom Z will be measured. The overall cost of the study is

$$NC(\mathbf{s}) = N + R\sum_{yw} s_{yw} E[N_{wy}(\theta)]$$

where R is the ratio of per-subject costs between the validation and main studies and $C(\mathbf{s})$ is the cost per main-study subject for a given sampling scheme. The asymptotic relative cost efficiency (relative to a "fully validated" design with X measured on everybody and no substudy, i.e., $\mathbf{s} \equiv 1$) can then be defined as

$$ARCE(\mathbf{s}|\theta) = \frac{V(1,\theta)C(1)}{V(\mathbf{s},\theta)C(\mathbf{s})}$$

The optimization then seeks to maximize this quantity with respect to the sampling fractions for a given choice of model parameters.

Using an approach like this, Holcroft and Spiegelman (1999) showed that a balanced design was nearly optimal across a broad range of parameters and provide a FORTRAN program to determine the best design for a given set of parameters. Although some further improvement is possible by optimal sampling of the four possible combinations of W and Y (Thomas 2007a), the optimal sample fractions depend on knowledge of the true model parameters θ, whereas a balanced design can be expected to perform reasonably well over a broad range of values. See also Reilly (1996) for a range of epidemiologic applications using pilot samples to estimate the quantities needed to determine the optimal design, Breslow and Chatterjee (1999) for a discussion of the design of two-phase studies using semiparametric methods and Holford and Stack (1995), Wong et al. (1999), Spiegelman et al. (2001) for designs using repeated measures or multiple surrogates when no gold standard is available; the latter also consider both internal and external validation study designs. Stram et al. (1995) consider a somewhat different optimization problem, the choice of the numbers of subjects and replicate measurements on each (effectively holding the main study size fixed); they conclude that for a reasonable range of variance and cost ratios for a diet validation study, four or five replicate measurements would suffice. The reader is also reminded of the discussion in Chapter 5 of the design of two-phase case-control studies where the aim was to collect information on other variables (confounders, modifiers, more detailed exposures) in the second stage.

Some examples of complex dosimetry systems

Frequently, exposure assessment entails elements of both classical and Berkson errors, sometimes shared, possibly differential. We call such exposure assessment protocols "complex dosimetry systems." We conclude this chapter with a narrative discussion of several examples.

Dose reconstruction for "downwinder" studies

Nevada Test Site downwinders

During the two decades following World War II, the United States exploded hundreds of nuclear weapons above ground at the Nevada Test Site (NTS) and the Pacific Proving Grounds in the Marshall Islands. These tests generated large amounts of fallout, which were dispersed widely in surrounding areas. For the NTS tests, much of this landed in western Utah, with smaller amounts elsewhere in the western United States and, for that matter, worldwide in even smaller quantities (NCI 1997). In addition to the military participants of the tests themselves, residents of some areas downwind received sufficient doses to the bone marrow (from external gamma radiation) and to the thyroid gland (primarily from internal radioiodine I^{131} contamination) that associations with leukemia and thyroid abnormalities (cancer, benign nodules, thyroiditis, hypothyroidism, etc.) might potentially be detectable (Gilbert et al. 1998). To investigate this possibility two studies were conducted, a population-based case-control study of leukemia in the entire state of Utah (Stevens et al. 1990) and a cohort study of children clinically examined for thyroid abnormalities (Kerber et al. 1993). Supporting these studies was an intensive dose reconstruction effort (Simon et al. 1995; Till et al. 1995), which traced the fallout from every NTS test, beginning with information about the yield of the bomb and decay of radionuclides, modeling its atmospheric transport and deposition on the ground, uptake by plants and grazing animals, transfer to milk, and distribution of milk and vegetables to consumers. This was combined with subjects' responses (for the thyroid study) to questions about consumption of milk from cows and backyard goats, leafy greens, and breastfeeding or (for the leukemia study) residential history records maintained by the Mormon church, to estimate individual doses. A unique feature of both dosimetry efforts was the use of Monte Carlo methods to assign not just individual doses but also uncertainty estimates for each dose assignment. Some elements of this complex dosimetry system were based on various kinds of measurements, others based on distributions of expert judgment. For each random draw for the various unknowns (including missing or uncertain questionnaire data), a new dose estimate could be produced for each of the study subjects, based on the

Table 11.2. Estimates of relative risk of thyroid neoplasms (N = 19) at 1 Gy (95% CI) in children downwind of the Nevada Test Site (Adapted from Mallick et al. 2002.)

Error model	Dose distribution	Dose–response model	
		Parametric	Semiparametric
No error		9.4 (4.5–13.8)	13.8 (2.5–18.9)
Classical	Normal	17.1 (8.5–24.5)	21.5 (9.4–36.7)
	Semiparametric	15.8 (7.7–23.2)	19.0 (8.0–32.7)
Berkson		7.9 (3.8–11.4)	10.0 (3.1–13.2)
Mixture	Normal	13.2 (5.0–23.1)	16.4 (2.2–34.8)
	Semiparametric	10.9 (2.6–22.6)	14.2 (1.7–33.6)

same input parameters. Over many such realizations, a distribution of dose estimates was obtained that properly reflected the various uncertainties, and was summarized by a geometric mean and geometric standard deviation (GSD) for each individual's dose estimate.

For the leukemia study, the average GSDs of the dose uncertainties was 1.15. Treating these as entirely classical error increased the slope estimate from 1.08 per rad (95% CI 0.26–1.98) to 1.22 (0.18–2.66) allowing only for uncertainties in estimates of deposition or 1.28 (0.26–2.88) allowing also for uncertainties in residence histories. Additional adjustment for shielding uncertainties using a Berkson error model increased it to 1.40 (0.36–3.20) (Thomas 1999). For the thyroid study, the uncertainties were much larger, an average GSD of 2.8. The semiparametric classical and Berkson error mixture model described earlier (Mallick et al. 2002) yielded the various estimates of RR at 1 Gy summarized in Table 11.2. As expected, adjustment for measurement error assuming it was entirely classical increased the relative risk estimate substantially. The semiparametric model suggested substantial skewness in the distribution of true doses, however, leading to somewhat more modest and probably more appropriate increases in adjusted relative risks. Risk estimates were actually reduced somewhat assuming the errors were entirely Berksonian, and the mixture model produced intermediate results. Treating the dose–response model semiparametrically yielded higher risks at 1 Gy under all measurement error models.

Hanford Thyroid Disease Study

The Hanford Nuclear Reservation in southeast part of the state of Washington was the site of plutonium production for the atomic bombs during World War II and the following decades. During this period, substantial emissions of radionuclides, notably I^{131}, were released from the plant and were deposited in local soil and rivers, eventually making it up the food chain in a manner similar to the NTS fallout. In response to public concern about the health risks experienced by the downwind population, a large retrospective cohort study, known as the Hanford Thyroid Disease Study (HTDS), was undertaken by investigators at the Fred Hutchison Cancer Research Center in Seattle, Washington (Davis et al. 2004; Hamilton et al. 2005; Kopecky et al. 2005).

Exposure assessment for the HTDS was based on a separate project known as the Hanford Environmental Dose Reconstruction (HEDR) Project (Shipler et al. 1996; Napier 2002; Tatham et al. 2002; Kopecky et al. 2004), using methods very similar to those described above for the NTS fallout studies. The dose–response analysis, however, went far beyond the incorporation of the GSDs of individual dose estimates, aiming to explore the problem of additive and multiplicative, shared and unshared, classical and Berkson-type errors, using the entire set of realizations of simulated doses. We denote by $\hat{\mathbf{X}}_s = (\hat{X}_{is}|\mathbf{W}_i)_{i=1,\ldots,n}$ the vector of all subjects' dose estimates from Monte Carlo sample $s = 1, \ldots, 100$, given the various inputs $\mathbf{W}_i$ to each individual's dose calculation. Stram and Kopecky (2003) and Kopecky et al. (2004) discussed the use of these dose distributions in a formal maximum likelihood framework for estimating the effects of dose on thyroid abnormalities, using a Monte Carlo approximation to the likelihood of the form

$$L(\beta) = \frac{1}{100} \sum_{s=1}^{100} \prod_{i=1}^{n} p_\beta(Y_1|\hat{X}_{is})$$

In particular, they addressed the influence of shared and unshared, multiplicative and additive errors on the bias and power of analyses using this likelihood, and concluded that the distribution of quantifiable variation in doses between individuals was sufficient to yield adequate power for detecting dose–response relationships expected on the basis of studies in other populations. Nevertheless, this conclusion is not without controversy, some authors (Hoffman et al. 2006) arguing that previous analyses had underestimated the uncertainties in various ways, notably by neglecting the effects of classical error.

Colorado Plateau uranium miners

Another radiation example—the Colorado Plateau uranium miners—illustrates a number of other points. Here, individual doses were assigned using the job-exposure matrix approach, in which exposure levels were assigned to mine-year combinations and linked to the job histories for the individual miners in the cohort obtained from company payroll records.

Although there are doubtless some uncertainties in the individual job histories, the major component of uncertainty here is in the assignment of radon daughter exposures to mines, as available measurements were very spotty; indeed, many mines lacked any measurements at all for several years. To address this problem, the original investigators developed an elaborate hierarchical system, geographically clustering mines into districts within regions within states (Lundin et al. 1971). Where an individual mine was lacking any measurements in a given year, an average of measurements in surrounding years was used. If the gap was too long, an average of measurements for the same year for other mines in the same district was used, or if these were inadequate, in the same region, or same state. Stram et al. (1999) later formalized this somewhat *ad hoc* approach into a multilevel hierarchical model, involving random effects in the means and slopes over time at the mine, district, region, and state levels, to derive an estimate of the "true" dose rate $X_{srdm}(t)$ for a given year t in a given mine m within a given district d, region r, and state s, which combined the available data for that combination with estimates of the posterior expectations of the relevant random effects, given all the data. This also yielded an estimate of the uncertainty of each exposure assignment. The resulting estimates and their uncertainties were then combined with the individual miners' work histories to yield a measurement-error-corrected dose history $x_i(t)$ for each miner, which were used to fit various exposure–time–response models. Typically, uncertainties were much larger in the early years, reflecting the paucity of measurements before the hazards were widely recognized, so this greater uncertainty had the potential to modify the effect of such temporal modifying factors as age at exposure, latency, and dose–rate/duration. In particular, the authors found that the strong modifying effect of dose–rate (a long low dose being more hazardous than a short intense one for the same total dose) was considerably attenuated, but not completely eliminated, after adjustment for dose uncertainties.

Use of biodosimetry in the atomic bomb survivors study

Cullings (2006) provide a historical review of the evolution of the atomic bomb dosimetry, beginning with the pioneering work of Jablon (1971) (see Box 11.1). As described previously, the effect of classical measurement error is generally to attenuate the slope of a dose–response relationships and possibly also to change its shape. One of the earliest measurement error analyses (Prentice 1982) using Jablon's model demonstrated that the apparent negative quadratic term in a linear–quadratic model for this cohort would be attenuated by allowance for multiplicative measurement errors, while the linear term would be increased. Similar results have been reported by several others in relation to the effects of measurement error on either the magnitude of the quadratic term (Jablon 1971; Gilbert 1984; Little 2000) or the evidence for a threshold (Little and Muirhead 1996; Hoel and Li 1998; 1998; 2004b). Other analyses of cancer mortality among the atomic bomb survivors indicated that correction for this bias would increase the excess relative risk per Gy by about 17% for leukemia and by about 10% for solid cancers, assuming measurement errors were multiplicative and lognormally distributed with a logarithmic standard deviation of about 40% (Pierce et al. 1990). But how is one to decide how big the error variance really is? For this one needs some external information, as the structural error model is not identifiable using data on Z and Y alone.

11.1 First description of the effects of errors in the physical dosimetry for the atomic bomb survivors

Jablon (1971) was the first to recognize the nonlinear relationship between true and estimated doses induced by measurement error. His quantitative model is of such importance as to merit a detailed description, but the nonmathematical reader can safely skip this passage.

Let D_{true} denote an individual's true distance from the hypocenter, D_{est} the estimated distance and H the height of the bomb at the time of detonation. Theoretical physics considerations give the following relationship between true dose X and "slant distance" $R = \sqrt{D_{true}^2 + H^2}$:

$$X = f(D_{true}) = \alpha e^{-R/L}/R^2$$

where the denominator corresponds to the usual inverse square law and the exponential term is the effect of absorption by air, with "relaxation distance" L estimated to be about 0.25 km. Below, it will be convenient to denote by $g(X)$ the inverse function of $f(D_{true})$, i.e., the distance D_{true} corresponding to a true dose X.

Jablon treats the exposed population as being roughly uniformly distributed over a 20 km circle around the hypocenter in Hiroshima and a linear strip in Nagasaki, so the prior distribution of distances $\pi(D_{true})$ is proportional to D_{true} in Hiroshima and uniform in Nagasaki. The probability of survival, based on previous work, was assumed to be $S(D_{true}) = 1/(1 + 2.33 D_{true}^{-4})$. Combining these relations, one can compute the posterior density of true doses given survival as

$$\Pr(X|S) = \pi[g(X)]S[g(X)]\frac{dg(X)}{dX}$$

Jablon treats the reported distances as lognormally distributed around the true locations, so the conditional probability given the dose Z calculated using estimated location is

$$\Pr(X|Z, S) \propto \Pr(X|S) \Pr[g(Z)|g(X)]$$

From this, he evaluated $E(X|Z)$ numerically and provided an approximate analytical expression for this relationship.

Jablon also considered other sources of error including the location of the hypocenter, the form of the "9-parameter" model used for shielding, evaluation of the parameters of the dosimetry and survival models, and rounding error. By expressing the various uncertainties on a log scale, he could combine them with the uncertainties in location, expressed in meters, and showed that the overall uncertainty corresponded to an effective uncertainty in location of 47 m in Hiroshima (of which 38 was due to location) and 62 m in Nagasaki (of which 50 were due to location).

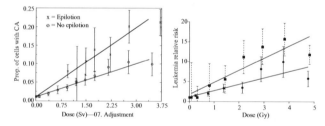

Figure 11.6. Dose–response for chromosomal aberrations (left) and acute leukemia (right) in the atomic bomb survivors by presence or absence of severe epilation. (Left figure Reprinted with permission from Sposto et al. 1991, right figure based on data from Neriishi et al. 1991.)

One possible source of such information is the joint dependence of multiple endpoints on dose, since one of the effects of measurement error is to induce a correlation between outcomes Y_1 and Y_2 conditional on Z, even if Y_1 and Y_2 are independent given X. Figure 11.6 illustrates two such analyses, one of the proportion of cells with chromosomal aberrations (Sposto et al. 1991), the other of leukemia mortality (Neriishi et al. 1991), each stratified by the presence or absence of severe epilation (the self-report of the loss of at least two-third of the hair on the head as a symptom of acute radiation sickness immediately after the bombing). In both cases, the dose–response is more than twice as steep in those reporting epilation than in those without.

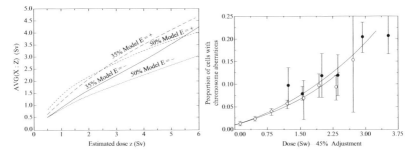

Figure 11.7. Predictions of a measurement error model: left, the conditional expectation of true given estimated doses for subjects with ($E = +$) and without ($E = -$) severe epilation, assuming 35% and 50% errors in a lognormal distribution of true dose; right, dose–response for chromosomal aberrations in these two groups using the measurement error corrected doses, assuming a 45% error model. (Reprinted with permission from Sposto et al. 1991.)

One explanation for this phenomenon is that at any given level of estimated dose (based only on location and shielding), those with epilation tended to have higher true doses than those without. Figure 11.7(left) illustrates this by plotting $E(X|Z, Y_1)$ as a function of Z for different assumptions about $\mathrm{var}(Z|X)$; at any given value of Z, those with epilation have substantially larger average X than those without. After correction for measurement errors, the two dose–response curves line up almost perfectly (Figure 11.7, right).

An alternative explanation is that the two endpoints are not really independent given true dose, that is, that there is inter-individual variation in radiosensitivity, epilation being an indicator that an individual is more radiosensitive. The structural error model is still not identifiable with data only on Y_1, Y_2, and Z, *unless* one makes the additional assumption of independence given X. Nevertheless, one can use such analyses to estimate the range of possible error variances that are compatible with the observed correlation—or conversely to estimate the possible correlation in radiosensitivity after allowance for various degrees of measurement error (Stram and Sposto 1991). Neriishi et al. (1991) took the former approach and showed that the apparent 2.4-fold difference in leukemia slopes without allowance for measurement error declined to 1.8 under the 35% error model (but still significant) and declined further to 1.6 under the 50% error model (no longer significant). Sposto et al. (1991) took the latter approach and found that the apparent difference in chromosome aberrations between epilation groups was minimized at a 46–50% error, depending upon the method used.

It is also worth commenting on the nonlinearity of the curves of $E(X|Z)$ in Figure 11.7 (left), as one of the effects of measurement error pointed out earlier is the change in the shape of a dose–response relationship.

If the true dose–response $\Pr(Y|X)$ is itself nonlinear, then calculation of $E[\Pr(Y|X)|Z]$ for the regression substitution method will involve the higher moments $E(X^n|Z)$. Furthermore, if the distribution of X is unknown or not conjugate with $Z|X$, then the calculation of $\Pr(X|Z)$ can be quite complex. The earlier papers (Pierce et al. 1990; 1991; 1992) treated $\Pr(X)$ has having a Weibull distribution, whereas more recently a semiparametric approach (Pierce and Kellerer 2004) has been developed (see the description of the method earlier in this chapter). Reassuringly, the results of these new analyses yield virtually the same estimates of the attenuation factors for leukemia and solid cancer mortality as those quoted above. These analyses also provide some further guidance as to the likely degree of measurement error. Their current recommendation (Pierce et al. 2007) involves a combination of 40% multiplicative errors and 20% additive errors, for an overall coefficient of variation of 44%. These authors also explain how the measurement errors are viewed as comprising a mixture of classical and Berkson error components, the former arising mainly from uncertainties about individuals' locations and shielding, the latter from the averaging done in assigning doses to subjects with similar locations and shielding.

Recall that the publicly available data for the atomic bomb survivor cohorts has been a grouped dataset, finely cross-classified by categories of dose, age, calendar time, gender, city, and a few other factors, with each cell of the tabulation containing the numbers of person-years and cancers of each type and the mean doses. An important advantage of the regression substitution approach is that such a tabulation can simply be augmented by including the estimates of $E(X^n|Z)$ for each cell so that other investigators could easily re-analyze the data with the measurement error corrected doses.

Like the correlation between two endpoints Y_1 and Y_2 given Z, measurement error can also lead to overdispersion of a continuous endpoint Y given Z. Thus, in analyzing stable chromosomal aberrations, one might treat the observed proportion of aberrant cells Y out of N cells scored as having an over-dispersed binomial distribution with mean $E(Y|X) = \mu_\beta(X)$ and variance

$$\text{var}(Y|X) = \frac{\mu_\beta(X)(1 - \mu_\beta(X))}{N} + \lambda \left(\frac{N-1}{N} \right) [\mu_\beta(X)]^2$$

where λ is an overdispersion parameter that can be estimated along with the dose–response parameters β by an appropriately weighted regression or generalized linear models (Otake and Prentice 1984; Stram and Mizuno 1989; Stram et al. 1993; Cologne et al. 1998). [This formula derives from a model in which for at any given dose each person's Y_i has a binomial distribution with parameter μ_i, and these person-specific μ_is have a beta distribution with variance λ; similar results apply for the Poisson distribution with a gamma distribution in individual rates (Otake and Prentice, 1984).] This parameter incorporates both natural biological variation, as well as dose measurement errors (if the uncorrected Z were used instead of X). Using measurement-error corrected doses $\hat{X}(Z, \sigma)$ with measurement error variance σ^2 and searching for the value of σ that minimizes λ would then provide an estimator of the measurement error variance. Analyses using this approach have yielded similar estimates as those based on other methods.

Ultimately, one might imagine using biomarkers B directly as biological dosimeters (alone or in combination with physical dosimetry), rather than simply as a means of estimating the measurement error variance in physical dosimetry. Given a calibration sample comprising a set of (B, X) measurements, one could estimate $\mu_\beta(B)$ as described above. Then in principle one should be able to estimate the dose for an individual with biomarker measurement B simply using the inverse of this function, $\hat{X}(B) = \mu_\beta^{-1}(B)$. The difficulty with this classical approach, however, is that the resulting dose estimate can be negative or nonexistent. To overcome these difficulties, Bender et al. (1988) introduced the posterior means estimator that entails integrating the likelihood of the biomarker measurement over the prior distributions of doses in the target population and of the model parameters. Cologne et al. (1998) provide a clear discussion of the statistical issues and simulation study of the use of this approach in epidemiologic dose–response analysis. The posterior means estimator with a correctly specified prior distribution of doses yielded nearly the same slope estimates as using physical dosimetry alone, with only slightly larger standard errors, whereas the classical estimator and the posterior means estimator with an incorrectly specified prior were both severely biased.

A subtlety somewhat unique to the atomic bomb survivor cohort is the interpretation of the reference population: when we speak of $\Pr(X)$, are we referring to the distribution of true doses among all exposed or of the survivors to 1950 who constitute the cohort being followed? The dosimetry system that produces the estimated doses Z uses only information on individuals' locations and shielding, together with knowledge of the physics of the energy distribution from the bombs; it does not aim to exploit any biological knowledge about the chances of survival in relation to dose. Obviously the survivors at any given level of estimated dose Z will tend to have had lower true doses than those who died before start of follow-up. (To put these numbers in perspective, the LD_{50} for whole-body irradiation is about 2.5–3 Gy.) Thus, if $\Pr(X)$ were to refer to the distribution of true doses among all exposed individuals, then one would have to include a term for the probability of survival given true dose. This would have to rely on external data about survival from acute radiation sickness and medium-term consequences, as the survivor cohort obviously contains no information to allow this to be estimated directly. Instead, all measurement error analyses of the atomic bomb survivor cohort have treated $\Pr(X)$ as the distribution of true doses *among the population of survivors*, without further correction for survivorship.

On-going work at the Radiation Effects Research Foundation aims to exploit a variety of novel biomarkers of dose, such as *in vivo* mutant T-cell frequencies (Hakoda et al. 1988), Glycophorin-A mutations, and tooth enamel electron spin resonance measurements. These are generally available on only a small subset of survivors, requiring a two-stage approach to the analysis, as discussed earlier. Approaches to the analysis of individual rather than grouped data are also being explored, so as to facilitate the use of biomarkers and incorporate spatial correlations in doses between individuals at similar locations.

Children's Health Study

A quite different example derives from the Children's Health Study (CHS) of air pollution, focusing on variation in individual exposures within communities. In addition to the central site measurements made continuously throughout the study, as used in the between-community comparisons discussed in Chapter 10, information about individual variability came from several sources: place of residence, time-activity patterns, household characteristics, traffic patterns, meteorology, and limited measurements at a subset of locations (schools and a sample of homes) in each community. Navidi and Lurman (1995) described the use of a "microenvironmental" approach to incorporating some of these data into a model for individual exposure assessment, focusing on time spent at school, at home, and outdoors and housing characteristics relevant to each pollutant (e.g., air conditioning for ozone, gas stoves for NO_2, and pets for particulates). Unfortunately, the extent to which true individual variability could be quantified by these methods was somewhat limited, so there was inadequate range of variation within communities to estimate exposure–response relationships at that level for most pollutants. However, analyses of between-community associations using the mean of personal exposures instead of the ambient levels showed somewhat stronger relationships than simply using ambient levels.

Subsequently, Gauderman et al. (2005) conducted a survey of NO_2 concentrations in 289 homes, selected to represent a range of high and low traffic exposures in 11 of the 12 communities. Using the mean of two 2-week measurements (one in summer, one in winter) for each home, a substantial range of variation was found within communities, and a significant within-community association with prevalent asthma was demonstrated for the subset of children included in this survey. For the cohort as a whole, McConnell (2005) demonstrated an association of asthma with residence near major roadways and with predicted pollutant concentrations based on an atmospheric dispersion model (Benson 1989). No direct pollution measurements were used in this analysis, however. These two types of information were later combined in a general Bayesian measurement error framework by Molitor et al. (2006), similar to that illustrated in Figure 11.1. Specifically, simple traffic metrics such as distance to the nearest freeway, the more complex dispersion model predictions, and the central-site long-term average concentrations, constituted a vector of predictors W for the unknown true exposures X. The available short-term measurements Z were then assumed to be lognormally distributed around the true long-term mean, with an adjustment for temporal variation based on the central-site data. For this analysis, the outcome Y was eight-year rates of change in lung function growth. The measurement error adjusted regression of Y on X for the substudy participants was somewhat larger and more significant than either the naïve regression of Y on W or on Z alone. It remains to be seen whether extending this model to children not in the substudy (for whom the predictors W would be available, but not Z) would yield similar improvements.

For this to be likely, however, it would be essential to exploit the spatial correlations in the exposure data, as discussed in Chapter 9. In a subsequent paper (Molitor et al. 2007), a spatial autocorrelation process was added to both the true exposures X and in the outcomes Y, separately at the within- and between-community levels. Although the between-community comparisons showed little spatial correlation in either component, the within-community comparisons substantially improved the predictive power of the model, particularly those for exposure. This analysis was also restricted to the participants in the substudy, but now a much better imputation of exposures to cohort members for whom no measurements are available would be possible, by "borrowing strength" from measurements at neighboring locations, appropriately weighted by the distance between them.

12 Multiple risk factors and interactions

Up to this point, we have been concerned mainly with one exposure factor at a time, although as we have seen in Chapter 6, even this can be quite complex, particularly if there are time-related modifying factors like latency or if exposure is extended over a long period at varying intensity. We now turn our attention to the analysis of multiple risk factors, say, two or more exposures or one exposure of primary interest together with other confounders or modifiers. For now, we set aside the complications discussed earlier involving time-dependent exposures and focus on the effects of two or more fixed covariates.

In Chapter 2, we distinguished between confounding and effect modification. To review, a confounder C is a variable that is related both to the exposure E and the outcome Y and thus has the potential to induce a spurious association between E and Y where no causal connection exists, or to bias the magnitude of a real association upwards or downwards. A modifier M, on the other hand, is a variable for which the magnitude of the association between E and Y differs at different levels of M; for example, there might be no association in one subgroup of M and a strong one in another subgroup. This phenomenon goes by various names, such as interaction or synergism, all meaning that the effect of two or more factors in combination is different from what would be predicted by each factor separately. We begin by addressing methods for dealing with confounders and then turn our attention to effect modification.

Frequently in environmental epidemiology, we are faced with exposures that are themselves complex mixtures of hazardous substances. Particulate air pollution, for example, comprises a broad range of particle types, varying in size and chemical composition, coexisting and reacting with other gaseous pollutants. Sorting out their separate contributions to a health effect, or even assessing the effect of the mixture as a whole where its composition varies from place to place, can be a formidable challenge. Often we are left with many different statistical models for the joint effects of multiple agents that fit the data reasonably well. What then are we to conclude about either the contribution of individual factors or their combination?

We conclude this chapter with a discussion of interactions between two or more exposures or between a single exposure and a genetic susceptibility factor. More complex pathways involving many genes and multiple exposure factors they interact with are discussed in terms of "empirical" models involving a combination of main effects and interactions. We will revisit this problem in the following chapter, where we take a more mechanistic approach.

At this point, our focus is on assessing the relationship between various risk factors and disease at the population level as the basis for risk assessment in guiding environmental regulation policy in Chapter 15. We will return to these issues in Chapter 16, where the focus shifts to the individual and we ask questions about how we should assess whether a particular disease in an individual with a history of exposure to an established risk factor was caused by that exposure or by something else.

Dealing with confounders

The four basic methods for controlling confounding—restriction, stratification, matching, and covariate adjustment—were introduced in Chapters 3 and 4. An association cannot by confounded by a categorical variable like sex if the analysis is restricted only to males or only to females. Stratification extends this basic principle by reasoning that if this is true for each gender separately, then the results of such sex-specific analyses can be combined across genders to obtain an unconfounded summary estimate. There are various ways of accomplishing this combination, such as the comparison of standardized rates, as discussed in Chapters 3 and 4. These techniques are collectively known as stratification, but it is worth emphasizing that this term implies more than just dividing the sample into strata; it is the *combination* of estimates across strata that is the key concept. For a continuous variable like age, one could create categories (say, five-year intervals) and stratify the analysis in this way, but depending upon how broad the categories are, some residual confounding could remain within categories. The same applies if there are several confounders to be controlled simultaneously.

By a similar reasoning, an association cannot be confounded if the groups being compared—exposed and unexposed in a cohort study, cases and controls in a case-control study—are individually matched on the confounders, so this technique is often preferred when there are multiple confounders to be controlled, particularly confounders measured on continuous scales.

A key advantage of stratification and matching is that no assumptions are required about the effects of the confounders. By defining separate strata for levels of each variable jointly, the association is free of the possibility of confounding across levels. But as the number of factors to be controlled increases, many strata may lack any subjects in one of the groups being compared (leaving those in the comparison group for that stratum uninformative); correspondingly, in a matched study, the prospects of finding an exact match may diminish as the matching criteria become more numerous or more restrictive. In this circumstance, one might turn instead to covariate adjustment using a multivariate model, such as logistic or Cox regression (Chapter 4). While these approaches have the advantage of great flexibility for adjusting for multiple confounders, they require stronger assumptions about the joint effects of these variables and the exposures of interest on disease.

The possibility of residual confounding can never be completely eliminated in any observational study. Although one may be able to control for all *measured* confounders with some degree of confidence, there may be still other variables that have not been measured or errors in the variables that have, which are not balanced across the comparison groups. Only randomization can assure balance in all potential confounders simultaneously (Greenland 1990), and even then, this guarantee applies only in expectation across hypothetical replications of a study, not in any single study sample. What *can* be guaranteed, however, is that a statistical estimation procedure based on a randomized design will yield an unbiased estimator, in the sense that across hypothetical replications, the average value of the estimate will equal the true value. Likewise, under the null hypothesis, a significance test will reject at the nominal level, meaning that a 5% significance test would reject the null hypothesis in 5% of replicate samples.

One of the most widely used methods of stratified analysis is the Mantel–Haenszel estimator of the odds ratio (OR). Consider the data layout in Table 12.1 for a hypothetical case-control study. Within any stratum s, the OR would be computed as $\psi_s = a_s d_s / b_s c_s$. The question is how best to combine these estimates across strata. An obvious approach would be to take a weighted average, weighting each contribution inversely by its variance, so that the most precise estimates receive the heaviest weight. Since the ORs themselves tend to have highly skewed distributions and their variances tend to be proportional to their magnitude, Woolf (1955) introduced a natural estimator by averaging their logarithms:

$$\ln(\hat{\psi}) = \frac{\sum_s \ln(\hat{\psi}_s)/V_s}{\sum_s 1/V_s}, \quad \text{where } V_s = \text{var}\left[\ln(\hat{\psi}_s)\right] = \frac{1}{a_s} + \frac{1}{b_s} + \frac{1}{c_s} + \frac{1}{d_s},$$

$$\text{with var}\left[\ln(\hat{\psi})\right] = \frac{1}{\sum_s 1/V_s}$$

One difficulty with this estimator, however, is that it is undefined if even one cell is empty. An ad hoc solution to this problem adds a small value (conventionally 1/2) to every cell, but this then yields an estimator that is no longer unbiased. In a classic paper, Mantel and Haenszel (1959)

Table 12.1. Hypothetical data for a case-control study illustrating a stratified analysis

Confounder stratum	Exposure status	Controls	Cases	Total
s = 1	Z = 0	a_1	c_1	v_1
	Z = 1	b_1	d_1	w_1
	Total	t_1	u_1	n_1
s = 2	Z = 0	a_2	c_2	v_2
	Z = 1	b_2	d_2	w_2
	Total	t_2	u_2	n_2
...	...	...	...	...
s = S	Z = 0	a_S	c_S	v_S
	Z = 1	b_S	d_S	w_S
	Total	t_S	u_S	n_S

introduced the simple estimator

$$\hat{\psi} = \frac{\sum_s a_s d_s / n_s}{\sum_s b_s c_s / n_s} = \frac{\sum_s G_s}{\sum_s H_s}$$

that does not have this problem. It can be shown that this estimator is unbiased and is nearly fully efficient relative to the maximum likelihood estimator (Robins et al. 1986). Various estimators of its variance have been suggested, the recommended one being

$$\mathrm{var}(\ln \hat{\psi}) = \frac{\sum_s G_s P_s}{2 \left(\sum_s G_s \right)^2} + \frac{\sum_s G_s Q_s + H_s Q_s}{2 \left(\sum_s G_s \right)^2} + \frac{\sum_s H_s Q_s}{2 \left(\sum_s H_s \right)^2}$$

where $P_s = (a_s + d_s)/n_s$ and $Q_s = (b_s + c_s)/n_s$. A relatively simple way of putting a confidence limit on the estimate is the test-based procedure (Miettinen 1985),

$$\widehat{\mathrm{OR}}_{\mathrm{MH}}^{1 \pm Z_{1-\alpha/2}/\chi_{\mathrm{MH}}}$$

which has the attractive property that the lower limit is exactly 1 if the test is exactly significant at any particular significance level α. The Mantel–Haenszel significance test takes a somewhat different form from the OR estimate,

$$\chi_{\mathrm{MH}}^2 = \frac{\left(\sum_s a_s - \sum_s E(a_s) \right)^2}{\sum_s V(a_s)} = \frac{\left(\sum_s a_s - \sum_s t_s v_s / n_s \right)^2}{\sum_s t_s u_s v_s w_s / n_s^2 (n_s - 1)}$$

and also has a chi square distribution on 1 df. (Analogous formulae are available for rate data with person-time denominators, see Rothman and Greenland (1998).

The analysis of matched case-control studies is based on the McNemar procedure for binary exposure variables (Chapter 3) and conditional logistic regression (Chapter 4).

Selection of confounders, model choice, and model averaging

Given the relative ease with which multivariate methods can be applied to control for multiple confounders, a more important issue is which of the many possible variables should be adjusted for. The basic principle is that for a variable to be a true confounder, it must be associated with exposure and a risk factor for disease conditional on exposure (i.e., not merely a reflection of a direct effect of exposure on disease). Failure to adjust for any such variable will lead to a biased estimate (upwards or downwards, depending upon whether the two associations are in the same or opposite directions), but adjustment for a variable that is not a confounder is also unwise. The downside of such unnecessary adjustments is not bias but an inflation of the variance and loss of power. Generally, adjustment for a variable that is an independent risk factor but unrelated to exposure has little effect, but the converse situation (matching or adjustment for a variable that is associated with exposure but not with disease) goes under the name of "overmatching" or "overadjustment," and can have a severe impact on variance and power.

Consider a hypothetical study of ambient air pollution and asthma. One might be tempted to match on neighborhood in an attempt to control for various nebulous aspects of socioeconomic status that are not easily measured or not even recognized. However, since individuals from the same neighborhood would have essentially the same exposure to ambient pollution, there would be little scope for comparison of pollution levels within neighborhoods, at least if central site measurements were to be used as the measure of exposure. If, in fact, whatever variables neighborhood is a surrogate for did have a causal effect on asthma risk, then failure to adjust for it would yield biased tests and estimates. In this circumstance, one would have to try to assess the variables that were etiologically important in the risk of asthma across neighborhoods and control for them instead.

How are these principles to be operationalized? Day et al. (1980) pointed out that by judicious selection of potential confounders, it may be possible to "explain away" any association. Clearly, an analysis strategy that aims to accomplish this (or the converse) by systematically searching for variables that would reduce (or increase) an association would be highly biased.

Since there is relatively little penalty associated with adjustment for variables that are true risk factors but may not be strongly associated with exposure, some have advocated a stepwise search for risk factors, irrespective of their association with exposure. However, stepwise variable selection procedures are well known to be unreliable for selecting the right set of variables, and tests or confidence limits based only on a final "best model" will fail to properly account for uncertainty about the selection of variables.

Another strategy would be to assess separately the relationships of each variable to exposure and to disease and only adjust for those that are associated with both (conditional on any other variables chosen). But in either of these approaches, exactly how are we to assess whether an association "exists" or not? The obvious choice—statistical significance—depends both on the magnitude of the association and on the sample size, so true confounders may fail to be detected in small samples or variables with only very weak observed associations might be deemed significant in very large samples, yet have little confounding effect. The choice of significance level in this context is quite arbitrary, but it is generally felt that a failure to adjust for a true confounder would be a more serious error than inclusion of a nonconfounder, so significance-based criteria should be much more liberal than might be considered appropriate for the exposure variables of primary interest.

These considerations have led some authors (see, e.g., Rothman and Greenland 1998) to advocate an approach that depends upon the strength of each variable's associations with both exposure and disease indirectly through their effects on the exposure–disease association of real interest. Using this "change in estimate" criterion, the usual recommendation is to adjust for any variable that leads to a change in the OR_{DE} of 10% or more, upwards or downwards, after adjusting for any other variables that meet this criterion. Although the 10% figure is also quite arbitrary, it has the merit of being independent of sample size and in general appears to perform quite well. By selecting variables independently of the direction of their confounding effects, the bias noted by Day et al. above does not arise.

Even the change in estimate criterion has some difficulties when it comes to inference about the model parameters, however, similar to the pitfalls in stepwise regression discussed above. In what order should variables be tested for their effect on the exposure–disease association conditional on other variables? Any how should the uncertainty about variable choice be reflected in the final confidence limits and significance tests? One attractive approach to this problem is "model averaging" (see Box 12.1) which in principle can be applied in either a frequentist or a Bayesian fashion, but (at least until recently) has been better developed in the Bayesian context (Madigan and Raftery 1994; Raftery et al. 1997; Hoeting et al. 1999;

Viallefont et al. 2001). Basically, one considers the full range of possible adjustment models and forms a weighted average of the parameters of interest across all of these alternative models, weighted by some measure of the fit of each model.

An example of the use of Bayesian model averaging (BMA) is provided by studies of the acute effects of particulate air pollution on mortality, with uncertainty about which which measure(s) of pollution to use and which of several weather variables to adjust for. Clyde et al. (2000) describe the BMA methodology in detail, with an application to time series data on mortality from Birmingham, AL, while Dominici et al. (2003b) provide an application of similar methods to data from Phoenix. With 8 particulate and 20 weather variables, Dominici et al. found the posterior probability π_m for model m (see Box 12.1) was 46% for the most likely model, with the top 25 models accounting for 87% of the total. The BMA estimate for an interquartile range of particulates was 1.028 with a 95% posterior credibility interval (CI) 1.000 − 1.045. In contrast, the single best model yielded a CI of 1.014 − 1.047, demonstrating the underestimation of uncertainty from failure to account for model uncertainty. The posterior probability of no particulate effect (the sum of πs over all models with none of the 8 particulate variables) was 0.052, or a posterior odds of 18:1. This compares with a prior probability of no particulate effect (assuming all models were equally likely a priori) of 0.747 (=Bin(0|8, 1/28)), or a prior odds of 0.339:1. The ratio of posterior to prior odds yields a Bayes Factor of 54, within the range Kass and Raftery would call "strong evidence."

BMA can be misused, however, particularly when applied to sets of highly correlated variables. For example, Koop and Tole (2004) used BMA to examine which of a number of highly correlated measures of particulates and their gaseous co-pollutants were responsible for the health effects. They concluded that there was so much uncertainty about model form that the causal constituent could not be determined with any confidence. Thomas et al. (2007a) replied that model averaging is more appropriately used to obtain a model with more stable predictive ability than to interpret individual regression coefficients and concluded that these data still supported a strong effect of air pollution overall, despite the uncertainty about which constituent(s) were responsible.

Multipollutant models have been widely considered in the air pollution literature, but mainly by exploring the fits of alternative single- and two-pollutant models in a descriptive manner (see, e.g., Samet et al. 2000a; Schwartz 2000a; Sarnat et al. 2001). MacLehose et al. (2007) compare hierarchical BMA approaches to this problem with several different parametric and nonparametric prior models for the effects of the individual components of a complex mixture. By using a hierarchical framework,

12.1 *Bayesian model averaging and hierarchical models*

Let β represent the adjusted log OR for the exposure–disease association of primary interest, let Θ represent a vector of regression coefficients for all P possible confounders, and let $m = 1, \ldots, M = 2^P$ index the set of all subsets of the P covariates, with corresponding estimates $\hat{\beta}_m, \hat{\Theta}_m$ and sampling variance $V_m = \text{var}(\hat{\beta}_m)$. Now let

$$\pi_m = \Pr(m|D) \propto \Pr(m) \int \Pr(D|m, \beta, \Theta) \Pr(\beta, \Theta) d\beta \, d\Theta$$

be the posterior probability of model m given the data D. Then the model-averaged estimator of β is given by $\hat{\beta} = \Sigma_m \hat{\beta}_m \pi_m$ with variance $\text{var}(\hat{\beta}) = \Sigma_m \pi_m [\hat{V}_m + (\hat{\beta}_m - \hat{\beta})^2]$. A good approximation to the posterior probability is based on the Bayesian Information Criterion, $BIC = 2 \ln L - P \ln(n)$, where n is the sample size. If all the models are *a priori* equally likely, then

$$\pi_m \cong \frac{\exp(-\text{BIC}_m/2)}{\Sigma_m \exp(-\text{BIC}_m/2)}$$

When the number of variables P is large, exhaustive enumeration of all possible models becomes computationally impractical, but Monte Carlo methods can be used to sample the subset of more likely models efficiently (George and McCulloch 1993; George and Foster 2000; Yi et al. 2003). Madigan and Raftery (1994) provide an efficient way of pruning the set of models to be considered based on the principle of "Occam's razor."

Results can be summarized in terms of the posterior probability that each variable is included in the model $P_p = \Sigma_m \pi_m I(\beta_{mp} \neq 0)$ and its posterior expectation, either marginally or conditional on being in the model. Inference on either the set of models or the set of variables can be accomplished using "Bayes factors" (Kass and Raftery 1995), as described in Chapter 4.

So far, we have assumed that all the variables are "exchangeable," in the sense that absent any specific information to the contrary, we cannot predict in advance that any particular variable is more or less likely to be relevant. The approach can be extended to a multilevel model by including "prior covariates" that could relate to either the prior probability that a variable is in the model or the prior expectation of its effect size. Such prior covariates do not constitute a declaration that such variables *are* more or less important or quantitatively how much more so they might be; they simply define "exchangeability classes" that *could* differ and within which we cannot make any further *a priori* distinctions. Putting such prior covariates into a hierarchical regression framework provides a flexible way of allowing for multiple types of prior knowledge or even continuous variables. Letting Z_p denote a vector of prior covariates for parameter β_p one specifies the second level model by a pair of regression equations

$$\text{logit } \Pr(\beta_p \neq 0) = \omega' Z_p$$

$$E(\beta_p | \beta_p \neq 0) = \varphi' Z_p$$

The entire system of equations can be fitted by maximizing the marginal likelihood, $L(\omega, \varphi) = \Pr(Y | X, Z)$, or by MCMC methods (Conti et al. 2003), providing estimates of πm, $\Pr(\beta_p \neq 0 | D, Z)$, and $E(\beta_p | \beta_p \neq 0, D, Z)$.

effect estimates "borrow strength" from other similar effects. In an application to retinal degeneration in the wives of applicators of 18 herbicides, only one exposure showed a significant effect by maximum likelihood, but it was substantially reduced and nonsignificant in all four hierarchical models. In an accompanying editorial, Thomas et al. (2007b) elaborate on the potential for incorporating prior covariates characterizing each of the specific constituents—their sources, chemical similarities, interactions with metabolizing enzymes, etc., and compare model averaging approaches with hierarchical models for estimating a single fully saturated model.

Where the focus is on a single exposure variable of primary interest with uncertainty about which variables to adjust for rather than uncertainty about all possible models, Crainiceanu et al. (2007) suggest a two-stage approach: first they identify a subset of possible models for predicting exposure and choose the one at which the deviance stabilizes; then, they force these variables into the disease model and explore the remaining variables for further improvement in the deviance. No model averaging is done. They show by simulation that their method yields consistent estimates of the adjusted relative risk, whereas BMA can lead to biased estimates of both the effect size and its uncertainty, partly because the averaging can include models that do not include variables needed to properly control for confounding, and partly because different adjustment models can yield exposure effects with differing interpretations.

The performance of these various alternative approaches has been studied by simulation in a series of papers (Greenland 1993; 1994a; 1997; 1999a; 2000; Steenland et al. 2000).

Testing for interactions

Epidemiologists have long recognized that most diseases result from a complex "web of causation" (MacMahon and Pugh 1970), whereby one or more external agents ("exposures") taken into the body initiate a disease process, the outcome of which could depend upon many host factors (age, genetic susceptibility, nutritional status, immune competence, etc.). Exposures may occur over an extended period of time with some cumulative effect, and exposure to multiple agents could have synergistic or antagonistic effects different from what might result from each separately. These general notions were formalized by Rothman (1976a) in a "sufficient component causes model," which postulates that disease can result from any of several sufficient causal constellations, each of which may comprise several components (e.g., exposure plus susceptibility plus timing) that are all necessary to make them a complete cause. This framework provides a useful way to think about exposure to multiple exposures or risk factors.

Table 12.2 illustrates the calculation of the ORs for two binary exposure variables A and B, first estimating a separate OR for each of the possible combinations of exposure to A and/or B, relative to those exposed to neither, then the conditional ORs for each factor specific to the level of the other (e.g., $\text{OR}_{A|B=0}$ or $\text{OR}_{A|B=1}$ and similarly for $\text{OR}_{B|A}$). Table 12.3 then shows the different ways such ORs might be displayed. The most complete presentation is that given in the top panel ("Unconditional"), which provides the effects of each combination of factors along with

Table 12.2. Calculation of joint and conditional ORs for two binary exposure factors

Exposures		Numbers of		Odds ratios		
A	B	Cases	Controls	AB	$A\|B$	$B\|A$
No	No	n_{00}	m_{00}	1	1	1
No	Yes	n_{01}	m_{01}	$\dfrac{n_{01}m_{00}}{n_{00}m_{01}}$	1	$\dfrac{n_{01}m_{00}}{n_{00}m_{01}}$
Yes	No	n_{10}	m_{10}	$\dfrac{n_{10}m_{00}}{n_{00}m_{10}}$	$\dfrac{n_{10}m_{00}}{n_{00}m_{10}}$	1
Yes	Yes	n_{11}	m_{11}	$\dfrac{n_{11}m_{00}}{n_{00}m_{11}}$	$\dfrac{n_{11}m_{01}}{n_{01}m_{11}}$	$\dfrac{n_{11}m_{10}}{n_{10}m_{11}}$

Table 12.3. Typical presentations of ORs for two binary exposure factors, together with an illustrative example (see text)

Exposure A	Exposure B		
	No	Yes	Total
Unconditional			
No	1	$OR_B = 2$	1
Yes	$OR_A = 0.5$	$OR_{AB} = 1.5$	$aOR_A = 0.67$
Total	1	$aOR_B = 2.3$	—
Conditional on A			
No	1	$OR_{B|A=0} = 2$	—
Yes	1	$OR_{B|A=1} = 3$	—
Conditional on B			
No	1	1	—
Yes	$OR_{A|B=0} = 0.5$	$OR_{A|B=1} = 0.75$	—

the marginal effects of each factor adjusted for the other (aOR_A and aOR_B). The presentation of conditional ORs—though readily showing how the effect of one variable might be modified by the other—fails to communicate the effect of the modifying variable itself in the absence of the other, since the latter OR is set to one. For example, suppose $OR_A = 0.5$, $OR_B = 2$, and $OR_{AB} = 1.5$. Then $OR_{A|B=0} = 0.5$ and $OR_{A|B=1} = 1.5/2 = 0.75$, apparently suggesting that A is always protective, whereas the effect of the two in combination clearly is deleterious. Conversely, $OR_{B|A=0} = 2$ and $OR_{B|A=1} = 1.5/0.5 = 3$, suggesting that the effect of B is always deleterious, whereas the risk in those exposed to B is actually smaller in the presence of A than in the absence of it.

Consider the hypothetical data shown in Table 12.4. In this example, it is clear that smoking is a much larger contributor to risk than is the environmental hazard. The two factors are not confounded, however, since in the population at risk, the prevalence of smokers is the same in the exposed and unexposed. The two effects on risk are multiplicative, individuals with both factors having a relative risk of 30, the product of the relative risks for smoking (10) and for the environmental hazard (3).

The multiplicative model just illustrated can be represented as

$$RR_{mult} = RR_E \times RR_S$$

where RR_E is the relative risk for exposed nonsmokers relative to unexposed nonsmokers, and RR_S is the relative risk for unexposed smokers relative to unexposed nonsmokers. Under this model, the effect of exposure is the same in both nonsmokers ($RR_{E|NS} = 3$) and smokers ($RR_{E|S} = 30/10 = 3$).

Table 12.4. Hypothetical data from a cohort study of some environmental hazard and tobacco smoking

Environmental hazard	Smoking habit	Number at risk	Cancer cases	Relative risk
No	Never	1000	10	1
No	Current	1000	100	10
Yes	Never	1000	30	3
Yes	Current	1000	300	30

An additive model takes the form

$$RR_{add} = 1 + (RR_E - 1) + (RR_S - 1)$$
$$= RR_E + RR_S - 1$$

in other words, the risk from exposure to both factors is the background risk plus the sum of the *additional* risks from each factor separately. Thus, in our hypothetical example, if the single-factor risks were the same as before, we would have expected a relative risk for exposed smokers of $1 + (10 - 1) + (3 - 1) = 12$, rather than 30 as above. Under this model, the excess relative risk (ERR $=$ RR $- 1$) for exposure is the same in non-smokers (ERR$_{E|NS} = 3 - 1 = 2$) and smokers (ERR$_{E|S} = 12 - 10 = 2$). Of course, the truth could also be less than additive (e.g., a joint *RR* of 11), greater than multiplicative (e.g., 50), or something in between (e.g., 20).

In practice, neither model is likely to be exactly correct, so one might want a more general family of models that would include both the additive and multiplicate forms as special cases, thereby providing an alternative against which to test the fit of each. The first such model to be proposed was an exponential mixture (Thomas 1981) of the form

$$RR(E, S) = RR_{add}^{1-\alpha} \times RR_{mult}^{\alpha}$$

which takes the additive form when $\alpha = 0$ and the multiplicative form when $\alpha = 1$. Other mixture models have been suggested (Breslow and Storer 1985; Moolgavkar and Venzon 1987; Lubin and Gaffey 1988), of which the most consistent under arbitrary recoding of the constituent sub-models appears to be one based on the Box–Cox transformation (Guerro and Johnson 1982) of the form

$$RR(E, S) = \begin{cases} \exp(\beta_1 E + \beta_2 S) & \text{if } \alpha = 0 \\ [1 + \alpha(\beta_1 E + \beta_2 S)]^{1/\alpha} & \text{if } \alpha \neq 0 \end{cases}$$

Unfortunately, by combining exponential-multiplicative and linear-additive models in this manner, comparisons of the form of joint effect of two factors and the shape of their separate exposure–response relationships are confounded if the variables are continuous.

Epidemiologists use the terms "effect modification" or "interaction" to describe any departure of the observed joint risk from what might be expected on the basis of a simple model involving only the effects of the separate factors. Any test or estimate of interaction is thus model specific, that is, one must specify which main effects model the observed data deviates from. For example, one could define a multiplicative interaction relative risk as

$$RR_{Int(mult)} = \frac{RR_{joint}}{RR_E \times RR_S}$$

or an additive interaction relative risk as

$$RR_{Int(add)} = RR_{joint} - RR_E - RR_S + 1$$

For the data illustrated in Table 12.2, $RR_{Int(mult)} = 1$ and $RR_{Int(add)} = 18$, indicating no departure from a multiplicative model but a large positive deviation from an additive model. Likewise, if the joint RR were 12, the multiplicative interaction RR would have been 0.4 and the additive interaction would have been 0, indicating a less-than-multiplicative joint effect and no departure from an additive model. These concepts have natural extensions to more than two risk factors, such as the inclusion of main effects and interactions in a logistic regression model (for testing departures from a multiplicative model). These parameters can be used to estimate the proportion of disease among exposed individuals that is attributable to the separate or joint action of each factor or other unknown factors, as will be explained in Chapter 16.

Confidence limits on these derived parameters can be derived from their asymptotic variances, in a similar manner to those for main effects described in Chapter 4. For example, the logarithm of the multiplicative interaction OR is simply

$$\ln(OR_{Int(Mult)}) = \ln(OR_{AB}) - \ln(OR_A) - \ln(OR_B)$$
$$= \ln(OR_{A|B=1}) - \ln(OR_{A|B=0})$$

But $var(\ln OR_{AB}) = 1/m_{11} + 1/m_{00} + 1/n_{11} + 1/n_{00}$, and similarly for the other terms, so

$$var(\ln OR_{Int(Mult)}) = 1/m_{00} + 1/m_{01} + 1/m_{10} + 1/m_{11}$$
$$+ 1/n_{00} + 1/n_{01} + 1/n_{10} + 1/n_{11}$$

In practice, one is generally looking at more than two simple binary variables at once, so it is more convenient to use logistic regression for the purpose of testing and estimating interaction effects, at least for testing departures from a multiplicative model. To see this, note that under a logistic model, $\text{logit} \Pr(Y = 1|A, B) = \beta_0 + \beta_1 A + \beta_2 B + \beta_3 A \times B$, the OR is $\text{OR}(A, B) = \exp(\beta_1 A + \beta_2 B + \beta_3 A \times B) = \text{OR}_A \times \text{OR}_B \times \text{OR}_{\text{Int(Mult)}}$. In other words, one can estimate the three ORs in Table 12.2 simply by including indicator variables for the main effects of each factor along with their product in a logistic regression (conditional or unconditional, depending upon whether the study design is matched or not). Likewise, the conditional ORs can be computed as $\text{OR}_{A|B} = \exp(\beta_1 + \beta_3 B)$ and similarly for $\text{OR}_{B|A}$. This regression framework extends naturally to incorporate adjustment for confounders and interactions amongst more than two variables, including higher-order interactions (e.g., three-way interactions $A \times B \times C$, etc.)

The terms "interaction," "effect modification," and "synergy" are often used interchangeably, but there is an extensive literature on this subject (Rothman 1974; 1976b; Koopman 1977; Miettinen 1982a; Rothman and Greenland 1998). "Synergy" implies a public health impact that is larger than the sum of the effects of each variable separately—in short, a greater-than-additive model (a less-than-additive effect would be called "antagonism"). "Interaction" in the statistical literature means any departure from a pure main effects model; this main effects model could take any form, but an important case is what is known as "intrinsic" interaction in the sense that it is not possible to represent the joint effect in terms of main effects alone on any scale (Tukey 1949). In the biological literature, interaction typically is used in a different sense to refer to a biological effect that depends jointly on two or more factors, such as an interaction between a substrate and an enzyme that metabolizes it or between an antigen and an antibody. Biological interactions can lead to effects at the population level that are nevertheless indistinguishable from a simple main effects model or could require a statistical interaction term; conversely, a statistical interaction might or might not imply the presence of a biological interaction. See (Siemiatycki and Thomas 1981) for further discussion of this distinction, with examples from multistage carcinogenesis as described in the next chapter.

For studies of gene–environment interactions, two alternative designs are available that do not require control subjects. The case-only design relies instead on an assumption that genotype and exposure are independently distributed in the population at risk, so that any association between the two factors among cases implies an interaction in risk, specifically a departure from a multiplicative model. The design cannot be used for testing main effects, however, and the assumption of gene–environment independence is not always tenable: some genes could affect behavior, directly or indirectly (e.g., through family history of a genetic disease with a modifiable environmental component). The case-parent-triad design instead estimates a genetic relative risk by comparing the genotypes of cases to the set of genotypes the case could have inherited, given the genotypes of the parents. By stratifying on the exposure of the case, differences in the genetic relative risk for exposed and unexposed cases would indicate a gene–environment interaction, without requiring any exposure information for the parents. This design requires a weaker assumption of gene–environment independence, conditional on parental genotype (i.e., within-family rather than between-family independence). For further discussion of these approaches see (Thomas 2004).

Multiple exposures, source apportionment, and complex mixtures

Frequently in environmental epidemiology, we are confronted with an exposure that is really a complex mixture of many constituents. "Air pollution," for example, comprises both gaseous pollutants like oxides of nitrogen (NO_x, including NO, NO_2, etc.), ozone (O_3), and various acid aerosols (HCl, H_2SO_4, N_2NO_3, etc.), as well as particulate matter of various sizes and chemical compositions (collectively known as PM, with sizes designated as coarse PM_{10}, fine $PM_{2.5}$, and ultrafine $PM_{0.25}$). These various components of the mixture can have different health effects, come from different sources, and have different spatial and temporal distributions. It is thus not just an academic matter to separate their effects, since different the public health goals (reducing mortality from specific causes, preventing asthma exacerbations, improving lung function) may require control of different constituents, and the regulations to accomplish these goals may require interventions directed at different sources of pollution.

Furthermore, it is possible that the various constituents may interact chemically or biologically. For example, NO and O_3 are highly reactive, yielding NO_2 and O_2; thus NO from fresh vehicle exhaust, which occurs in high concentrations near highways, tends to deplete the background concentrations of O_3, leading to lower concentration of the latter near highways. Depending upon which pollutant has the stronger health effects, one might thus see apparently protective or deleterious effects from living close to major highways. It is also possible that long-term exposure to O_3 sets up a chronic inflammatory process that make an individual more sensitive to a short-term insult with PM or vice versa.

Unlike an experimental study, where it would be possible to expose animals or cell cultures to various pollutants separately and in combination in a balanced factorial design and observe their joint effects and interactions directly, the epidemiologist is limited to observing these effects within the range of combinations that are available naturally in different settings. Typically, the various pollutant concentrations will tend to be highly correlated with each other. In Southern California, for example, O_3 is generated by the action of intense sunlight on other pollutants and tends to be only moderately correlated with them. PM, NO_2, and acids, however, tend to be highly correlated with each other, both spatially and temporally, with correlations in long-term average concentrations across various locations typically of the order of 0.8–0.9 or higher. In this circumstance, simply putting all the pollutants into a multiple regression model is unlikely to be rewarding, as their multicollinearity will lead to highly unstable coefficients. Before exploring various statistical approaches to addressing uncertainty about the selection of variables, we consider two other approaches to this problem.

The first is known as "source apportionment" (Schauer et al. 1996; Zheng et al. 2002). Using chemical mass balance techniques and knowledge of the chemical profiles of specific sources (e.g., road dust, gasoline and diesel vehicles, wood burning, cattle manure, manufacturing plants, etc.), one can use the measured distribution of chemical constituents in a complex mixture to reconstruct the proportion of the mixture that derived from the various sources. In some situations, the availability of spatially resolved data on specific chemical constituents, together with meteorological data and atmospheric dispersion modeling techniques, may permit the identification of specific point sources as well. These estimates of source contributions may be much less correlated across different locations than the concentrations themselves, and thus more amenable to multiple regression modeling of their health effects. Such an analysis also yields a direct estimate of the burden of health decrements attributable to specific pollution sources, which is the ultimate goal of environmental regulations. Nikolov et al. (2006) illustrate use of a Bayesian structural equations approach to analyzing the dependence of heart rate variability on sources of particulates in the Boston area.

A quite different approach is known as "genetic fingerprinting" (Rothman et al. 2001; Brennan 2002; Gant and Zhang 2005; Hunter 2005; Kraft and Hunter 2005). Here the basic idea is that different constituents of a complex mixture are metabolized by different enzymes, leading to different health effects. As noted by Hunter,

The finding of an interaction between exposure to a complex mixture and a specific variant of a metabolic gene 'points the finger' at the substrates of the gene as the causal components of the complex mixture.

By identifying the specific genes encoding the enzymes relevant for specific constituents that appear to most strongly modify the exposure–response relationship for the mixture, one can indirectly identify the constituent that may be responsible for the observed health effect (Thomas 2007b).

The section on "Selection of confounders" above introduced the idea of model averaging as a means of dealing with uncertainty about model form. There, the focus was on a particular covariate and one was interested in obtaining an estimate of its coefficient and its variance by averaging across the space of all possible models that included that variable. In a similar manner, this approach can be applied to estimate the posterior distribution for whole sets of variables jointly to address questions of model selection. Extending this approach to models incorporating interaction terms produces a space of possible models that is much larger: if there are P variables available, then there can be $2^P - 1$ possible main effects and interactions and hence $2^{2^P - 1}$ possible models. This number can be reduced substantially by restricting to the subset of models in which, for any interaction effect in the model, all the constituent main effects and lower-order interactions are also included (Chipman 1996). (In the log-linear models literature, this is known as the set of "hierarchical" models, but this is not to be confused with the usage of that term to describe multi-level models adopted here.) To incorporate prior covariates in a multi-level model, one might score the possible interactions on the basis of whether they appeared in the same biological pathway. The result would be a set of estimates for each term in the model and posterior probabilities or Bayes factors for effects and for models. In particular, one might be interested in assessing the posterior probability that any particular variable had *either* a main effect or contributed to one or more interaction terms, simply by summing the posterior probabilities over all models to which that variable contributed. See Conti et al. (2003) for a discussion of this approach to metabolic pathways, where the covariate vector comprises a set of main effects and interactions (up to third order) between two exposures (well done red meat and tobacco smoking) and six genes involved in metabolizing the polycyclic aromatic hydrocarbons and heterocyclic amines they produce through two distinct pathways.

Examples of interactions

Environment x environment

We conclude this chapter with some specific examples. First, we consider the interaction between radon and tobacco smoking, an issue that is of particular relevance for assessing the effect of domestic radon in both case-control and ecologic studies, as well as for establishing compensation policies for smoking and nonsmoking uranium miners with lung cancer. Bcause the risks from domestic radon are relatively small, case-control studies have limited power for distinguishing alternative models for this interaction (Lubin and Boice 1997; Krewski et al. 2005b; Krewski et al. 2006). More informative data comes from various miner cohorts, including the Colorado plateau uranium miners. Whittemore and McMillan (1983) found that both categorical and linear models for cumulative smoking and radon strongly rejected an additive joint effect in favor of a multiplicative one. The exponential mixture of linear main effects models of the form described earlier produced an estimate of the interaction parameter $\alpha = 0.94$. However, such mixture parameters can have highly skewed confidence limits (Moolgavkar and Venzon 1987). A subsequent reanalysis of these data (Lubin and Gaffey 1988) yielded an estimate of the interaction parameter of $\alpha = 0.4$, apparently closer to additivity than multiplicativity, but the likelihood ratio test rejected the additive model ($\chi_1^2 = 9.8$) and not the multiplicative ($\chi_1^2 = 1.1$). A linear mixture showed an even more skewed likelihood, with $\alpha = 0.1$ (apparently nearly additive) but with likelihood ratio tests that again rejected the additive but not the multiplicative model. More comprehensive reanalyses of the 11 available uranium miner cohorts by the National Academy of Sciences BEIR VI Committee (Lubin et al. 1994; Lubin et al. 1995; NAS 1999) have generally found the joint effect to be intermediate between additive and multiplicative.

When the temporal sequence of the two exposures was considered, exposure to radon followed by smoking produced a significantly more-than-multiplicative effect, whereas the reverse sequence produced a significantly less-than-multiplicative effect (Thomas et al. 1994), suggesting that smoking may act as a promoter of radon-initiated cells. (Similar analyses using the mechanistic two-stage clonal expansion model of carcinogenesis will be described in the following chapter.) These findings are compatible with an experimental study in rats (Gray et al. 1986), which found that if tobacco smoke exposure was applied after radon exposure, four times as many cancers occurred as when the same total doses were applied in the reverse sequence.

Residential studies have generally low power for testing for radon-smoking interaction effects, but a joint analysis of the five case-control studies with smoking data yielded a summary RR estimate of 1.18 (0.8–1.6) at 150 Bq/m^3 in nonsmokers, not larger than the RR of 1.24 (1.0–1.5) in all subjects ignoring smoking (Lubin and Boice 1997), as would be expected under an additive model. More recent re-analyses (Krewski et al. 2005b; Krewski et al. 2006) also found no evidence of different relative risks in nonsmokers and smokers or subgroups thereof.

Gene × environment

As an example of gene–environment interactions, we turn to the Children's Health Study and consider interactions between various air pollutants

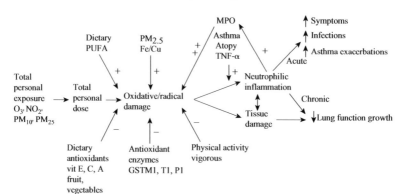

Figure 12.1. Hypothesized representation of the oxidative stress and inflammatory pathways for the respiratory effects of air pollution and the various genes involved. (Reprinted with permission from Gilliland et al. 1999.)

and the genes that have been postulated to be involved in oxidative stress and inflammatory pathways (Figure 12.1). Gilliland et al. (1999) postulated that

respiratory effects in children from exposure to gaseous air pollutants (O_3, NO_2, acids) and particulates (PM_{10} and $PM_{2.5}$) result from chronically increased oxidative stress, alterations in immune regulation, and repeated pathologic inflammatory responses that overcome lung defenses to disrupt the normal regulatory and repair processes. In this theoretical framework, the effects of O_3, NO_2, PM_{10}, and $PM_{2.5}$ are mediated by complex, interacting, and self-enhancing processes of oxidative, radical, and enzymatic attack on the [respiratory epithelial lining fluid] RELF, epithelial cells, and macrophages. These processes are coupled to a persistent inflammatory response that produces tissue damage, decreased ventilatory capacity, increased airway reactivity, decreased macrophage clearance, and altered immune functions. The inflammatory response, if not properly regulated, may produce tissue damage from the activity of secreted proteases, oxidants, and radicals. Inactivation of protease inhibitors by oxidant air pollutants may further enhance the damage from proteases released during neutrophilic inflammation.

They went on to characterize the specific genes that were involved in these processes, many of which have now been tested in a substantial proportion of CHS subjects. The first reports (Gilliland et al. 2002a; Gilliland et al. 2002b; Gilliland et al. 2002c) focused on the glutathione-S-transferase superfamily of genes involved in detoxification of oxidant species. Variants in *GSTP1* were found to have a main effect on the frequency and severity of respiratory illnesses, and both *GSTP1* and *GSTM1* on lung function growth. Furthermore, *GSTM1* was found to interact with maternal smoking in pregnancy: among children with the *GSTM1* null genotype, there were strong associations of various asthma and wheezing endpoints with maternal smoking, but none among those with functioning *GSTM1* genotypes.

To test this hypothesis experimentally, a randomized crossover chamber study (Gilliland et al. 2004) was conducted, in which subjects with sensitivity to ragweed pollen were challenged with allergen alone and allergen plus diesel exhaust particles in random order. Those with the *GSTM1* null genotype had significantly larger changes in IgE and histamine following diesel particles than those with a functioning genotype, and those with both the *GSTM1* null and *GSTP1* I/I genotypes at codon 105 had even larger responses.

Asthma has also been found to be associated with Intracellular Adhesion Molecule-1 haplotypes (Li et al. 2005b) and with an interaction between ozone and Tumor Necrosis Factor (Li et al. 2006). Dietary anti-oxidant intake has also been found to have protective effects (Gilliland et al. 2003). So far, most of these various factors have been explored one-at-a-time or in pairwise combinations, but there is a need for a more comprehensive treatment. Interactions amongst 20 genes were explored using a novel "Focused Interaction Testing Framework" (Millstein et al. 2006), which identified a three-way interaction between the genes *NQO1*, *MPO*, and *CAT*, all involved in the oxidative stress pathway in whites and Hispanics and replicated in African-Americans and Asian-Americans. Asthma is doubtless a complex disease with a variety of genes controlling immune and airway responses (e.g., *CD14*) interacting with a broad range of endotoxins and other allergens (Martinez 2007a, b).

Another study is examining the risk of second breast cancers in women treated for a first breast cancer in relation to radiation dose to the untreated breast and the *ATM* gene, which plays a central role in the repair of double-strand breaks that can be induced by radiation. This study uses a counter-matched design (Chapter 5) in which each case-control triplet (comprising one second cancer and two unilateral cases matched on age at diagnosis of the first, survival to the time of the second cancer, and center) contain two women treated with radiotherapy and one without. This design ensures that all triplets will be discordant for radiation and therefore greatly improves the power for testing both the main effect of radiation and gene–radiation interactions, while having negligible effect on the power for genetic main effects. Preliminary results suggest that both radiation and *ATM* genotype have relatively weak effects on their own (at least overall, although somewhat stronger in younger women with longer latency). However, in combination their effects were much stronger—an interaction *RR* of 3.6 (95% CI 1.3–6.8) for rare variants classified as "likely deleterious" overall, rising to 6.0 (1.2–29) in women under age 45 at diagnosis of the first cancer and 9.1 (2.0–42) in women with at least 5 years latency (J.L. Bernstein et al., Unpublished data).

Complex pathways

Polycyclic aromatic hydrocarbons (PAHs) and heterocyclic amines (HCAs), have been postulated to be involved in the etiology of colorectal polyps and cancer. Both PAHs and HCAs are contained in tobacco smoke and in well-done red meat and each is metabolized by different genes, some having the effect of activating these chemicals to potent carcinogens, others involved in detoxifying them. Complex biological pathways, such as these call for a comprehensive analysis of interaction effects (Thomas 2005b). The hierarchical modeling approach described in Box 12.1 provides an empirical framework for accomplishing this (Conti et al. 2003). The following chapter will discuss more mechanistic approaches.

Ultimately, it will be essential to incorporate various markers of the internal workings of a postulated pathway, perhaps in the form of biomarker measurements of intermediate metabolites, external bioinformatic knowledge about the structure and parameters the network, or toxicologic assays of the biological effects of the agents under study. For example, in a multi-city study of air pollution, one might apply stored particulate samples from each city to cell cultures with a range of genes experimentally knocked down to assess their genotype-specific biological activities, and then incorporate these measurements directly into the analysis of G×E interactions in epidemiologic data (Thomas 2007b). See Thomas (2005); Conti et al. (2007); Parl et al. (2008); Thomas et al. (2008) for further discussion about approaches to incorporating biomarkers and other forms of biological knowledge into pathway-driven analyses.

13 Mechanistic models

The empiric models discussed heretofore have not explicitly incorporated knowledge of specific biological mechanisms, although in some cases, the form of the dose–response model may be motivated by theoretical considerations, such as the linear-quadratic cell-killing model used in radiobiology (see Chapter 6). In some instances, however, the underlying disease process may be well enough understood to allow it to be described mathematically. Probably the greatest activity along these lines has been in the field of cancer epidemiology. Two models in particular have dominated this field: the multistage model of Armitage and Doll (1954) and the two-event clonal-expansion model of Moolgavkar and Knudson (1981). For thorough reviews of this literature, see Whittemore and Keller (1978), Peto (1977), Moolgavkar (1986), and Thomas (1988); here, we merely sketch some of the basic ideas.

A few other biological systems have been modeled mathematically, such as the insulin–glucose metabolism pathway (Bergman et al. 2003), but few have attempted to incorporate environmental exposures. Important exceptions are physiologically based pharmacokinetic (PBPK) models for the metabolism of toxic agents. While we begin by treating the external exposure at a given time as the relevant dose (as might be appropriate for, say, ionizing radiation), later in this chapter we will consider models that account for the metabolic activation, detoxification, and physical movement of agents between compartments of the body before the active agent reaches the tissue. This will provide a framework for allowing for the modifying effects of various metabolic genes, providing a more mechanistic basis for Gene×Environment interactions than the purely descriptive models of the previous chapter.

Stochastic models of carcinogenesis

Armitage–Doll multistage model

The Armitage–Doll multistage model postulates that cancer arises from a single cell that undergoes a sequence of K heritable changes (point mutations, chromosomal rearrangements, insertions, deletions, changes in methylation, etc.), in a particular sequence. Suppose first that the rate of each of these mutations is constant over time, μ_k, $k = 1, \ldots, K$

$$N \xrightarrow{\mu_1} I_1 \xrightarrow{\mu_2} I_2 \xrightarrow{\mu_3} \cdots \xrightarrow{\mu_{K-1}} I_{K-1} \xrightarrow{\mu_K} M$$

Figure 13.1. Schematic representation of the Armitage–Doll multistage model of carcinogenesis.

(Figure 13.1) and suppose further that the rate of one or more of these changes depends on exposure to carcinogens. From these simple assumptions, one can then show that the hazard rate for the incidence of cancer (or more precisely, the appearance of the first truly malignant cell) following continuous exposure at a constant rate X is approximately $\lambda(t, X) = \alpha t^{K-1} \prod_{k=1}^{K} (1 + \beta_k X)$, where β_k is the slope of the relative mutation rate for the kth step of the process per unit X. Thus, the hazard has a power-function dependence on age and a polynomial dependence on exposure rate with order equal to the number of dose-dependent stages. The model does not require any inherent variation in the stage-specific mutation rates (other than their implicit dependence on possibly time-varying carcinogen exposures) to produce a dramatic increase in cancer rates with age. Thus, the there is no independent aging effect, say due to loss of immune competence (Peto et al. 1985).

The multistage model also implies that two carcinogens X_1 and X_2 would produce an additive effect if they act additively at the same stage (i.e., if $\mu_k = \mu_{k0}(1 + \beta_{k1} X_1 + \beta_{k2} X_2)$ and a multiplicative effect if they act at different stages. If instead exposure is instantaneous with intensity $X(u)$ at age u, its effect is modified by the age at and time since exposure: if it acts at a single stage k, then the excess relative risk at time t is approximately proportional to

$$Z_k(t) = X(u) \frac{u^{k-1}(t-u)^{K-k-1}}{t^{K-1}}$$

and for an extended exposure at varying dose rates, the excess relative risk is obtained by integrating this expression over u (Whittemore 1977; Day and Brown 1980). Analogous expressions are available for time-dependent exposures to multiple agents acting at multiple stages (Thomas 1983b). For example, if two stages k_1 and k_2 are sensitive, then the resulting risk involves terms of the same form for the effect of each stage separately, plus an additional term for the interaction between exposures at time u and v of the form

$$t^{1-K} \int_0^t \int_0^v X(u)X(v)u^{k_1-1}(v-u)^{k_2-k_1-1}(t-v)^{k_1-1} \, du \, dv$$

illustrating a case of nonadditivity as discussed under extended exposure histories in Chapter 6. These expressions are only approximations to the far more complex exact solution of the stochastic differential equations

(Moolgavkar 1978); the approximate expressions given above are valid only when the mutation rates are all small.

Pierce and Mendelsohn (1999) and Pierce and Vaeth (2003) have developed an alternative formulation of the multistage model with *all* stages but the last equally sensitive, as might be plausible if, say, all mutation rates were equally dependent on exposure. They demonstrated that in this case, the absolute excess rates for an instantaneous exposure would depend only on attained age, not age at exposure or latency as in the case where only a single stage is dose dependent. This prediction is generally consistent with the pattern observed in the atomic bomb survivor cohort, as described in greater detail in the section on Genomic Instability below.

The multistage model has also been fitted to epidemiologic data on a wide variety of cancers and exposure factors, including lung cancer in relation to arsenic (Brown and Chu 1983a,b), coke oven emissions (Dong et al. 1988), and asbestos and smoking jointly (Thomas 1983b), as well as leukemia and benzene (Crump et al. 1987) and solid cancers and radiation among the atomic bomb survivors (Thomas 1990; Little et al. 1992; Heidenreich et al. 2002a) and uranium miners (Thomas 1990; Heidenreich et al. 2002a; Little et al. 2002), as well as studies of radon-exposed rats (Heidenreich et al. 2000). Particularly notable are a series of papers on applications to data on tobacco smoking (Brown and Chu 1987; Freedman and Navidi 1989; 1990). Chapter 6 described various empiric models for smoking. For a constant intensity exposure z from age t_0 to t_1, the multistage model with only a single sensitive stage k out of K would predict a hazard rate of

$$\lambda(t) = \lambda_0 t^{K-1} + \lambda_k X \int_{t_0}^{t_1} u^{k-1}(t-u)^{K-k-1}\, du$$

$$= \lambda_0 t^{K-1} + \lambda_k X \sum_{j=0}^{K-k-1} \binom{K-k-1}{j} t^j \frac{\left(t_1^{K-j-1} - t_0^{K-j-1}\right)}{K-j+1}$$

with similar but more complex expressions if multiple stages were sensitive. All these expressions are simply polynomials in smoking intensity X and age at initiation, age at cessation, and attained age. Thomas (1982; 1988) and the previously cited papers illustrate the predictions of this model for the dependence of risk on these various factors.

Several authors (Crump et al. 1976; Peto 1977; Hoel 1979; Day and Brown 1980; Portier and Hoel 1983) have discussed the implications of multistage models for cancer risk assessment, which we will revisit in Chapter 15. In particular, the "linearized multistage model" (Crump 1984; 1996) for lifetime risk $R(X)$ data as a function of a constant applied dose rate z in animal carcinogenity experiments has the form $R(X) = 1 - \exp(-\beta_0 - \Sigma_k \beta_k X^k)$, with no temporal modifiers; in most

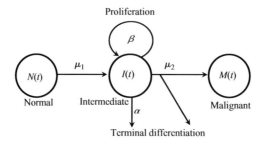

Figure 13.2. Schematic representation of the Moolgavkar–Knudson TSCE model.

applications only the linear or linear-quadratic terms are included and agencies often rely on some upper confidence limits on the predicted risk for setting safety standards.

Moolgavkar–Knudson two-stage clonal-expansion model

The Moolgavkar–Knudson two-stage clonal expansion (TSCE) model (Moolgavkar and Venzon 1979) postulates that cancer results from a clone of cells from which one descendent has undergone two mutational events at rates $\mu_1[X(t)]$ and $\mu_2[X(t)]$, either or both of which may depend on exposure to carcinogens (Figure 13.2). Suppose further that the clone of intermediate cells is subject to a birth-and-death process (birth rate α, death rate β) with net proliferation rate $\rho[X(t)] = \alpha - \beta$ that may also depend on carcinogen exposures. (Here, "birth" refers to the division of a stem cell into two stem cells and "death" refers to terminal differentiation or apoptosis; the normal division of a stem cell into another stem cell plus a terminally differentiated cell has no impact on the total number of stem cells at risk and can be ignored.) The number of normal stem cells at risk $N(t)$ varies with age, depending on the rate of development of the target tissue. Finally, in genetically susceptible individuals (carriers), all cells carry the first mutation at birth.

An approximate expression for the resulting incidence rate at age t is then

$$\lambda(t, X) = \mu_2[X(t)] \int_0^t N(u)\mu_1[X(u)] \exp\left\{\int_u^t \rho[X(v)]\,\mathrm{d}v\right\}\,\mathrm{d}u$$

for noncarriers, representing the expected number of cells with the first mutation being created at time u, multiplied by the expected number of descendents of each at time t, integrated over u, and then multiplied by the second mutation rate at time t. For carriers, this reduces to

$$\lambda(t, X) = \mu_2[X(t)]N(0) \exp\left\{\int_0^t \rho[X(v)]\,\mathrm{d}v\right\}$$

Again, note that these expressions are only approximate solutions to the stochastic process (Moolgavkar et al. 1988), the validity of which depends even more strongly than in the Armitage–Doll model upon the mutation rates being small, because with fewer mutational events, these rates must be much higher to yield comparable cancer rates.

There have been many applications of the TSCE model to various carcinogen exposures, including smoking (Moolgavkar et al. 1989; Heidenreich et al. 2002b), radon (Moolgavkar et al. 1990; 1993; Luebeck et al. 1999; Curtis et al. 2001; Heidenreich and Paretzke 2001; Little et al. 2002; Heidenreich et al. 2004a,b), the atomic bomb survivors (Little 1996; Heidenreich et al. 1997; 2002a; Kai et al. 1997; Pierce 2003; Jacob and Jacob 2004), plutonium in Mayak workers (Jacob et al. 2005; Jacob et al. 2007), bone cancer in radium and plutonium in animals and humans (Bijwaard et al. 2002; 2004; Bijwaard and Dekkers 2007), cadmium (Stayner et al. 1995), and arsenic, radon, and smoking jointly (Hazelton et al. 2001). As an example, the original application to the cohort of U.S. uranium miners (Moolgavkar et al. 1993) found effects of radon on both the first mutation and net proliferation rate of intermediate cells ("promotion"), similar to earlier analyses of data on radon-exposed rats (Moolgavkar et al. 1990; Heidenreich et al. 2000; Bijwaard et al. 2001). Both analyses showed that the "inverse dose rate effect" found in the various descriptive analyses (Chapter 6) could be accounted by this promotion (although some may be simply the result of their assumed nonlinear dependence of the various rates on smoking and radon concentrations). Their analysis of the miner data assumed that radon and smoking contributed additively to the mutation rates, but nevertheless yielded predicted cumulative hazard rates that showed a synergistic effect—greater than additive, but submultiplicative. A later application to a cohort of Chinese tin miners (Hazelton et al. 2001) exposed to arsenic as well as radon and tobacco also found interactive effects of the three exposures: attributable risks of 20% for the interactive effect of arsenic and tobacco, 11% for arsenic and radon, 10% for radon and tobacco, and 9% for the three-way interaction, in addition to main effects, with only 9% remaining for background factors. Although all three exposures had effects on growth and death rates of intermediate cells, they had quite different effects on the *net* proliferation rates.

The model has also been fitted to population incidence data (with no exposure information) for breast (Moolgavkar et al. 1980) and colon cancers (Moolgavkar and Luebeck 1992). The latter compared the fit of the Armitage–Doll multistage model with those of the TSCE model and an extension involving three mutational events. They concluded that all three models fit the data equally well, but estimates of mutation rates from the three-stage model were more consistent with experimentally observed values. A later analysis of SEER data (Luebeck and Moolgavkar 2002), however, found that two rare events followed by a common one with clonal expansion leading to an adenomatous polyp, in turn followed by only one more rare event leading to malignant transformation, provided a good fit. Whereas the two rare events could be homozygous mutations in the *APC* gene, the common event is more likely to represent a repositioning within the colon crypt. They concluded that there is no need to invoke genomic instability (see next section). See also Little (1995a) for further discussion of the need for more than two stages in such models.

Few of these reports have provided any formal assessment of goodness of fit, focusing instead on comparisons between alternative models. This can be done, however, by grouping the subjects in various ways and comparing the numbers of observed and predicted cases; for example, Moolgavkar et al. (1993) grouped uranium miners by the temporal sequence of their radon and smoking exposure histories and reported good agreement with the predictions of their two-stage model.

Variants incorporating genomic instability and DNA repair

Despite the successes of the Armitage–Doll and Moolgavkar–Knudson models, both are somewhat unsatisfying. In the multistage model, the number of events has generally been estimated at about 5–7 for most epithelial cancers, to account for the steep age dependency in cancer incidence rates as a function of age. The specific mutational events have not been identified, however, and it seems unlikely that there would be a single pathway to cancer requiring these specific mutations to occur in the same sequence on all occasions. Furthermore, it is not clear why only one or two of these events would be related to carcinogen exposure, as is generally required to account for the approximate linearity of most dose–response relationships. The best observational evidence in support of the multiple mutational events comes from the colorectal cancer work of Fearon and Vogelstein (1990), who described specific chromosomal changes typically associated with the progression from normal epithelium, through hyperplasia, metaplasia, adenoma, to carcinoma and metastases. Nevertheless, many authors have questioned the need for as many as 5–7 stages when experimental biologists tend to recognize only initiation, promotion, and progression. By adding clonal expansion, the Moolgavkar–Knudson model can describe the age pattern with only two mutational events; for colorectal cancer, they found two- or three-mutation models with clonal expansion described the data equally well (Moolgavkar and Luebeck, 1992).

In all these models, the mutation rates from one stage to the next and (in the models involving clonal expansion) the birth and death rates are assumed to be homogeneous across individuals and across cells within an individual. It seems more plausible that there would be some heterogeneity in these rates and that in a competitive process, the more aggressive cells would eventually tend to predominate. Evidence in support of this hypothesis comes from experiments in which colorectal tumors have been microdissected and the spectrum of mutations in different regions used to infer the age of the tumor (Tsao et al. 2000). In this experiment, the initial event is assumed to be loss of mismatch repair (MMR) capability (e.g., mutation or loss of *MLH1* or *MSH2*) leading to accumulation of further mutations at a much accelerated rate. The data for those analyses was obtained from the distribution microsatellite alleles in noncoding regions: the length of the terminal expansion period is estimated using the variance in allele lengths within loci across a tumor, whereas the interval from loss of MMR to the start of the terminal expansion is estimated using the variance of the differences between the most common alleles and the germline, using coalescent theory. It seems reasonable to assume that a similar process would be happening to functionally important genes, where each additional mutation could have an influence on the rate of subsequent mutations and/or the kinetics of these intermediate lesions.

These considerations suggest an alternative model of carcinogenesis incorporating the following features:

- An unlimited number of possible paths to a fully malignant cancer clone, each characterized by a sequence of mutational events, but with neither the number of such events nor the specific changes being fixed.
- Each cell being at risk of a mutational, a birth, or a death event at independent time-invariant Poisson rates that are specific to that particular cell type and homogeneous for all cells of that type.
- Each successive mutation altering the mutation, birth, and death rates of the resulting daughter cells; the distribution of these rates may become increasingly shifted towards more aggressive values with each successive stage and over time for any given stage, solely as a consequence of natural selection, without the need to postulate any systematic effect of mutation per se in that direction.

Figure 13.3 provides a schematic representation of the model. Let $N_k(t)$ represent the total number of cells which have accumulated k mutations, where N_0 represents the number of normal stem cells. For simplicity, we assume that N_0 is constant over adult life, but in principle this could be modeled deterministically based on knowledge of the growth of the target organ as done by Moolgavkar, et al. (1980) for breast cancer, for example. Now let N_{km} represent the number of cells with k mutations of specific types m. Let the rate of mutation per cell from type km be denoted μ_{km} and let the birth and death rates of type km cells be denoted α_{km} and

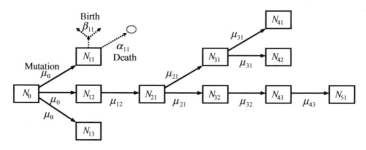

Figure 13.3. Schematic representation of a branching process model for carcinogenesis.

β_{km}, respectively. Then the possible transitions have rates are given by

$$\Pr\left(N_{km} \rightarrow N_{km} - 1, N_{k+1,m'} = 1\right) = \mu_{km} N_{km}(t)\, dt$$

$$\Pr\left(N_{km} \rightarrow N_{km} + 1\right) = \alpha_{km} N_{km}(t)\, dt$$

$$\Pr\left(N_{km} \rightarrow N_{km} - 1\right) = \beta_{km} N_{km}(t)\, dt$$

The cells generated by a mutation inherit the properties of their parent, with some random perturbations that might be given by (say) gamma distributions,

$$\mu_{k+1,m'} = \mu_{km} \Gamma(m_1, m_2)$$

$$\alpha_{k+1,m'} = \alpha_{km} \Gamma(a_1, a_2)$$

$$\beta_{k+1,m'} = \beta_{km} \Gamma(b_1, b_2)$$

In full generality, this model appears intractable, but a useful simplification was found by Pierce and Vaeth (2003) by reducing the past history of mutational states s_r after event r of a cell to a vector of states $s = \{s_1, \ldots, s_r, \ldots\}$. They then assumed that the hazard rate for the next transition rate depends in some arbitrary fashion on that history and upon exposure at that moment, but not intrinsically upon age. Then the hazard rate for all cells at risk at time t can be written as

$$\lambda_{Z(t),s}(t) = \{\lambda_{1,s_1}[1 + \beta z(t)], \ldots, \lambda_{r,s_r}[1 + \beta z(t)], \ldots\}$$

where $Z(t) = \{z(u)\}_{u \leq t}$ represents the entire past history of exposure. Their crucial observation was that if one then transformed the age scale from t to $t' = t + \beta X(t)$, where $X(t) = \int_0^t x(u)\, du$ is cumulative dose, the transformed rates became simply $\lambda_{Z(t),s}{}'(t') = \lambda_s(t)$. In other words, the age-transformed rates do not depend upon exposure, so that whatever complex dependency the baseline rates of cancer $\mu_0(t)$ may have on the mutational history of the population of cells, the rates $\mu_z(t)$ among exposed individuals will have exactly the same form on this age-transformed scale. Thus, one can write the relative risk for cancer as

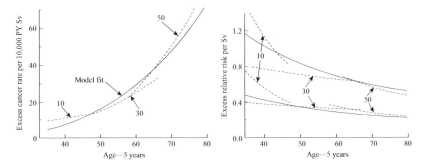

Figure 13.4. Excess absolute (left panel) and relative (right panel) rates of solid cancers, excluding thyroid and breast. The dotted lines show the observed effects at 1 Gy for different ages at exposure (separately for males and females on the left panel). The solid lines are the model fits under the attained-age dependent absolute risk model (left panel) and the conventional age-at-exposure dependent relative risk model (right panel). (Reproduced with permission from Pierce and Mendelsohn 1999.)

simply

$$\mathrm{RR}(t) = \frac{\mu_Z(t)}{\mu_0(t)} = \frac{\mu_0(t + \beta X(t))}{\mu_0(t)}[1 + \beta x(t)] \qquad (13.1)$$

In particular, under the multistage model, for which $\mu_0(t) \propto t^{k-1}$, this leads to the expression

$$\mathrm{RR}(t) = \left\{ 1 + \beta \frac{X(t)}{t} \right\}^{k-1} [1 + \beta x(t)]$$

the term in square brackets being trivial in most cases. Although the term in curly brackets has a polynomial dependence on dose, for doses and ages yielding RRs less than about 3, a linear approximation is adequate, so that $\mathrm{RR}(t) = 1 + [\beta(k-1)/t]X(t)$, that is, a slope coefficient that is inversely proportional to attained age. As an empirical description of dose–time–response relationships, this result holds quite generally without relying on a specific choice of the number of stages, including for broad groupings of cancers that may involve different numbers of stages. This basic idea had been previously developed by Pierce and Mendelssohn (1999), leading to a reinterpretation of the temporal modifying factors discussed in Chapter 6 putting more emphasis on attained age rather than age at or time since exposure. As shown in Figure 13.4, the model (left panel) fits the observed absolute excess rates relatively well, showing only slight dependence on age at exposure, whereas the conventional description in terms of age-at-exposure dependent relative risks (right panel) does not fit anywhere nearly as well.

Little et al. (1997; 1999) have pointed out, however, that the fit of the absolute excess risk model is not as good when extended to the full age range and to all solid cancers (not excluding the hormone-dependent cancers as done by Pierce et al.). Little (personal communication) compared the fit of the Pierce–Mendelsohn model with a general relative risk model allowing modification by attained age (but not age at exposure or latency) and a generalized Armitage–Doll multistage model (Little et al. 1992) and found the fits were indistinguishable with the Pierce and Mendelsohn restrictions, but the multistage model fitted significantly better without these restrictions.

To illustrate the generality of the model, Pierce and Vaeth applied Eq. (13.1) to data from the atomic bomb survivors, uranium miners, and the American Cancer Society cohort to examine the effect of cessation of smoking. All three provided good fits to the data with only a single free parameter β (beyond the arbitrary dependence of baseline rates $\mu_0(t)$ on age).

Equation (13.1) is very general, applying both to instantaneous and extended exposures. It implicitly allows for differential proliferation or death of cells having different mutational histories and does not require any predetermined number or sequence of mutations as the Armitage–Doll model does. Simple generalizations allow it also to incorporate age-dependent mutation rates or the possibility that an increment of exposure might confer some long-lasting instability to a cell, resulting in transition rates that depend upon not the current state of the cell but on how long it has been in that state. However, comparing the fits of the Pierce–Vaeth and several other variants of the multistage and two-stage models to the atomic bomb survivor data, Heidenreich et al. (2002a) concluded that they all yielded similar fits to the data, yet produced very different predictions of the dependence of excess risk on age at and time since exposure, calling into question Pierce and Vaeth's claim that excess risks depended only on attained age. See Pierce (2003) for a rejoinder to their specific criticism and general comments on the aims of mechanistic modeling.

Other models that explicitly allow for destabilizing mutations have been developed by Nowak et al. (2002) and Little and Wright (2003). The latter is a generalization of the TSCE model that allows for an arbitrary number of stages and levels of genomic instability. When applied to the U.S. population incidence data for colorectal cancer, the best fit was obtained with five stages and two levels of instability. Comparison of the predicted excess risks following an instantaneous exposure with the observed patterns in the atomic bomb survivor data suggested that radiation might act on one of the early destabilizing mutations (Little and Wright 2003). However, a subsequent reanalysis of the U.S. data (Little and Li 2007) found that the two-stage clonal expansion model, which does not involve genomic instability, fitted the data about as well as the Nowak et al. and Little and

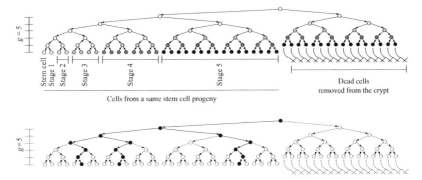

Figure 13.5. Schematic representation of the coalescent model of carcinogenesis. The upper panel shows the ancestry of all cell in a colon crypt derived from a single stem cell, including (right) those that died. The lower panel shows the ancestry of only the subset of cells in the tissue sample. (Reproduced with permission from Nicolas et al. 2007.)

Wright two-stage genomic instability models and significantly better than the Little and Wright models with three or five stages.

The models described above are closely related to coalescent theory that is widely used in population genetics to describe the common ancestry of populations of organisms. The same mathematical models can be used to describe the ancestry of a clonal population of cells derived from a common stem cell (Figure 13.5). A general introduction to coalescent models in genetics is provided in the companion volume (Thomas 2004), with references to the technical details elsewhere, but an example of its application to clonal cell populations in colon cancer is provided by Nicholas et al. (2007). The experimental data derives from microdissections of colon tumors, like Tsao et al. (1999) data described above, but here using DNA methylation as a marker of genetic changes between lineages. Thus, the distribution of methylation states across cells within the same tumor specimen can provide information about the age of the tumor and its evolutionary history. Formal fitting of the coalescent model using Bayesian methods allowed estimation of the number of stem cells in a colon crypt. The authors found that this was at least 8, with posterior mode between 15 and 20, and concluded that both hyper- and hypo-methylated cells coexist in the same crypt in a synergistic manner.

Models of bystander effects

Most of the models we have considered so far effectively treat initiating events as occurring independently across the exposed cells. Accumulating evidence suggests, however, that this is not the case: cells that are not directly hit by a quantum of radiation can nevertheless undergo a transformation as a result of damage to neighboring cells. This phenomenon is called the "bystander effect." The mechanisms are not well understood, but are thought to relate to some combination of inter-cellular gap–junction communication and diffusion of signaling molecules (Hamada et al. 2007; Morgan and Sowa 2007). In addition, directly hit cells can exhibit an adaptive response that may contribute to the bystander effect through the production of reactive oxygen or nitrogen species (ROS/RNS) (Matsumoto et al. 2007; Tapio and Jacob 2007), as well as various other effects (induced radioresistance, genomic instability, epigenetic changes, etc.). Experimental evidence for these phenomena derives from studies in which various biological responses are compared across *in vitro* cultures irradiated in different ways: by coculturing irradiated and nonirradiated cells; using very low-fluence alpha beams so that most cells are not hit; using charged-particle microbeams; or by transferring culture medium from irradiated to nonirradiated cells. Such experiments have typically revealed a steeper dose–response at very low doses when only some cells are irradiated than when all are, followed by a plateau at higher doses, as well as responses in cultures that have not been irradiated at all but had irradiated cells (or filtered medium from irradiated cultures) transplanted into them.

In an attempt to describe these observations mathematically, Brenner et al. (2001) postulated a model in which, for each directly hit cell, k neighboring cells would be exposed to a bystander stimulus and some small proportion σ of hypersensitive cells would undergo an "all or none" transformation independent of the dose. They derive an expression for the proportion of surviving cells following a heterogeneous exposure of the form:

$$\mathrm{TF} = vq\overline{N} + \sigma\left(1 - e^{-k\overline{N}}\right)e^{q\overline{N}}$$

(Figure 13.6), where TF denotes the fraction of cells transformed, $\overline{N}$ the average number of α particle traversals per cell (proportional to dose) and the other adjustable parameters are the probability q of a single cell surviving and the rate v of transformation in response to a direct hit.

The model also allows for extended exposures at low-dose rates, allowing the population of hypersensitive cells to be replenished (Brenner and Sachs 2002). Applying this verson of the model to data on radon from 11 miner cohorts, they demonstrated that it could explain both the downward curvilinearity as well as the inverse dose–rate effect described in Chapter 6, and estimated that in the range of domestic radon concentrations, the majority of the excess would be due to bystander, rather than direct effects (Brenner and Sachs 2003).

These results are controversial, however. Little and Wakeford (2001) did a combined analysis of the uranium miner and residential radon datasets and concluded that the ratio of low- to high-dose slopes was only 2.4–4.0, much smaller than predicted by the Brenner and Sachs model, and

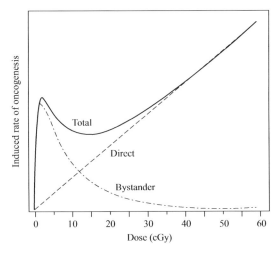

Figure 13.6. Extrapolation to very low doses of the prediction of the model for bystander effects: experimental data derive from doses in the range of 10–80 cGy that would be expected to deliver on average 1–8 alpha particle traversals per cell. (Reproduced with permission from Brenner et al. 2001.)

consequently that the number of cells k subject to bystander effects in their model would be less than 1 with an upper confidence limit of 7; they argued that this would be incompatible with the large number of cells expected to be in neighborhoods that have been experimentally demonstrated to be affected, and thus that the bystander effect was unlikely to contribute much to low-dose risks. Little (2004a) also reanalyzed the 11 uranium miner datasets with a more sophisticated fitting method, accounting for latency and the covariance between parameters, and found that the miner data were equally well fitted by a simple linear relative risk model with adjustments for age at exposure and attained age. Little et al. (2005) subsequently proposed a more complex stochastic model that takes account of spatial location and repopulation and showed that it produces similar downward curvilinearity of the dose–response and the enhancing effect of protraction.

Human data—or even *in vitro* experimental data—showing an actual downturn in risk like that portrayed in Figure 13.6 are elusive. One exception is a report of chromosomal translocations in individuals exposed *in utero* to the atomic bombs (Ohtaki et al. 2004). This study showed a surprisingly weak association with dose overall (in light of other studies of cancer in individuals subject to fetal irradiation), *except* at very low doses (<100 mSv) where there was a significant elevation in frequency. Their Figure 4 (not shown) is remarkably similar to Figure 13.6, and is fitted to essentially the same form of relationship as proposed by Brenner and Sachs ($Y = \alpha + \beta X e^{-\gamma Z} + \delta X$). The authors do not interpret their results in terms of bystander effects, since here the dominant exposure would be to gamma rays, not densely ionizing particles, but instead they consider the pool of progenitor cells as containing two subpopulations, one highly sensitive to translocations but also to cell killing (the $\beta X e^{-\gamma X}$ term), the other less sensitive and not subject to cell killing (the δX term).

Bystander effects can also be protective (Mothersill and Seymour 2004). A rather complex mathematical model aims to integrate bystander effects and adaptive response with genomic instability (Scott 2004). The basic idea entails modeling the activation of one or more apoptosis signaling pathways initiated by release of TGF-β by a transformed cell, leading a complex cascade of ROS/RNS secretions by the undamaged neighboring cells, ultimately resulting in selective elimination of the transformed cells. The mathematical model is fitted using Bayesian Markov chain Monte Carlo (MCMC) methods. A qualitative model based on ideas from chaos theory that incorporates both beneficial and deleterious bystander effects has also been suggested (Mothersill and Seymour 2003), but appears not to be been developed quantitatively.

For an authoritative discussion of the biology of genomic instability and its relation to bystander effects, see a recent United Nations report (UNSCEAR 2006).

Microdosimetry models

The concept of dose that we have relied upon throughout this book actually represents a statistical expectation of insults received across a *population* of targets, for example, individual cells or DNA molecules. In modeling the carcinogenesis process, we are really concerned with the expected biological responses of *individual* cells. If this cellular-level dose–response is nonlinear, then the organ's dose–response will not be simply a function of the average dose but will entail the variance and possibly the higher moments of its distribution. A crucial biological observation (Crowther 1924) was that after low-dose exposure to X-rays a few phages or viruses in a culture would be inactivated, while on average the culture could tolerate very large doses. The resolution of this apparent paradox was that they received very different doses from the same exposure, rather than that their biological responses to the same dose varied. In radiobiology, this realization has given rise to the field of "microdosimetry," which aims to characterize the spatial distribution of energy deposited by a quantum of radiation in biologically relevant targets across its path via a stochastic theory for the transfer of discrete quanta of energy (Rossi 1959). It thus depends critically upon the volume of the presumed target—an entire cell, the nucleus, a chromosome, a single base-pair, or at the opposite extreme, a neighborhood of cells subject to bystander effects.

Early work (Kellerer and Rossi 1972; 1978) established the "theory of dual radiation action" in which radiation was thought to produce sub-lesions in at a rate proportional to average ("absorbed") dose, which could then interact within some bounded region of uniform sensitivity to produce lesions with biological effects on mutation or survival. The dose–response at the cellular level would then be a linear-quadratic function of dose, the linear term representing the probabilities of a single inactivating event (e.g., a double strand break or an interaction between a sub-lesion and an undamaged chromosome) and the quadratic term representing two interacting events to the same site (e.g., two single-strand breaks with no intervening repair). This idea—along with consideration of cell killing as a exponential survival distribution—leads to the standard model that has been used in most epidemiologic analyses of radiation effects, $\lambda(t, Z) = \lambda_0(t)[1 + (\beta_1 Z + \beta_2 Z^2) \exp(-\beta_3 Z - \beta_4 Z^2)]$ (Chapter 6).

Goodhead (2006) provides a historical account of the development of the field of microdosimetry and its current state based on Monte Carlo simulation of every interaction along the path of a charged particle and their secondary ionizations and excitations (see Box 13.1). The present paradigm entails "clustered" damage to DNA from multiple ionizations within a single track, with the biological effectiveness depending upon the complexity of the damage. A spectrum of target volumes thus appears to be relevant for different endpoints, ranging from 3 to 10 nm for initial sub-lesions, to $0.1 - 0.5\ \mu$m for subsequent interactions leading to lesions, to $\sim 10\ \mu$m for intracellular adaptive responses or several mm for inter-cellular bystander effects.

13.1 *Stochastic theory of microdosimetry*

Bardies and Pihet (2000) provide a more quantitative treatment of the standard stochastic theory. Let z denote the specific energy deposited by a single interaction and let n represent the number of interactions per unit volume, assumed to have a Poisson distribution with mean λ. The dose distribution is thus a bivariate function $f(z, n)$ with absorbed dose $\overline{D}$ being its overall mean

$$\overline{D} = \sum_{n=1}^{\infty} n \int_0^{\infty} z f(z, n)\, \mathrm{d}z = E(n|\lambda)E(z|n = 1) = \lambda \overline{z}_F$$

where $\overline{z}_F = \int z f(z, 1)\, \mathrm{d}z$ is the mean energy deposited per interaction. Also relevant is the proportion $g(z) = z f(z, 1)/\overline{z}_F$ of the total energy density deposited at specific energy z, with mean

$$\bar{z}_D = \int z g(z)\, dz = (\bar{z}_F)^{-1} \int z^2 f(z,1)\, dz$$

$$= E^2(z|n=1) + \mathrm{var}(z|n=1)$$

Then if the biological effect of each interaction is proportional to z^2, the overall biological effect can be shown to be proportional to $\bar{z}_D \bar{D} + \bar{D}^2$. On the other hand, the quantity $\bar{z}_F$ is the relevant one if the biological effect is all-or-none, depending only the number of interactions, not on the dose, and hence proportional to $\bar{z}_F E(n) = \bar{z}_F \lambda$. As described earlier, this would be the appropriate way to model the response of bystander cells.

In situations with a nonuniform distribution of dose (e.g., from internally deposited radionuclides), the marginal distribution of absorbed dose is more complex

$$f(D|\lambda) = \sum_{n=1}^{\infty} \frac{e^{-\lambda}\lambda^n}{n!} f\left(\frac{D}{n}, n\right)$$

where the joint probability densities $f(z, n)$ are obtained from the single-track densities $f(z, 1)$ by successive convolution

$$f(z,n) = \int_0^z f(z-x, n-1) f(x, 1)\, dx$$

Roesch (1977) showed how to overcome the computational difficulty of these multiple convolutions using Fourier transforms. More detailed stochastic modeling involves track-structure calculations based on Monte Carlo simulation or analytic approximations to this process.

To date, there seems to have been little effort made to marry the fields of stochastic modeling of the carcinogenesis process with microdosimetry considerations, perhaps in part because the evidence for these phenomena derives mainly from in vitro experiments rather than epidemiology, but this will likely change with the increasing use of modern molecular epidemiology tools and biomarkers. Kellerer (1996) concludes,

Microdosimetric have become indispensable to radiation therapy and to radiation protection.... [R]isk estimates are still linked to epidemiological investigations rather than radiobiological studies.... Microdosimetry has served as an important heuristic guide in radiation biology, but rarely as a tool for truly quantitative analyses. More precise radiobiological data and techniques are required. In cellular studies, the microdosimetric analyses may appear more sophisticated and complex than the incomplete biological data would seem to warrant. Fairly crude analyses are usually adequate, and this includes the use of LET with all its limitations. *With advancing techniques of molecular biology the situation is certain to change; sophisticated mathematical tools and the use of microdosimetric data will become essential.* [emphasis added]

An example of this marriage of molecular biology and microdosimetry is the use of chromosomal aberration data to estimate the relative biological effectiveness of high-energy gamma rays and neutrons for the atomic bomb survivors (Sasaki et al. 2006).

Mechanistic models for noncancer endpoints

PBPK models

Pathway-based epidemiologic models generally entail some combination of the following elements (Figure 13.7):

- Measured inputs in the form of genotypes G at loci thought to be relevant to the hypothesized pathway and exposures E to their environmental substrates.

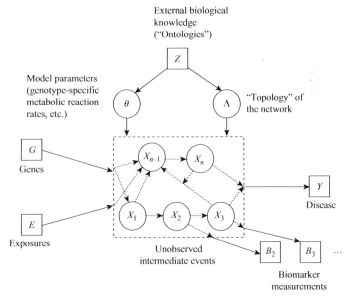

Figure 13.7. Schematic overview of a pathway-based model for the relationship between genes, exposures, disease, and biomarkers: boxes represent measured quantities, circles represent unmeasured latent variables or parameters to be estimated. (Reproduced with permission from Parl et al. 2008.)

- A measured outcome phenotype Y—in the case of cancer, a disease outcome and the age at diagnosis or censoring, although more generally, this could be a continuous trait or vector of traits, possibly measured longitudinally
- Possibly, additional biomarker measurements B of intermediate states in the postulated pathway
- Some model for the underlying unobservable pathogenic process, represented by a series of latent variables X, structured by some postulated topology Λ and involving a vector of parameters θ, typically representing rates of the intermediate steps and their dependence on genotypes;
- Some external knowledge Z about the structure of the model and the parameters.

Within this general paradigm, an investigator has many specific tools that could be used to represent a model, ranging from purely exploratory methods like Classification and Regression Trees (Cook et al. 2004) or Neural Networks (Chakraborty et al. 2005), to highly parametric PBPK models, with hypothesis-driven statistical approaches like hierarchical Bayes (Conti et al. 2003) or logic regression (Kooperberg and Ruczinski 2005) somewhere in-between. Although exploratory methods can be useful for detecting subtle patterns in complex multi-dimensional datasets (Hoh and Ott 2003; Moore 2003; Cook et al. 2004), they generally do not attempt to incorporate prior knowledge about the structure of a pathway, so we will not consider them further here. Instead, we will focus on mechanistic and empirical methods that in one way or another allow an investigator to formally incorporate biological knowledge (or beliefs) about specific pathways hypothesized to be relevant to the disease and causal factors under study.

Let $\theta_i = (\theta_{i1}, \theta_{i2}, \ldots, \theta_{iK})$ denote the vector of metabolic rate parameters (e.g., activation rates λ and detoxification rates μ in a linear kinetic model or $V_{\max}$ and k_m for a Michaelis–Menton kinetic model, described below) involving K reactions specific to individual i with a vector genotypes $G_i = (G_{i1}, G_{i2}, \ldots, G_{iK})$ at the relevant loci, and let X_{iK} denote the predicted final metabolite concentration in this process. The relationships among these variables are depicted in Figure 13.8.

Here X_{ik} denotes the steady-state solution for metabolite k in subject i from the system of differential equations forming the PBPK model, represented by triangles to indicate deterministic nodes in a graphical model. For example, one might use a system of first-order linear kinetic equations of the form

$$\frac{dX_{ik}}{dt} = \lambda_{i,k-1} X_{i,k-1} - (\lambda_{ik} + \mu_{ik}) X_{ik}$$

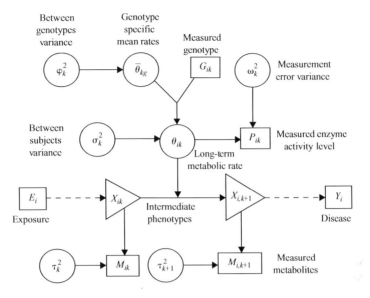

Figure 13.8. Schematic representation of one intermediate step in a PBPK model for intermediate metabolite concentrations in a metabolic model. Boxes represent measured quantities, circles latent variables or model parameters, and triangles represent deterministic quantities given by the steady-state solution to the differential equations in the PBPK model.

in which the rate of change of the concentration of the kth metabolite is the rate at which it is generated from its predecessor minus the rate at which it is either transformed into its successor or eliminated. This system has the steady-state solution (setting $dX/dt = 0$),

$$X_{ik} = X_{i,k-1} \left(\frac{\lambda_{i,k-1}}{\lambda_{ik} + \mu_{ik}} \right)$$

Stringing along several such linear reactions in series still preserves the proportionality of the final metabolite concentration to the input concentration, with a proportionality coefficient that is simply the product of the various fractions in parentheses. However, nonlinear reactions or networks involving feedback loops require more complex treatment. For example, a reaction catalyzed by an enzyme that is in limited supply can saturate, leading to the Michaelis–Menten expression for the multiplier of the substrate concentration in the reaction rate

$$\lambda_k = \frac{V_{\max}^{(k)}}{k_m^{(k)} + X_k}$$

This would be substituted for λ_{ik} or μ_{ik} in the differential equations above, depending upon whether it was an activation or detoxification reaction that is saturated. Either V_{max} or k_m for a given reaction could depend upon a person's genotype at the relevant locus, V_{max} being the maximum reaction rate when fully saturated and k_m being the substrate concentration at which the reaction rate is one-half its maximum. The reaction rate at low concentrations is thus simply V_{max}/k_m. Solution to the system of differential equations gives the time course of each metabolite, for any specified input concentration history. For dose–response modeling of some ultimate disease outcome, one might use the time-weighted average of the metabolite concentration or its peak concentration as the relevant tissue dose, depending upon the postulated biological effect. For most epidemiologic purposes, however, such detailed knowledge of exposure histories on a time-scale comparable to the metabolic processes (minutes or hours) is not available, so the equilibrium solution in relation to average exposure is sufficient. See (Moolgavkar et al. 1999b) for a review of pharmacokinetic and receptor-binding models.

The inputs to this system represented in Figure 13.8 comprise an individual's exposures E_i and genotypes G_i, which in turn influence that person's rate parameters θ_i. For example, one might assume that there is some person-to-person variability in these rates among people with the same genotype because of various other unmeasured characteristics, and adopt a statistical model for this unobserved variability, such as a lognormal distribution with logarithmic mean $\bar{\theta}_k$ and logarithmic standard deviation σ_k. Ideally, these genotype-specific population means and standard deviations would have prior distributions that were informed by laboratory assays of experimentally measured rates in appropriate model systems.

One might also have short-term biomarker measurements, such as "metabolomic" measurements of intermediate metabolite concentrations M_{ik} or "proteomic" measurements of enzyme activity levels P_{ik}. These also might be assumed to be lognormally distributed around their respective unobserved long-term average values with logarithmic standard deviations τ_k and ω_k, respectively. Finally, one might assume a logistic or proportional hazards model for disease risk as a function of the final metabolite concentration, for example, logit $\Pr(Y_i = 1|X_{iK}) = \beta_0 + \beta_1 X_{iK}$. An MCMC approach to fitting this entire combination of deterministic and stochastic models was described by Cortessis and Thomas (2003).

Their method was applied to data on colorectal polyps in relation to two environmental exposures, consumption of well-done red meat and tobacco smoking. Both of these are sources of two classes of known carcinogens (Figure 13.9), polycyclic aromatic hydrocarbons (PAHs, such as benzo(a)pyrene) and heterocyclic amines (HCAs, such as MeIQx). These substances are in turn converted to more potent carcinogens through a series of activating enzymes (CYP1A2, NAT1, and NAT2 for HCA, CYP1A1, and EPHX1 for PAHs) and detoxifying enzymes (UDP-G for HCA, various GSTs such as GSTM3 for PAHs). The results demonstrated

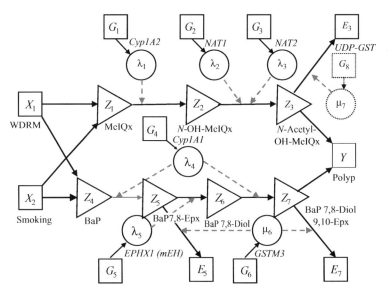

Figure 13.9. Directed acyclic graph representing the model for metabolism of well done red meat (WDRM, X_1) and tobacco smoking (X_2) through a series of heterocyclic amines (Z_1–Z_3) and polycyclic aromatic hydrocarbons (Z_4–Z_7) under the action of various metabolic genes (G_1–G_6) determining the relevant enzyme activation rates λ and detoxification rates μ. (Adapted with permission from Cortessis and Thomas 2003b and Conti et al. 2003.)

a stronger effect of smoking than of well-done red meat through both pathways, and a stronger effect of the HCA than the PAH pathway on polyp risk. Such inferences are based essentially on the pattern of inter-actions of the various substrates with the relevant genes on the different pathways.

General comments

As in any other form of statistical modeling, the analyst should be cautious in interpretation. An pointed out by Jansen (2003),

So, the modeling of the interplay of many genes—which is the aim of complex systems biology—is not without danger. Any model can be wrong (almost by definition), *but particularly complex (overparameterized) models have much flexibility to hide their lack of biological relevance.* (emphasis added).

A good fit to a particular model does not of course establish the truth of the model. Instead the value of models, whether descriptive or mechanistic, lies in their ability to organize a range of hypotheses into a systematic framework in which simpler models can be tested against more complex alternatives. The usefulness of the multistage model of carcinogenesis, for example, lies not in our belief that it is a completely accurate description of the process but rather in its ability to distinguish whether a carcinogen appears to act early or late in the process or at more than one stage. Similarly, the importance of the Moolgavkar–Knudson model lies in its ability to test whether a carcinogen acts as an "initiator" (i.e., on the mutation rates) or a "promoter" (i.e., on proliferation rates). Such inferences can be valuable, even if the model itself is an incomplete description of the process, as must always be the case.

Although mechanistic models do make some testable predictions about such things as the shape of the dose–response relationship and the modifying effects of time-related variables, testing such patterns against epidemiologic data tends to provide only very weak evidence in support of the alternative models and only within the context of all the other assumptions involved. Generally, comparisons of alternative models (or specific submodels) can only be accomplished by direct fitting, and visualization of the fit to complex epidemiologic datasets can be challenging. Any mechanistic interpretations of model fits should therefore consider carefully the robustness of these conclusions to possible misspecification of other parts of the model.

Intervention studies

Need for Assessment of Public Health Interventions

In 1970, the U.S. Congress enacted the Clean Air Act, empowering the Environmental Protection Agency (EPA) to set standards for certain categories of "criteria" air pollutants. The 1990 Clean Air Act Amendment further required EPA to report periodically to Congress on the costs and benefits of this program. The first such report was published in 1997 and provided a retrospective assessment of the program from 1970 to 1990 (EPA 1997). In 1999, EPA published its second report (EPA 1999), a prospective assessment of projected costs and benefits from 1990 to 2010. Although in that report it announced its intent to provide updates every two years, this does not seem to have happened; to date, only an "analytical blueprint" for future reports has been made publicly available (http://www.epa.gov/oar/sect812/). Nevertheless, the 1999 report provides arguably the most cogent case in support of the past and projected future accomplishments of the Clean Air Act. Table 14.1 summarizes the key findings of the report, showing a four-fold ratio of benefits to costs. Setting aside for the moment the controversial issue of the monetary valuation of health benefits, it is clear that the dominant benefit assessed by the EPA is for mortality, based on an estimate of 23,000 avoided cases over age 30 per year (95% CI 14,000–32,000). (This calculation does not take into account the amount of life shortening, however, a point we will return to later.) Congress subsequently directed the EPA to commission a report by the National Academy of Sciences (NAS 2002), which criticized many details about the EPA's approach, but endorsed their main conclusions.

The U.S. EPA is not alone in making such claims. In the same year, a panel of the World Health Organization Ministerial Conference on Environment and Health (Kunzli et al. 2000) performed a health impact analysis of air pollution in Austria, Switzerland, and France, concluding that 6% of all deaths were attributable to air pollution, about half of that excess being due to pollution from motor vehicles. In addition, they estimated that more than 25,000 chronic bronchitis cases in adults, 290,000 episodes of bronchitis in children, half a million asthma attacks, and 16 million person-days of restricted activities could be attributed to pollution from mobile sources. They estimated that the economic costs of these effects added up to 1.7% of the gross domestic product of the three countries.

Table 14.1. Summary of monetized costs and benefits of the U.S. Clean Air Act, projected to 2000 and 2010 in billions of 1990 dollars, mean and 95% CI. (Data from EPA 1999.)

	2000	2010
Direct costs	$19	$27
NAAQS	$8.6	$14.5
Mobile sources	$7.4	$9.0
Hazardous air pollutants	$0.8	$0.8
Acid deposition	$2.3	$2.0
Permits	$0.3	$0.3
Direct benefits	$71 (16, 160)	$110 (26, 270)
Mortality	$63	$100 (14, 250)
Chronic illness		$5.8 (0.4, 18)
Hospitalization	$5.1	$0.5 (0.0, 1.2)
Minor illness		$1.5 (1.0, 2.2)
Welfare	$3	$4.8 (3.4, 6.1)
Total benefits—costs	$52 (−3, 140)	$83 (−1, 240)
Benefit-cost ratio	3.7 (0.8, 8.4)	4.1 (1.0, 10)

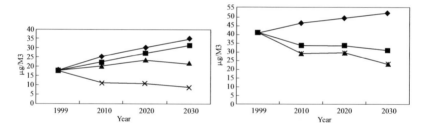

Figure 14.1. Projected changes in PM$_{10}$ (left) and SO$_2$ (right) concentrations in Beijing under "business as usual" scenario (diamonds), clean energy consumption (squares), same plus energy efficiency program (triangles), and same plus green transportation. (Reproduced with permission from Pan et al. 2007.)

In China, Pan et al. (2007) evaluated the potential benefits from several proposed strategies for controlling air pollution in Beijing. Figure 14.1 shows the projected changes in future levels of particulates and SO$_2$ under the various scenarios considered. Using these estimates, they projected decreases of 39–287 acute excess deaths per year and 462–3242 chronic excess deaths from the reductions in PM$_{10}$ under the different scenarios, and decreases of 400–554 short-term excess deaths from the reductions in SO$_2$ by the year 2030.

Such estimates are of course based on epidemiologic exposure–response relationships inferred from observational data on groups with different exposure histories. How likely is it that *changes* in exposure would really yield the benefits that are forecast by such methods? This is essentially a question about "counterfactual inference" (as will be discussed further in Chapter 16), that is, "what would the outcomes of the *same* epidemiologic study subjects have been had they been exposed to different histories of air pollution?" Without delving into the fundamentals of causal inference at this point, this chapter will address approaches to answering this question based on "natural experiments" in which reductions—planned or fortuitous—in some exposure have occurred and it has been possible to observe directly the changes in health endpoints that followed. Of course, the changes in outcomes may not actually have been *caused* by the changes in exposure; other factors could have changed concurrently. Only a randomized controlled trial could really answer the question of causality (Greenland 1990), but this is not generally feasible for most widespread environmental exposures. Changes in air pollution levels, for example, require large-scale government regulation and it is not generally feasible to randomly assign groups to receive the intervention or not. (Needless to say, it is certainly not possible to randomize the assignment of *individuals* in the population to environments!) Nevertheless, well designed "quasi-experimental" or observational designs can provide evidence of a benefit that would be convincing to policy makers weighing alternative control strategies.

The problem of evaluating the benefits of various programs has been considered for decades in the social science literature and has generally gone under the name of "intervention research" (Weiss 1972). A pioneering book by Campbell and Stanley (1963) described a range of study designs for program evaluation where randomized trials are not feasible. We review some of those approaches that may be suitable for evaluation of environmental health interventions below. Recently, the term "accountability" has been introduced to describe research aimed at assessing the performance of environmental regulatory policies (HEI Accountability Working Group 2003), from regulatory action to changes in emissions, ambient air quality, human exposure, and ultimately health response. Here, we focus primarily on assessing the end result of this chain—the net effect of a regulatory intervention on a population health response—without attempting to decompose it into its constituent steps. In particular, regulators may be interested in the effects of different control programs. For example, the Beijing analysis described earlier shows how different strategies can have differing effects on particulates and SO_2 and hence on health effects. An intervention that controls the wrong source or the wrong pollutant could have no little or no effect on health.

The ultimate use of such data in cost-benefit analysis entails economics approaches that are beyond the scope of this book. See, for example,

Chapter 6 of the EPA cost/benefit report (EPA 1999) and the critiques in (NAS 2002) and (Krupnick and Morgenstern 2002).

Some "natural" and planned experiments

To set the stage for the methodological discussion that follows, we begin with a capsule summary of a number of examples of studies of the effects of relatively short-term changes in air pollution levels due to either intentional government interventions or fortuitously due to labor strikes or the like. This story could, of course, begin with the 1952 London fog episode, or even the 1930 Meuse Valley and 1948 Donora PA ones described in Chapter 8. Obviously if increases in air pollution could cause large increases in mortality, does not it follow logically that decreases in pollution should cause decreases in mortality? Perhaps, but this reasoning makes a number of assumptions than can be difficult to test. Is the exposure–response linear? Perhaps the current levels are already at the "no effect level," so that further reductions would have no additional benefit. (The available epidemiologic data seem to contradict the notion that the current levels have no adverse effects, however.) Perhaps short-term fluctuations do not translate into long-term effects due to "harvesting" (Chapter 8). (Again, the epidemiologic evidence seems to suggest chronic effects are even larger than those predicted by the acute effects studies.) Perhaps homeostasic mechanisms lead to essentially the same equilibrium response to whatever the long-term level of exposure is. Perhaps variation between individuals in their sensitivity to air pollution (due to age, genetics, nutritional status, underlying health conditions, etc.) leads to different associations at the individual and population levels. And so on. Hence, it behooves us to examine whether reductions in exposure *below* currently established regulatory limits will really produce a benefit of the size predicted by the available epidemiologic data.

Observational epidemiologic studies can provide some evidence about benefits of environmental improvements. An example is provided by a study of CHS subjects who moved from their communities to areas with better or worse pollution (Avol et al. 2001). The study showed a significant improvement in lung function among those moving to areas with lower PM_{10} and a worsening for moving toward higher PM_{10}, but no correlation with changes in ozone and only a weak, nonsignificant correlation with changes in NO_2. Such changes in exposure were not, however, the result of the kinds of systematic interventions described in the rest of this chapter, nor did they affect whole populations but self-selected individuals, and so could be subject to selection bias (e.g., children with respiratory problems being more likely to move than those in the same communities without).

Thus, they do not constitute true "intervention" studies, but do point to the kinds of improvements that might be anticipated from population-wide interventions.

Government interventions

As a result of a switch from oil to coal during the 1980, Dublin experienced a deterioration of air quality, with peaks in air pollution being associated with increased respiratory deaths. In response, the government banned the sale of coal within the city on September 1, 1990, leading to a 70% decline in particulate concentrations (as measured by "black smoke"). Clancy et al. (2002) undertook an "interrupted time-series" analysis (Webster et al. 2002) (essentially regressing the death rates on an indicator variable for before or after the ban, adjusting for the seasonal and long-term trends, weather, respiratory epidemics, and death rates in the rest of Ireland, as described further below), in order to estimate the decline attributable specifically to the ban. The time-series of mortality (Figure 14.2) mirrors the declines in black smoke; SO_2 concentrations also declined, but more gradually during this period, with no abrupt change detectable in 1990. Overall age-adjusted death rates

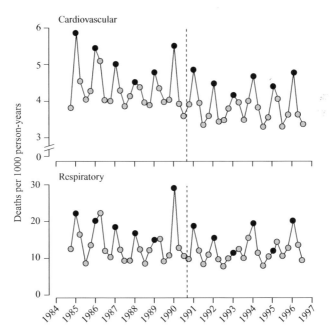

Figure 14.2. Cardiovascular and respiratory mortality in Dublin before and after the ban on coal sales (solid dots indicate winter). (Reproduced with permission from Clancy et al. 2002.)

dropped 58% for cardiovascular and 22% for respiratory causes between 1984–90 and 1990–96, but some of this could be due to other changes during the period. The time-series analysis yielded estimates of −10.3% and −15.5% change for the two groups of causes, after removing the overall time trends and adjusting for other factors ($p < 0.0001$ for both). These changes were larger than predicted by previous time-series analyses.

At virtually the same time (July 1990), Hong Kong undertook a similar intervention, in this case ordering all power plants and motor vehicles to use fuel with low sulphur content. This led to an immediate reduction in SO_2 concentrations in the polluted area by 80% and a 38% reduction in the sulphate concentrations of particles. Two subsequent epidemiologic publications took different approaches to evaluate the health effects of this intervention. Hedley et al. (2002) applied time-series methods similar to those used in Dublin to examine the effects on mortality and reported a 3.9% reduction in respiratory deaths and a 2.0% reduction in cardio-vascular deaths. They also calculated a gain in life expectancy of 20 days in females and 41 days in males.

Peters et al. (1996) focused instead on respiratory symptoms in children using the before/after exposed/control group design described in greater detail below. In the control district, there was no change in pollution during this period. Before the intervention, cough and sore throat were 22% more common, and wheezing 35% more common, in the polluted than the control districts. After the intervention, there was a greater decline in the prevalence of symptoms in the polluted than in the control district.

Since some health effects may not be reversible, one might anticipate a time lag between the introduction of an intervention and a demonstrable health benefit. Roosli et al. (2005) combined information from a meta-analysis of five studies of infant mortality and estimates of an exponential decay parameter in a dynamic model derived from the Utah Valley and Dublin studies to estimate the years of life lost (YLL). They estimated that 39% of the benefit occurred in the same year and 80% within five years of the intervention, but the overall YLL (42,400 for the Swiss population, 95% CI 22,600–63,600, 4% of which was due to infant mortality) was not particularly sensitive to the choice of decay parameter.

Typically, health improvements may be due to a range of policies rather than any single action. Two Swiss studies illustrate different approaches to assessing effectiveness in such a situation. The SCARPOL study (Bayer-Oglesby et al. 2005) used three cross-sectional surveys to evaluate the improvement in various childhood respiratory conditions between 1992 and 2001, a period during which air pollution levels declined considerably due to a range of factors, but the design of this study did not allow them to determine the effects of any specific changes. In contrast, SAPALDIA—a cohort study of lung function changes in adults (Downs et al. 2007) over roughly the same period—found a significant inverse association between changes in lung function and changes in PM$_{10}$ across individuals. Not surprisingly, however, the biggest improvement in air pollution occurred in the communities with the highest baseline levels. This correlation was so high that it was not possible to separate the effects of the reduction in air pollution from a continuing effect of past exposures. The investigators also did not describe the effects of any copollutants.

Labor strikes

In the 1960s, copper smelters were the source of most of the sulfate emissions in the southwestern United States. Between July 1967 and April 1968, a nationwide copper smelter strike produced a particularly large reduction in sulfate particles—about 60% in the states of Utah, Nevada, Arizona, and New Mexico—providing an opportunity for a before–during–after comparison using Poisson regression time-series methods (Pope et al. 2007). Over the four-state area, total mortality decreased by 2.5% (95% CI 1.1–4.0) during this period, adjusting for temporal trends, total mortality in the seven bordering states, and national cause-specific mortality from influenza/pneumonia, cardiovascular disease, and other respiratory disease.

Two decades later, a 10-month strike at a steel mill in Provo, Utah, produced a 2–3-fold reduction in the winter-time level of PM$_{10}$ compared with the year before and the year after. A corresponding 2–3-fold reduction in children's hospitalizations for pneumonia, pleurisy, bronchitis, and asthma was observed during winter months of the strike compared to the winters before and after, with somewhat smaller reductions among adults (Pope 1989).

German reunification

Before reunification, many cities of East Germany were among the most polluted parts of Europe. The rapid economic and regulatory changes following reunification led to a dramatic decline in particulate and SO$_2$ concentrations, providing a unique opportunity for evaluation by comparison with corresponding trends in West Germany, where no such abrupt changes occurred. For example, one study (Frye et al. 2003) reported a decline in TSP from 79 to 25 $\mu g/m^3$ in three East German communities between 1992 and 1999 and a decline in SO$_2$ from 113 to 6 $\mu g/m^3$. Another study (Heinrich et al. 2002) found similar declines in these same pollutants, but an increase in nucleation mode (10–30 nm) particles during this same period. Several studies, mainly of children's respiratory symptoms and lung function have been reported (Kramer et al. 1999; Heinrich et al. 2000; 2002; Frye et al. 2003), most employing some form of before-after design with one or more West German control groups. These various studies found consistently larger improvement in various health indicators in East than in West Germany during this period, suggesting that at least some adverse effects are reversible.

Olympic games

The 1996 Atlanta Olympic Games provided an opportunity to examine the transient effect of a 23% decline in traffic and a 28% decline in ozone concentrations. Friedman et al. (2001) used a before–during–after design (with no external control group) to examine childhood asthma acute care events. They found a significant 42% decline based on claims in the Georgia Medicaid files (with somewhat smaller changes using other data sources). These changes were based on a comparison of the 17-day Olympic period to a baseline that combined the four weeks before and four weeks following the Olympics. Unfortunately, the investigators did not report the two baseline rates separately, so it is not possible to assess how rapidly asthma rates returned to baseline as traffic and ozone concentrations reverted to normal.

Similar studies were conducted before, during, and after the Beijing 2008 Olympics, but using a panel study design instead of aggregate population records. A panel of 131 medical residents at the First Hospital of Peking University were enrolled and studied twice in the two months before, twice during, and twice in the two months following the Olympics (along with panels of 40 retires and 50 children). A broad suite of biomarkers (of inflammation, autonomic tone, endothelial function, platelet function, oxidative stress, etc.) were measured on each occasion. In addition, interactions with various candidate genes (e.g., *GSTM1*) will be studied.

As preparation for the Olympics, the government conducted a pilot test of their proposed traffic restrictions during a 4-day period in August 2007, permitting cars to drive only on even or odd days based on their license plate numbers. Air pollution levels did not decline appreciably until after the restriction period, perhaps because of concurrent changes in weather conditions. However, a previously established panel of 40 elderly persons was studied during the period and associations between continuously monitored heart rate variability and concurrent PM_{10} levels were found (Dr. Wei Huang, personal communication).

Study designs for program evaluation

Randomized experimental designs

We begin this discussion of study designs with the ideal—albeit seldom, if ever, feasible—design, the double-blind randomized controlled trial. Widely used in clinical medicine and—more relevant here—in prevention research (e.g., Prentice et al. 2005), the basic idea is that subjects are individually randomly assigned to the treatments being compared (possibly within strata of predictive factors), and then followed in an identical manner to observe their outcomes. To avoid bias, neither the participants nor the researchers are aware of the treatment assignment until the codes are revealed during the final analysis at the end of the trial. This design is the gold standard for causal inference, as the groups are guaranteed to be equivalent *in expectation* on all potential confounding factors, known or unknown. Of course, the play of chance could mean that some risk factor is out of balance between the groups. But lack of balance could occur in either direction and is likely to be balanced by other risk factors that are out of balance in the opposite direction. Hence, significance tests and estimates are guaranteed to have the right statistical properties (nominal test size, unbiased estimators, nominal confidence interval coverage, etc.) across hypothetical replications of the study. (As an aside it is worth noting that imbalance on known risk factors for which data have been collected can always be adjusted for in the analysis, but as the unadjusted analysis has, asymptotically at least, the correct statistical properties, it is not obvious that adjustment helps and it has the potential to create bias or lead to less efficient estimators in small samples or nonlinear models. The question whether this is a desirable procedure has been highly controversial; see, for example, Fisher 1935; Gail et al. 1988; Senn 1989; Robinson and Jewell 1991; Gail et al. 1996; Greenland et al. 1999b).

An important variant of the randomized comparative trial is a crossover trial. Like the observational case-crossover design described in Chapter 5, this design relies on comparisons *within* rather than *between* subjects. Each subject receives both treatments and their responses following each treatment are compared. (Note the similarity to the counterfactual approach to causal inference mentioned earlier: here, rather than having to *infer* what an individual's response to the counterfactual treatment would have been by assuming it would be similar to those observed on that treatment, these responses can be *observed* directly in a crossover trial.) To avoid contamination effects of one treatment by another, the order in which subjects receive the treatments is assigned at random, and there is usually a "washout" period between application of the different treatments. In addition to ensuring exact comparability of the treatment comparisons, the design can be more powerful than a group comparison because the treatment effect is tested against a within-subject rather than a between-subject residual variance, which will typically be smaller (Chapter 7). However, the design is only feasible for effects that occur soon after treatment begins and do not persist long after treatment ends. Thus, the design is not much use for studying the long-term effects of therapeutic or preventive interventions, or mortality.

Individual versus group randomization

One fundamental problem with using the randomized controlled trial approach to evaluate an *environmental* intervention is that such "treatments" typically affect whole populations—for example, an intervention aimed at lowering ambient air pollution levels—so that individuals within that population are more or less equally affected. Hence, it becomes impossible to randomly assign individuals to receive the intervention or not (or to different interventions). Of course, it still might be possible to mimic the treatment in a controlled laboratory setting (e.g., a chamber study), but this would not achieve the goal of finding out whether the intervention works on population scale, and again, would only be feasible for studying short-term effects. Although individual randomization is not possible, it may be possible to randomize larger population groups to alternative treatments. For example, whole cities might be allocated at random to receive an intervention or not. Such a strategy has been widely used to evaluate various social programs, such as advertising campaigns aimed at smoking cessation or prevention of drug abuse, but no examples come to mind for its application to government regulatory interventions to promote environmental health. In part, this may be because regulatory agencies are typically constrained to apply regulations uniformly across their entire jurisdiction. Of course, individual states have some discretion to enact their own regulations and could in principle agree to participate in such a group randomized experiment, but this presumes a level of political cooperation that is probably unrealistic to expect.

Nevertheless, it is worth considering in the abstract how such a study might be performed, if only for comparison with the kinds of inferences that can be obtained with the various nonrandomized designs discussed below. Suppose two treatments are to be compared, say an intervention to reduce some air pollutant (like the ban on coal sales or the use of high sulphur fuels described earlier) versus "business as usual." Suppose further that a reasonable number of administrative districts are available to participate in the study—at least twice the number of treatments to be compared to allow for replicate observations, but hopefully many more. These units are then assigned at random to receive the intervention or not, possibly within strata of potential confounding factors (geographic regions, demographic, or socioeconomic distributions, etc.). Aggregate measures of health status after the intervention are then compared between the intervention and control districts, for example, mortality, hospitalization, school absence rates, and so on. Alternatively, rather than relying on routinely gathered population health statistics, one might enroll panels of individuals in each district and study their health outcomes (symptoms, biomarkers, etc.) in greater detail. In either case, one must remember that the "effective sample size" is not the total number of individuals but rather the number of groups being compared. Of course, the number of individuals within each group will affect the statistical precision of the estimated outcome for that group, but the precision of the intervention effect estimate depends much more strongly on the number of groups. A swap in the treatment assignment of one pair of groups could have a very large effect on the outcome of the comparison if only a few groups are being compared, no matter how precisely each group's outcome is measured!

As with the randomized controlled trial of individuals, a group randomized trial has the property that the estimated treatment effect and significance test have their nominal properties guaranteed by randomization. This is true even if by chance some risk factors turn out—by an accident of randomization—to be out of balance between intervention and control groups. But recall that this is a claim about the statistical properties of tests and estimators over many hypothetical replications of the trial, not a claim that any particular trial got the right answer. Significance tests can only estimate the probability that the observed result could have occurred by chance had the allocation of sampling units to treatments been different. Because the number of sampling units is typically small, effects need to be much larger to be deemed statistically significant than if *individuals* had been assigned at random; confidence intervals on the estimated treatment effect will also tend to be much wider in a group randomized trial than in an individual trial as a result of the small number of groups being compared.

As in an individual-based trial, if one of more confounding factors turns out to be differently distributed between the treatment groups, it is also possible to adjust for it in the analysis. But the appropriateness of such adjustments is likely to be more controversial because of the smaller number of degrees of freedom available, so that asymptotic arguments about the validity of the procedure may not be applicable.

Staggered intervention design

Suppose it is not feasible to randomize individuals or groups to treatments. We now consider a range of nonrandomized designs, dubbed "quasi-experimental" by Campbell and Stanley (1963) in their classic text. The first such design presumes all sampling units ("districts") will eventually receive the intervention, but it will be introduced at different times across districts. Ideally, the timing of application would be assigned at random across districts, but this may not be feasible and is not essential to the validity of the design (but it helps). Health outcomes may be improving over time for reasons completely unrelated to the intervention, and some of these changes could occur abruptly at times that just happened to coincide with the intervention. But it would be quite unlikely that such changes would occur at different times in different districts in such a way as to line

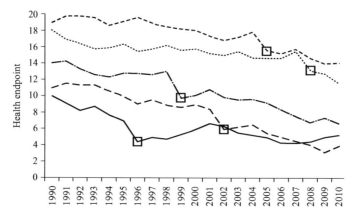

Figure 14.3. Hypothetical example of data from a staggered intervention study; dates of intervention for each city are marked by squares.

up exactly with the timing of the intervention in each place. Hence, one would be looking for a pattern of changes across districts like that shown in Figure 14.3, where there is a noticeable improvement in the outcome that coincides with the introduction of the intervention in each district. The analysis of such data might use a time-series model of the form

$$E[Y_i(t)] = a_i + s(t) + \beta Z_i(t) + e_i(t)$$

where $Y_i(t)$ denotes the aggregate measure of health outcome in district i at time t, $s(t)$ denotes a smooth function of time (assumed to be the same across cities), $Z_i(t)$ an indicator function for whether the intervention is in effect in city i at time t, and a_i and $e_i(t)$ are random error terms. Here the size of the treatment effect is given by β.

Before–after design with aggregate data (interrupted time-series study)

The "interrupted time-series" design (Webster et al. 2002) is essentially the same as that described in the previous section, but with only a single sampling unit available. The analysis would use the same time-series model (without the city-specific terms) and would estimate the treatment effect by the magnitude of the change before and after the intervention, after adjusting for long-term and seasonal trends and possibly other confounding factors (weather, influenza outbreaks, etc.). Because there is no built-in replication like in the staggered intervention design, confidence in a causal interpretation of an observed effect is much weaker. Multiple independent studies of this type may provide such replication, but without the advantage of a joint analysis of all the data using consistent data collection methods and analysis techniques. The Dublin and Hong Kong mortality studies cited earlier are examples of this kind of analysis.

Before-after design with individual data (panel study)

Panel studies were introduced in Chapter 7, and are well suited to evaluation research when it is desirable to examine the effects of the intervention with individual rather than aggregate data, so that subtler outcomes that may not be routinely available or intermediate variables like biomarkers can be studied. The design and statistical analysis is essentially the same as described earlier, with the main comparison being of *changes* in outcomes *within individuals* following the intervention, rather than correlations in daily fluctuations in outcomes with daily fluctuations in exposure. As with the interrupted time-series design for aggregate data, however, causal interpretation of any observed changes as an effect of the intervention is limited by there being only a single group, making it impossible to be certain that the observed changes were not due to some other factor that changed at the same time.

Before–after design with controls

Some of the weakness of the simple before–after designs can be remedied by the inclusion of an unexposed control group. Here, by "unexposed" we are referring to the *intervention*, not the environmental factor itself. The control district might be as exposed as the targeted group before or after the intervention. In either event, one would be looking for a change in the outcome of the targeted group following the intervention that was larger than the corresponding change in the control group. This is, in effect, a test of group × time interaction in a two-way analysis of variance. Equivalently, the design can also be seen as separate comparisons of treated versus control groups before and after the intervention. Depending upon whether the control group's exposure was more like the treated group's before or after, one might then be looking for a difference at one time and not at the other (Figure 14.4). The Hong Kong panel study of childhood respiratory symptoms described earlier is an example of this type of study.

As with a randomized trial, an investigator has the option of selecting controls to match the characteristics of the intervention group, either individually or by groups. Of course, only characteristics that have already been identified as potential risk factors and for which data are readily available at the time of control selection can be matched for. Furthermore, the number of characteristics that can be effectively matched on is limited, particularly in a group-matched setting. If whole cities are being chosen as controls for experimental cities, for example, there may be only a limited number to choose between; one can then only hope to select cities that are as similar as possible to their experimental counterparts on

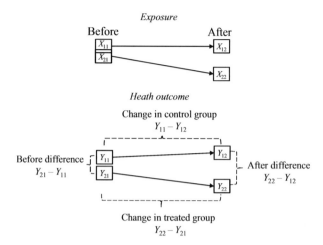

Figure 14.4. Hypothetical data on exposure and health outcomes in a before–after design with controls. The first subscript refers to the group, the second to the time period. The treatment effect is estimated by the difference in changes over time between treated and control groups $(Y_{22} - Y_{21}) - (Y_{12} - Y_{11})$, or equivalently by the change in between-group differences before vs. after $(Y_{22} - Y_{12}) - (Y_{21} - Y_{11})$, both quantities being equal to $Y_{22} - Y_{21} - Y_{12} + Y_{11}$.

a few relevant characteristics. With individual matching, one has greater scope for matching closely on multiple characteristics. But matching on pre-study outcomes can be counterproductive due to the phenomenon of regression-to-the-mean (Tu and Gilthorpe 2007): subjects with high initial values may be only randomly high and a follow-up measurement would be expected to be lower, with or without the intervention.

Before–during–after design

In some cases, the intervention is only temporary, as in the planned reductions in air pollution during the 2008 Olympic Games in Beijing. In this case, it is possible to look both for an improvement in short-term health indicators during the intervention and a return to previous levels of health sometime after the intervention ceases, and to study the length of time any beneficial effect persists. Although it is possible that some confounding factors would rise and fall in parallel with the targeted exposure and thereby account for any observed effect, a causal interpretation as an effect of the intervention itself is much more plausible than in a simple before–after comparison. Such an inference would be even stronger if it were possible to study a comparable control group not exposed to the intervention but subject to the same changes in other factors, and if no such pattern of changes in outcomes were observed in that group.

After-only design with controls

The previous designs all presume enough advance planning to allow an assessment the health status of a population before the intervention occurs. The least informative situation occurs when an *ex post facto* analysis is required with no basis for historical comparison. In this situation, all that can be done is a comparison of groups exposed and not exposed to the intervention, hoping that they would have been comparable before it, but with no solid basis for testing this assumption. Having multiple exposed and control groups to provide some replication and controlling for potential confounders by matching the groups in the design or by adjustment in the analysis will somewhat enhance one's confidence in the validity of the comparison. Different types of control groups may also be helpful for addressing different potential sources of bias (dubbed the "patched up design" by Campbell and Stanley), particularly if the results from different groups are similar, but conflicting results can be difficult to interpret.

Routine surveillance

While the assessment of the effects of specific interventions is best done by carefully designed, targeted, prospective studies, there is also a place for routine surveillance of exposures and health indicators, if only to build a long-term basis for historical comparisons. Major health events like mortality and hospitalizations are routinely compiled by government agencies, and have been the mainstay of time-series analyses like those discussed in Chapter 8. Data on subtler—but arguably more sensitive—endpoints like asthma attacks or absenteeism are harder to come by. Longitudinal measurements of certain biomarkers might be even more informative. Agencies charged with evaluating the costs and benefits of environmental regulations might do well to consider establishing more such surveillance systems long in advance of new interventions, so as to support future prospective studies. For example, the U.S. Centers for Disease Prevention and Control plans to develop a national environmental public health tracking network to identify environmental hazards, track exposure to them, and assess their health effects (http://www.cdc.gov/nceh/tracking/).

The Health Effects Institute Accountability Working Group (2003) report provides an excellent historical account of attempts to evaluate the health benefits of air pollution interventions and a thoughtful discussion of research needs, as background to a subsequent Request for Applications (HEI 2004). They call for further efforts to catalogue available national, state, and local databases of health indicators and potential confounding factors, and encourage further research into the feasibility of using available biomarkers and developing novel ones for health-impact assessments. Amongst the suggestions contained in the 2004 RFA are the following:

- *Prospective and retrospective studies to evaluate such targets as:* the heavy-duty low sulfur fuel rule; PM2.5 and O3 NAAQS state implementation plans; EPA's air toxics control plan; Tier II regulations for light and medium duty vehicles and fuels; California's diesel emissions reductions program; and various local initiatives.

- *Methods development:* techniques using data from surveillance systems; to measure time-varying confounding factors; surrogate markers of health effects; syntheses of preexisting studies; causal models; validation of model-based predictions, etc.

Statistical analysis issues

VSL versus VSLY approaches

As the various cost-benefit analyses discussed at the beginning of this chapter have shown, the most important benefit from reduced air pollution is reduction in mortality. Setting aside for the moment the question of how to assign a monetary value to a life saved, there is an important statistical question about how the mortality benefit should be calculated. The two competing approaches are generally known as the "value of a statistical life (VSL)" and the "value of a statistical life year (VSLY)." Both approaches require an assessment of the number of deaths prevented or the total number of years of life gained by changes in exposure before the economic analysis (assigning a value to such quantities) can begin. Although this may seem at face value like a relatively straight-forward application of the risk assessment techniques described in the following chapter, there is an important subtlety. Deaths can never truly be prevented, only delayed: sooner or later, we will all die! But it makes a big difference whether it is only postponed by a few days until one dies of the same cause one would have anyway (the harvesting hypothesis discussed in Chapter 8), or entirely avoided at some "premature" age until one eventually dies of another cause unrelated to exposure. Time-series methods count all excess deaths, but are silent about the length of life lost, whereas cohort studies reflect the change in long-term death rates due to exposure, implicitly involving the life expectancy. For the purpose of computing the benefit of reduced exposure, the person-years approach is obviously preferable but was not used by EPA as the primary basis of their cost-benefit analysis because of inadequacies in the data needed for the VSLY approach. Chapter 6 and Appendices D and H of their report (EPA 1999) provide an extensive discussion of the issues involved and their rationale for relying instead on the VSL approach, but it misses the following point. Assuming there is some (possibly very large) heterogeneity in individual risk, it is not possible to compute the number of deaths attributable to exposure without making additional untestable assumptions. Under somewhat weaker assumptions, however, the years of life lost attributable to exposure can be estimated. We will return to this point in Chapter 16.

Policy issues

Cost-benefit analysis

Economic estimation of the costs of an intervention is beyond the expertise of this author and the scope of this book, as is assessment of effect of the intervention on exposure distributions, which is usually the first step in assessing the effects on health. Since the dominant benefit appears to be mortality, it is worth a few words, however, about the sensitive question of the monetary value of a premature death or life-year-lost. That it is necessary—if somewhat distasteful—to assign such a value is obvious from the logic of a cost-benefit calculation. Only by expressing costs and benefits in comparable units can a ratio be computed as a dimensionless quantity. How then should one go about it?

For decades, the dominant paradigm in economics for assigning value to human life has been the "willingness to pay" or "willingness to accept compensation" approach. The value of material goods is established in the marketplace by how much people are willing to pay for them. So, the reasoning goes, one can establish the value of a premature death avoided or of an additional year of life by surveying individuals and asking them how much they would be willing to pay to avoid premature death. Numerous such surveys have been conducted, with a range of estimates from $0.6 to $13.5 million per premature death avoided, averaging about $4.8 million (the figure adopted by EPA). Dividing this figure by the population life expectancy yields a corresponding average monetary value of $137,000 per year of life saved. This figure almost certainly varies by age, however, not to mention an individual's personal perception of their quality of life remaining. Such factors are more difficult to estimate, however, so are ignored in the EPA's benefit assessment using the VSLY approach. In principle, however, what one really seeks is an integral over all ages of the *change* in the probability of surviving to each subsequent age under different exposure scenarios, multiplied by the value assigned to a year of life at that age, that is,

$$VSLY = \int_0^\infty f_1(t) \int_0^\infty [S_0(u|t) - S_1(u|t)] V(u) \, du \, dt$$

where $f_1(t)$ is the age at death distribution given current exposures, $S_x(u|t)$ is the probability of surviving to age u given survival to age t under exposure scenario x, and $V(u)$ is the value assigned to a year of life at age u.

Even if one were willing to assign values in this way, however, the fundamental difficulty with this calculation is that there is no way to estimate the counterfactual probabilities $S_x(u|t)$ in a heterogeneous population. Individuals dying at any given age in the presence of exposure are presumably the more sensitive, and their remaining expectation of life, had they not been exposed and not died as a result, would hardly be expected to be similar to the general population at that age. We will explore this issue further in Chapter 16.

These statistical issues aside, the whole rationale for cost-benefit calculations as a basis of public policy is highly controversial. See, for example, Hammitt and Graham (1999); Krupnick (2002); Kaiser (2003) for policy perspectives on the valuation of human life. For example, John Graham, the former director of the Office of Information and Regulatory Affairs (OIRA) in the White House Office of Management and Budget, has been a strong advocate of the VSLY approach, counting deaths at younger ages more heavily because they would entail more years of life lost. In response, EPA proposed adjusting the calculation of benefits by counting deaths above age 65 at 37% of the value assigned to deaths at younger ages. This provoked intense criticism from senior citizens and others, calling it a "senior death discount," so the EPA abandoned the proposal (E. Shogren, *Los Angeles Times*, May 8, 2003).

The social–political context of intervention research

The foregoing discussion should provide some idea of the highly politicized arena in which regulatory decisions are made and their impacts evaluated. Numerous vested interests and highly vocal advocacy groups are involved. Opinions are highly polarized. Decisions are made not just on the basis of science but on numerous social, political, and economic considerations. Nevertheless, science has an important role to play and scientists should contribute their expertise. But science operates in a culture that values objectivity. This can be difficult to maintain in the rough-and-tumble world of evaluation research, with its conflicting pressures from funding agencies, program staff, and interest groups (Weiss 1972). It may not be easy for scientists who are used to an ivory tower environment to maintain their objectivity in the conduct of their research and in reporting their findings dispassionately without shading their results to reflect their personal biases or to please their audience. But such objectivity is exactly what is needed to inform public policy.

This highly polarized policy-making environment is illustrated by the experience of the investigators conducting the Eurpoean health impact assessment for air pollution (Kunzli et al. 2000) described at the beginning of this chapter. The study was originally commissioned by the Swiss government as part of a comprehensive assessment of the costs of traffic (accidents, noise, air pollution, and property damage) to provide a basis for setting tariffs for truck traffic through Switzerland. Not surprisingly, the resulting ten-fold increase in these fees provoked an outcry, but nevertheless other countries became interested and the Swiss government lobbied for a second study. Ultimately, France and Austria joined in the second study, but Germany did not, as some of its scientific advisors insisted that reliance on cohort studies led to overestimates of the risk and that only the acute effects (time-series) studies should be used. Ironically, it is now widely agreed that the latter underestimate the risks (see the concluding portion of Chapter 8).

After reviewing the accumulated evidence about a range of health effects from different pollutants in southern California from the CHS, Kunzli et al. (2003) discussed two general strategies for improving the public health impact of air pollution: reducing emissions; and reducing individual exposures. The former include technological improvements (to vehicle design and fuels), urban design (limiting sprawl and building bicycle paths), and behavior (encouraging carpooling and forbidding idling of school buses). The latter also include technological improvements (air filtering in schools), urban design (separating schools from roadways), and behavioral (reducing outdoor activity when pollution is high). They go on to discuss some of the tensions between these alternative strategies. For example, school air conditioning requires energy which could lead to higher ambient pollution levels, whereas encouraging walking to school could increase individual exposures, and moving to less polluted communities in the suburbs will increase pollution from commuting. Such trade-offs can end up pitting one interest group against another—beyond the usual environmental quality versus economic costs debates—calling for a visionary approach to setting public health regulatory policy.

15 Risk assessment

> ...a continuing concern for methods,
> and especially the dissection of risk assessment,
> that would do credit to a Talmudic scholar
> and that threatens at times to bury
> all that is good and beautiful in epidemiology
> under an avalanche of mathematical trivia and neologisms.
> —(Stallones 1980)

Environmental epidemiology forms one the most important pillars for
setting environmental regulations. Policy makers distinguish two types of
activities, risk assessment and risk management. Risk assessment refers
to the scientific activity of combining evidence from all relevant sources
to assess the likely health consequences of various possible regulatory
strategies and the uncertainty about these predictions. Risk management
combines this scientific advice with consideration of economic, legal,
political, and other considerations to arrive at policy recommendations.
Risk managers frequently rely on the "precautionary principle," which
is perhaps best described in the Rio Declaration on Environment and
Development:

Where there are threats of serious or irreversible damage, lack of full scientific cer-
tainty shall not be used as a reason for postponing cost-effective measures to prevent
environmental degradation.

The application of the principle has been hotly debated, however; see
Renn (2007) for an overview of a series of talking points on this debate.
Risk management is beyond the scope of this book, however, so this
chapter will focus primarily on the use of epidemiologic evidence in the
risk assessment part of this process.

Of course, risk assessment often relies on more than just human data.
Epidemiologic data may be quite limited, or even if good human data are
available, there could be highly relevant experimental data from toxicological studies in animals, cell cultures (e.g., mutagenicity assays), exposure
assessment, or other sources. Combination of evidence across such disparate fields is more an art than a science, but we will review at least some
of the formal statistical methods that have been brought to bear on this
task later in this chapter.

Within the field of risk assessment, a further distinction is commonly
made between assessing the causality of an exposure–response relationship and quantifying the magnitude of the effect. Again, the former is

largely a judgment call, particularly where evidence from different fields are to be synthesized. Traditionally, epidemiologists have relied on criteria like those laid down by Sir Austin Bradford-Hill (1965) for evaluating the causality of the smoking and lung cancer association (see Chapter 2). Various expert bodies such as the International Agency for Research on Cancer (IARC), the Institute of Medicine (IOM) of the U.S. National Academy of Sciences, the U.S. Environmental Protection Agency (EPA), the National Toxicology Program (NTP), amongst others, have developed qualitative classifications of the strength of the scientific evidence. Although the specific terminology and definitions vary, these are broadly classified as "sufficient," "limited," or "inadequate" for humans and animals, supplemented with supporting evidence about mechanisms, chemical similarity to established carcinogens, or other relevant data. For example, IARC summarizes the evidence relating to cancer risk into five main categories: carcinogenic to humans; probably carcinogenic to humans; possibly carcinogenic to humans; not classifiable as to carcinogenicity in humans; and probably not carcinogenic in humans. The first three categories require at least direct observations of cancer in humans or laboratory animals and most agencies require multiple studies to establish a positive categorization. A recent IOM Committee on compensation for veterans (IOM 2007), reviewing the proliferation of criteria for judging evidence recommended the following simplified system: sufficient; equipoise and above; below equipoise; and against. Because this activity is largely a qualitative judgment, we will not consider it further in this chapter and will turn our attention to risk assessment.

Risk assessment from epidemiologic data

In general, the process of risk assessment entails three steps: a meta-analysis of the world literature to derive exposure–response relationships, including any modifying factors and their uncertainties; exposure assessment to describe the relationship between various exposure scenarios of potential regulatory interest and the resulting distribution of exposures or doses to human populations; and combination of the two to compute the predicted population risk from the various regulatory scenarios. We now consider each of these steps in turn.

Estimation of exposure–response relationships

Most of the earlier chapters of this book have been concerned with the analysis of epidemiologic study data to infer exposure–time–response relationships. But so far, we have considered only the analysis of a single study, where the raw data is available to the analyst. The task of the meta-analyst usually begins with the summaries of such analyses as reported in the scientific literature, without further access to the raw data. Of course, in some instances it has been possible for an analyst or an expert body to assemble the raw data from all the important studies and perform a combined analysis. Although a much more ambitious undertaking, such "mega-analyses" tend to be much more informative, as they allow all the studies to be analyzed in a comprehensive manner, using a consistent set of variable definitions and models (at least within the limitations of the data available from the different studies).

Risk estimates might be derived from purely empirical exposure-response models like those described in Chapter 6 or from mechanistic models like those in Chapter 13. The U.S. and California EPAs, for example, rely heavily on the linearized multistage model of carcinogenesis (Crump 1984) described in Chapter 13 for deriving risk estimates and confidence limits from animal carcinogenesis studies, as well as epidemiologic data.

The National Academy of Sciences Committee on the Biological Effects of Ionizing Radiation (BEIR IV) (NAS 1999) performed mega-analyses of all the major cohort studies of radon exposure among uranium miners and case-control studies of domestic radon exposures. Their analysis framework entailed an empirical form of exposure–time–response model for the miner studies that allowed for different slope coefficients across studies, while keeping other parts of the model (the modifying effects of age, latency, and smoking, for example) that were less well estimated common across studies. Thus, a typical model from their analysis might take the form

$$\lambda(t, s, Z) = \lambda_s(t)[1 + b_s Z(t) \exp(f(t)]$$

where s denotes study, $Z(t)$ denotes some empirically derived measure of latency-weighted cumulative radon exposure (as discussed in Chapter 6), and $f(t)$ some empirically derived modifying effect of attained age. The focus of the analysis is thus primarily on the estimation of the relative risk coefficients b_s and their within- and between-study variabilities.

Meta-analysis

Ideally, the estimation of an overall relative risk coefficient would be based on a mega-analysis of all the original data, but frequently all that is available is published analysis summaries of estimates and their standard errors, requiring a meta-analysis. Letting $V_s = \text{var}(\hat{b}_s)$, denote the sampling variability of the estimate $\hat{b}_s$ from study s, an obvious estimator of the average relative risk coefficient would be the simple variance-weighted average $\bar{\beta} = \Sigma_s(\hat{b}_s/V_s)/\Sigma_s(1/V_s)$ with naïve variance $\text{var}(\bar{\beta}) = 1/\Sigma_s(1/V_s)$. However, this estimator assumes that all studies are estimating the same parameter β_0. More likely, there is real heterogeneity in the "true" parameter values β_s being estimated by each study, due to various aspects of the populations studied and methodologies used. The random effects model assumes that these true β_s are normally distributed around β_0 with variance σ^2 and the study estimates $\hat{b}_s$ are in turn normally distributed around their respective β_s with variance V_s. Then the maximum likelihood estimate of the average risk coefficient β_0 is given by

$$\hat{\beta}_0 = \frac{\Sigma_s \hat{b}_s / (\hat{\sigma}^2 + V_s)}{\Sigma_s 1 / (\hat{\sigma}^2 + V_s)}$$

with model-based (inverse information) variance $\text{var}(\hat{\beta}_0) = [\Sigma_s 1/(\hat{\sigma}^2 + V_s)]^{-1}$, where the method of moments estimator of the between-populations variance of the true βs is given by $\hat{\sigma}^2 = (S-1)^{-1}\Sigma_{s=1}^{S}[(\hat{b}_s - \hat{\beta}_0)^2 - V_s]$ if positive, otherwise zero. (One could also use the more precise maximum likelihood estimator of σ^2, but this requires an iterative search.) The assumption that the true β_s are normally distributed may be questionable, in which case one might want to use the robust variance estimator

$$\text{var}(\hat{\beta}_0) = \frac{\Sigma_s(\hat{b}_s - \hat{\beta}_0)^2/(\hat{\sigma}^2 + V_s)^2}{[\Sigma_s 1/(\hat{\sigma}^2 + V_s)]^2}$$

While this estimator does not require the normality assumption, it may be poorly estimated if the number of studies is small.

More importantly, the meta-analyst should seek to understand *why* studies differ in their estimated risk coefficients rather than simply to obtain an average of them. This can be done using the technique of meta-regression (Greenland 1994b; Greenland and O'Rourke 2001), in which the analyst constructs a vector Z_s of covariates describing various aspects of the populations (e.g., their age distributions or geographic locations) and the study methodology (e.g., participation rates or subjective quality scores) that could potentially account for differences in results. The assumption of a common distribution of study-specific parameters β_s is then replaced by a regression model of the form $\beta_s \sim N(Z_s'\mu, \sigma^2)$. In addition to providing insight into discrepant results, the model provides a basis for estimating the risk for the specific population of interest to the regulator that might have been derived from a hypothetical ideal study (provided this did not require extrapolation too far beyond the range of the data!).

Here, for expository purposes, we have assumed that the sole parameter of interest is the relative risk coefficient β, but in reality, there are additional uncertainties in the parameters of the latency weighting in $Z(t)$ and the attained age modifying function $f(t)$, not to mention the basic form of the model and possibly other components. Furthermore, we have assumed that the parameter of interest is normally distributed. Such uncertainties might be better accounted for by deriving the likelihood of the full parameter vectors Θ_m for model m and the marginal likelihoods L_m or posterior probabilities π_m of each model, as described in Chapter 12.

Having derived a model from the miner data, the BEIR VI Committee then proceeded to perform a similar mega-analysis of the case–control studies of *domestic* radon and to compare the two models. Figure 15.1 provides a comparison of the two estimates, showing that, although the estimates from the low-dose residential studies individually have great uncertainty and as well as between-study variability, overall all the results are compatible with the risk estimates predicted by the high-dose miner studies and, indeed, are more compatible with those estimates than with an estimate of no effect. They also are clearly inconsistent with the inverse relationships suggested by the ecologic studies, which are potentially subject to the various forms of "ecologic" bias discussed in Chapter 10.

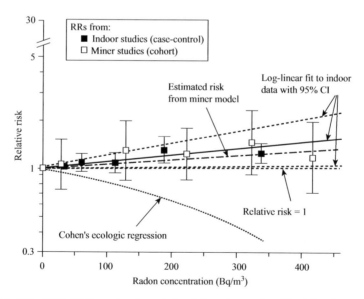

Figure 15.1. BEIR VI Committee's analysis of the miner and residential radon studies (Reproduced with permission from NAS 1999).

A key element of exposure–response analysis is, of course, establishing the basic form of the relationships—whether linear, loglinear, or involving some threshold or saturation—and how individuals might differ in their personal sensitivities to exposure across a population. A risk assessment will usually take such issues into consideration, perhaps using techniques such as those described in Chapter 6, including a range of parametric or flexible models (splines, etc.). Since regulators are generally interested in very low excess risks, it is commonplace to rely on the principle of *low-dose linearity*. The argument goes as follows (Crump et al. 1976): suppose the true form of the relation is $r(Z)$ such that $r'(0) > 0$ (i.e., any exposure beyond zero produces some increase in risk); then the (unknown) true dose–response can be approximated by a Taylor series as

$$r(Z) = r(0) + Zr'(0) + e(Z)$$

where $e(Z)$ denotes terms that become vanishingly small as Z goes to zero. The rationale for the assumption that $r'(0) > 0$ is that zero exposure is never truly attainable, so that the measured exposure is always in addition to some background level and it is implausible that a small additional exposure would not lead to some increase in risk. Thus, in the limit of very small exposures, any dose–response relationship should tend to linearity. Of course, the utility of this theoretical result depends upon how far from the linear range the observable data are. It certainly does not follow that linear extrapolation from the entirety of the observable data will lead to a valid estimate of the true low-dose slope; indeed, the two could differ by several orders of magnitude (Portier and Hoel 1983).

Inter-individual variability

Any risk assessment should, of course, incorporate what is known about measurable modifying factors (e.g., age, sex, sensitive subpopulations like asthmatics, etc.). But what about possible variation in sensitivity between individuals due to unknown factors, like polygenic background variation? First, suppose that the dose–response relationship were truly linear, but each person i had their own slope coefficient β_i, with these slopes having some distribution across the population, say $f(\beta)$. Then it follows that the population dose–response would still be approximately linear, $R(Z) = 1 + \bar{\beta}Z$ with slope $\bar{\beta} = E(\beta_i) = \int \beta f(\beta)\, d\beta$ equal to the population mean of these individual slopes. This approximation, however, depends upon the risks being small enough that there is no differential survival that would lead to premature elimination of the most sensitive members of the population. More precisely, the distribution of slopes among survivors at age t exposed to some level Z is given by

$$P(\beta|Z,t) \propto S(t|Z,\beta) f(\beta) = \exp\left[-\Lambda_0(t) r(Z|\beta)\right] f(\beta)$$

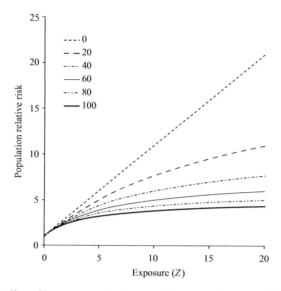

Figure 15.2. Effect of heterogeneity in slope coefficients in a linear model for individual risk on the population average relative risk as a function of age: gamma distributed slopes with mean 1 and variance 0.25 and a constant baseline hazard rate $\lambda_0 = 0.01$.

and hence the population dose–response would become

$$R(Z,t) = \int r(Z|\beta)\,P(\beta|Z,t)\,d\beta = \int (1+\beta Z)e^{-\Lambda_0(t)(1+\beta Z)}f(\beta)\,d\beta$$

The effect of this "survival of the fittest" effect is to leave the population at risk at the highest exposure levels increasingly selected with advancing age in favor of the less sensitive individuals, leading to an increasing downturn with age in the population dose–response relationship, as illustrated in Figure 15.2. This integral can be evaluated in closed form when β has either a normal or a gamma distribution, leading to

$$R(Z,t) = 1 + \mu Z - \Lambda_0(t)\sigma^2 Z^2$$

or

$$R(Z,t) = 1 + \frac{\mu Z}{1 + \Lambda_0(t)\sigma^2 Z/\mu}$$

respectively. The normal case includes some probability that β is negative, so that for sufficiently large risks, the dose–response can actually decline with increasing exposure, but this scenario is somewhat implausible. Figure 15.2 demonstrates the case for gamma-distributed individual slopes, which are necessarily positive.

A more important situation arises when the dose–response is nonlinear, with differences between individuals in the shape of the relationship. There are two important cases that are amenable to closed form solution: a probit model and a threshold model. In both cases, let us assume the risks are low enough that differential survival over time can be ignored.

For the probit model, let us first assume $\Pr(Y = 1|Z) = \Phi(\alpha_i + \beta Z)$, that is, the relative risk per unit dose is a constant across the population, but the level of risk varies between individuals. Furthermore, let us assume that the intercept terms α_i are normally distributed with mean μ_α and variance σ_α^2. Then it can be shown that the average population dose–response takes the form

$$\Pr(\gamma = 1|Z) = \Phi\left(\frac{\mu_\alpha + \beta Z}{\sqrt{1 + \sigma_\alpha^2}}\right)$$

Thus, the average population dose–response has the same basic form, but the slope is attenuated by the factor in the denominator [Figure 15.3(a)]. On the other hand, if the intercept is constant, but the slopes vary between individuals, then the average population dose–response becomes

$$\Pr(Y = 1|Z) = \Phi\left(\frac{\alpha + \mu_\beta Z}{\sqrt{1 + \sigma_\beta^2 Z^2}}\right)$$

In this case, the shape of the average population dose–response can be quite different from the individual ones, being more attenuated at high doses, but similar to that of an average individual at low doses [Figure 15.3(b)].

A second special case is the linear threshold model mentioned in Chapter 6. As in Eq. (6.1), we assume an individual's dose response is given by $r_i(Z) = 1 + \beta(Z - \tau_i)I(Z > \tau_i)$, that is, there is no increased risk until exposure exceeds an individual's personal threshold τ_i, at which point

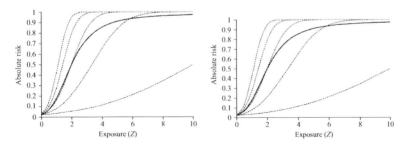

Figure 15.3. Effect of heterogeneity in intercepts (left panel) and slopes (right panel) on the population average dose response in a probit model; dotted lines represent ± 2, ± 1 or 0 SDs in individual intercepts or slopes; solid line represents the population average risk.

each additional increment of exposure has the same effect as for everyone else. Since both Z and τ are nonnegative, it makes sense to assume τ has a truncated normal, lognormal, or gamma distribution. The resulting population dose–response relationships are shown in Figure 15.4. All these curves have zero slope at zero exposure, but the transition to a linear dose–response is more gradual than for any individual, the location of the change point being determined by μ and the curvature at that point by σ^2.

Although we have focused in this discussion on the effect of heterogeneity on the *average* population risk, regulators are frequently concerned with the risk to the most sensitive individuals or subpopulations. Absent knowledge of specific modifying factors that can be taken into account in the epidemiologic analysis, this becomes a somewhat speculative exercise, and regulators usually fall back on arbitrarily chosen "safety factors" that are little more than guesses. For example, the U.S. EPA aims to base standards not on the average risk but on the risk to the 99th percentile of the distribution of susceptible individuals. In principle, the models described above could be used to estimate the variance parameters σ^2, and the fitted dose–response evaluated at some percentile of the distribution of the random effects, but in most circumstances, this is unlikely to be rewarding. First, even for a correctly-specified model, these variances are likely to be poorly estimated from epidemiologic data. More importantly, the estimate of variability is critically dependent upon the form of the assumed model for individual risk, which is not directly observable. For example, if the true dose–response were linear-quadratic with no heterogeneity across subjects, but one fitted a linear spline model with heterogeneity in thresholds to any finite dataset, one would likely obtain a nonzero estimate of the mean and variance of the thresholds, but the fitted upper bound on the distribution of risks across individuals would be meaningless. As we

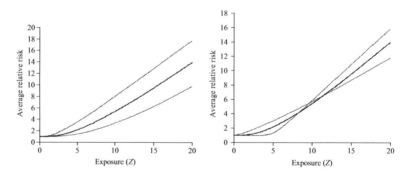

Figure 15.4. Effect of heterogeneity in thresholds in a linear spline model with a lognormal distribution of thresholds: left panel, varying the geometric mean $\mu = 2, 5, 10$ holding the logarithmic standard deviation (LSD) fixed at 1; right panel, varying LSD $= 0.25, 1, 4$ holding μ fixed at 5.

have seen, heterogeneity in slopes, intercepts, or thresholds have different effects on the population distribution of risks and there is little *a priori* basis for choosing between alternative models.

Exposure assessment

The next step of the process entails an evaluation of the distribution across the population of doses to the target organ that might result from various alternative regulatory scenarios. (As in Chapter 6, we note that in some situations it may be impossible to estimate distributions of actual dose and the analysis proceeds in terms of exposure instead, particularly if only exposure–response rather than dose–response data are available. Nevertheless, in what follows, we use the term "dose" interchangeably with "exposure.") We assume that $Z_i(t)$ in the previous section represents dose as a function of time for a given individual i in the population, and let Z represent some proposed regulatory limit on population exposures. Such limits could take many different forms, such as an upper bound on the instantaneous ambient concentration or a target for the long-term average ambient concentration. Limits could also be imposed on the emissions from particular sources or on some summary of the population distribution of personal exposures. The latter tend to be more difficult to implement. At issue also is whether the purpose of the regulation is to control the population average risk (or exposure) or the maximum risk to any individual. As discussed above, risk models might provide some basis for distinguishing between the risks per unit exposure across sensitive subgroups, in which case one might be concerned with controlling the population average risk or the risk to the most susceptible individual. Individuals doubtless do vary in terms of their sensitivities to any given exposure, but often the specific modifying factors are unknown and even the range of variability across individuals may be impossible to quantify. On the other hand, the distribution of individual *exposures* under some regulatory scenario might be more readily quantified than individual *risks*, so the estimation of doses to the most heavily exposed individuals is an important part of the process. Risk models from the epidemiologic data might be expressed in terms of exposure (in terms of external concentrations) or in terms of doses to particular organs, which might depend on such modifying factors as ventilation rates, activity levels, age, gender, or other factors. The task of the exposure assessor is to provide such translations.

This task is more difficult when the dose–response data derives from toxicologic experiments in animals. Then various adjustments must be made to translate the applied doses in the experimental situation to some kind of equivalent dose to humans. This may entail consideration of such factors as body size, metabolism, and relative lifespan. Such adjustments are often little more than guesswork, although some general principles have become conventional in the risk assessment community. Depending upon the context, for example, exposures might be computed in terms dose per body weight or per surface area and expressed as a proportion of normal lifespan rather than in units of calendar time. Crump et al. (1989) compared five such scaling, all previously shown to be highly correlated ($p < 0.001$) with human cancer potency estimates (Allen et al. 1988), on 23 chemicals for which both human and animal risk estimates were available and concluded that risk per unit body weight came closest to equality, with a range of ratios from 0.36 to 1.6. (Ideally one would prefer the index with the smallest *variance* in ratios, provided one had a suitable database for deriving a correction factor for the *mean* ratio; the Bayesian approach to evaluating the carcinogenity of plutonium described later in this chapter is an example of such an approach. On this basis, the California EPA cancer risk assessment guidelines specify the use of a surface area scaling instead.) Since toxicology experiments are generally performed at the maximum tolerated dose and perhaps a few lower doses, the possibility of nonlinearities due to saturation effects or biological effects that do not occur at all at lower doses must be considered (Hoel et al. 1988). Compartmental pharmacokinetic modeling, as discussed in Chapter 13 might be used to allow for such phenomena. For an extensive review of the methods used by the U.S. and California EPAs for risk assessment based on human and experimental data, see (OEHHA 2002). See also Crump et al. (1976); Hoel (1979); Hoel et al. (1983); Portier and Hoel (1983); Bailer and Hoel (1989); Roberts et al. (2001) and the IARC monograph (Moolgavkar et al. 1999a) for further discussion. Zeise et al. (2002) provide a comprehensive discussion of current methodological challenges and extensive references. The concluding section of this chapter describes a Bayesian approach to combining human and animal *risk* estimates, given a choice of a suitable dose metric. Ultimately the U.S. National Toxicology Program aims to rely more on mechanism-based biological observations than the traditional two-year rodent experiment for carcinogen risk assessment (Bucher and Portier 2004), but that time appears to be some ways away still.

Risk quantification

The final step entails computation of such quantities as the population average lifetime risk of disease $P(Z)$ resulting from various exposure scenarios Z. These are often summarized in a single number, such as the attributable number AN of cases due to exposure or the "unit risk estimate"

$$R_1 = [P(Z) - P(Z_0)]/(Z - Z_0) \tag{15.1}$$

where Z_0 represents some baseline exposure scenario (e.g., no exposure). Provided Z is sufficiently small, it is then reasonable to assume that the excess risk from some other scenario Z' would be proportional to the unit risk, that is, $P(Z') - P(Z_0) = R_1(Z' - Z_0)$. In the remainder of this section, we describe the calculation of this lifetime risk.

Let $\lambda(t, Z)$ represent some final risk model derived from the synthesis of the literature, as described in the first step above, and let $Z(t, Z)$ denote the expected exposure at time t under regulatory scenario Z, as described in the second step above. To compute the lifetime risk, we use lifetable calculations illustrated in Table 15.1 to compute the survival function

$$S(t, Z) = \exp\{-\int_0^t (\lambda[t, Z(t, Z)] + \mu(t))\, dt\}$$

Table 15.1. Lifetable risk assessment for exposure to 0.1 WL of radon daughters over a lifetime based on the BEIR VI concentration model [(NAS 1999), Table A-4, p. 151). See Table 3.1 for the baseline risks of lung cancer incidence and competing cause mortality

Age	Excess relative risk	Hazard rates for lung cancer		Probability of lung cancer		Life expectancy	
		Baseline	Exposed	Exposed	Attributable	Exposed	Lost
0–4	0.000	0.0	0.0	0.00000	0.00000	72.297	**−0.131**
5–9	0.019	0.0	0.0	0.00000	0.00000	67.326	−0.131
10–14	0.056	0.0	0.0	0.00000	0.00000	62.388	−0.131
15–19	0.089	0.2	0.2	0.00001	0.00000	57.458	−0.131
20–24	0.118	0.2	0.2	0.00001	0.00000	52.545	−0.131
25–29	0.142	0.4	0.5	0.00002	0.00000	47.666	−0.131
30–34	0.161	1.5	1.7	0.00008	0.00001	42.823	−0.131
35–39	0.180	4.9	5.8	0.00028	0.00004	38.018	−0.130
40–44	0.199	14.2	17.0	0.00080	0.00013	33.260	−0.130
45–49	0.218	37.6	45.8	0.00211	0.00038	28.563	−0.130
50–54	0.237	84.5	104.5	0.00467	0.00089	23.957	−0.128
55–59	0.146	162.1	185.7	0.00790	0.00100	19.491	−0.123
60–64	0.157	276.7	320.0	0.01257	0.00170	15.240	−0.114
65–69	0.085	404.3	438.7	0.01527	0.00120	11.311	−0.098
70–74	0.091	514.0	560.6	0.01625	0.00135	7.832	−0.078
75–79	0.030	572.1	589.2	0.01302	0.00038	4.933	−0.055
80–84	0.032	553.4	570.9	0.00838	0.00026	2.724	−0.034
85–89	0.033	446.5	461.4	0.00344	0.00011	1.256	−0.019
90+	0.035	446.5	462.1	0.00238	0.00008	0.509	−0.011
Total				**0.08718**	**0.00754**		

where $\mu(t)$ denotes the risk of death from competing causes. The lifetime risk is then computed as

$$P(\mathcal{Z}) = \int_0^\infty \lambda[t, Z(t, \mathcal{Z})] S(t, \mathcal{Z}) \, dt$$

and the attributable number is simply $AN(\mathcal{Z}) = [P(\mathcal{Z}) - P(\mathcal{Z}_0)]N$, where N is the total population size. Alternatively one might use the current age distribution $N(t)$ of the population to compute the attributable number as

$$AN(\mathcal{Z}) = \int_0^\infty \{\lambda[t, Z(t, \mathcal{Z})] - \lambda[t, Z(t, \mathcal{Z}_0)]\} N(t) \, dt$$

In the event that the effect of some proposed regulatory strategy would be instead to yield a probability distribution $p[Z(t, \mathcal{Z})]$ of population exposures, then one would obtain the unit risk or attributable number by integrating these quantities over that distribution.

The validity of these calculations depends upon an assumption of independence of competing risks, that is, that individuals dying of the cause of interest are no more or less likely to die of other causes than other individuals. Unfortunately, with only a single cause of death observable, this assumption is inherently untestable, although it seems likely that some causes of death are indeed dependent. Nevertheless, the assumption is commonly made for lack of a viable alternative.

Now it is tempting to define the unit risk as the average across subjects of the unit risks from Eq. (15.1), but this conflates two countervailing tendencies: on the one hand, at any given age, the hazard rate for death is increased by exposure, but also the probability of surviving to that age is decreased. It has therefore been argued (Thomas et al. 1992a) that a more appropriate measure of excess risk is to integrate the excess risk at each age attributable to exposure *against the survival distribution expected in the absence of exposure*, that is,

$$R_1 = \int_0^\infty \{\lambda[t, Z(t, \mathcal{Z})] - \lambda[t, Z(t, \mathcal{Z}_0)]\} S(t, \mathcal{Z}_0) \, dt / (\mathcal{Z} - \mathcal{Z}_0)$$

A disadvantage of any of these measures of excess risk is that they take no account of *when* the excess deaths occur. Two agents might cause the same total number of deaths, but one might act at a younger age than the other, leading to greater loss of life expectancy. Thus, an alternative measure of impact is the *expected years of life lost* (EYLL), computed as

$$EYLL = \int_0^\infty (S(t, \mathcal{Z}_0) - S(t, \mathcal{Z})) \, dt$$

$$= \int_0^\infty t \, (\lambda(t, \mathcal{Z}_0) S(t, \mathcal{Z}_0) - \lambda(t, \mathcal{Z}) S(t, \mathcal{Z})) \, dt$$

(This equality is readily demonstrated by integration by parts.)

These calculations are illustrated in Table 15.1 for a hypothetical lifetime of exposure to domestic radon at a concentration of 0.1 WL. The baseline incidence rates of lung cancer and death rates from competing causes are the same as those used in introducing the lifetable method in Table 3.1. Here, we compare those baseline calculations with those under the exposed scenario. The excess relative risks are given by the BEIR VI Committee's (NAS 1999) preferred model involving modifying effects of latency, concentration, and attained age, which takes the form

$$ \text{ERR}[t, Z(\cdot)] = \beta \varphi_t (\theta_1 Z_{5-14} + \theta_2 Z_{15-24} + \theta_3 Z_{25+}) \gamma_{\bar{Z}} $$

where $\beta = 0.0744$, φ_t is the modifying effect of attained age (1.00 for ages <55, 0.57 for ages 55–64, 0.29 for ages 65–74, and 0.09 over age 75), θ is the modifying effect of latency in the indicated periods prior to t ($\theta_1 = 1.00$, $\theta_2 = 0.78$, $\theta_3 = 0.51$), and $\gamma_{\bar{Z}}$ is the modifying effect of average concentration (1.0 for concentrations less than 0.5 WL, declining for higher concentrations). Here, we evaluate the risk for an average concentration of 0.1 WL. The column headed "Lung cancer incidence, exposed" is simply the baseline rates multiplied by $1 + \text{ERR}$ derived from this model at the indicated attained ages. "Probability of lung cancer" in the exposed is computed in the same way as described in Chapter 3, using the hazard rates for competing risks given in Table 3.1. The column headed "Attributable risk" of lung cancer is $[\lambda(t, Z) - \lambda(t, 0)]S(t, Z)$. Thus, the lifetime risk of lung cancer in the exposed is 8.718%, compared with 8.017% for the general population, for a lifetime attributable risk of radiation-induced lung cancer of 0.754%. (Note that this is not simply the difference of the two lifetime risks, 0.701%, because the exposed and unexposed hazard rates are integrated over the *same* survival distribution, that among the exposed population.) The life expectancy of the exposed population is 72.297 years, compared with 72.427 for the unexposed population, or a loss of life expectancy of 0.131 years. Thus, we would report the "unit risk" estimate as 7.5% per WL and the unit loss of life expectancy (LLE) as 1.3 years per WL.

As before, of course, if these risks vary across individuals, either because of variation in the distribution of personal exposures under a given scenario or because of variation in individual modifying factors, then the unit risk would have to be computed by taking a population average of this quantity across individuals.

Recall the discussion of some of the reasons for the huge difference between acute and chronic effects of air pollution discussed in Chapter 8. Whether exposure effects are assessed on the basis of excess deaths or loss of life expectancy can thus make a big difference in risk assessment. In fact, as we will see in the following chapter, the former is not estimable without making unverifiable assumptions about population homogeneity and biological mechanisms, although loss of life expectancy can be estimated under less restrictive assumptions.

Another quantity frequently used in the risk assessment literature is "Disability Adjusted Life Years (DALYs)," which adds weights to each year of life reflecting a subjective assessment of quality of life following a chronic illness or injury. See Murray and Lopez (1997); Gold et al. (2002); Pruss et al. (2002); Steenland and Armstrong (2006) for details of methods for calculating DALYs and Chapter 17 for their application to estimating the global burden of environmentally caused disease. Essentially, DALY is the years of life lost (YLL) due to premature mortality (as described above), *plus* a weighted sum of years of life lived with disability. Typical weights range from 0.81 for late stage cancer (any site) to 0.10 for untreated asthma or 0.06 with treatment.

Uncertainty analysis

Uncertainty analysis entails consideration of multiple components of uncertainty, including the sampling variability in the relative risk coefficient of primary interest, other parameters of the model (e.g., individual or temporal modifiers), between-study variability, and model form. Although in principle, one could use the delta method to compute the variance of R_1, this is difficult for a multi-parameter model, requiring the full covariance matrix of all the parameters. Furthermore, the distributions of individual parameters are often markedly skewed, so relying on asymptotic normality can be misleading. A simpler method is to randomly sample parameter values from the full likelihood of all the parameters and repeat the lifetable calculation for each sampled parameter vector, then tabulate the distribution of the resulting unit risk estimates. Such an approach was used by the NAS BEIR V committee (NAS 1990; Thomas et al. 1992a).

Given the inevitable uncertainty about the true model form, one might ask how one should estimate the relative risk (or any other epidemiologic effect parameter derived from it) and an "honest" confidence interval for it that allows for the possibility of model specification error. There is an extensive statistical literature on this question [see, e.g., Leamur (1978) for a review], but in practice the problem is frequently ignored. All too often, an investigator conducts a number of different analyses and ends up reporting only a single best-fitting model, or the one he or she has the strongest belief in, and reports confidence limits on the parameters of that model as if it were the "true" model. Sometimes, an investigator may acknowledge this uncertainty about model form by reporting a range of alternative models in the spirit of "sensitivity analyses," but this can leave the reader with a dilemma about which specific set of estimates to use, particularly if several models fit the data more or less equally well, yet yield different estimates and confidence intervals. There are, however, a number of formal approaches to this problem, such as the Bayes model averaging method described in Chapter 12.

Combination of evidence from different disciplines

As noted at the outset, epidemiologic evidence from humans is often limited or subject to various potential biases. An example comes from the National Academy's BEIR IV report (NAS 1988), in which an estimate of the carcinogenicity of plutonium in humans was sought. The available human data is very limited—zero bone cancers among individuals occupationally exposed in the Manhattan Project out of 449 and 324 person-year (PY)-rad, for Pu^{238} and Pu^{239}, respectively, excluding five years latency. Hence, the Committee adopted an empirical Bayes approach (DuMouchel and Harris 1983) that relied on the assumption that the ratio of carcinogenic potencies of plutonium to various other radionuclides would be roughly constant across species. Since there is substantially more human data about the carcinogenicity of various isotopes of radium and extensive animal data about plutonium, radium, and other radionuclides, it was possible to estimate the risk of plutonium in humans from a combined analysis, including an uncertainty analysis that incorporated the variability in the ratios of relative carcinogenicities across species. Specifically, the committee adopted a hierarchical relative risk model of the form

$$\lambda_{sr}(Z) = \lambda_{0s}(1 + \beta_{sr} Z)$$

$$\log(\beta_{sr}) = \mu + \alpha_s + \gamma_r + \varepsilon_{sr}$$

$$\alpha_s \sim N(0, \sigma_S^2), \quad \gamma_r \sim N(0, \sigma_R^2), \quad \text{and} \quad \varepsilon_{sr} \sim N(0, \sigma_E^2)$$

where s indexes species and r indexes radionuclides. The basic assumption is thus that the ratio of slope coefficients β_{sr} between radionuclides are approximately constant across species (or vice-versa), with some random deviations with logarithmic standard deviation σ_E. The model was fitted to data from four human studies of two isotopes of radium and to six dog and five rat studies of two isotopes of plutonium and one of radium. Note that the extremely sparse human data on plutonium were not used in the first stage of this analysis. On this basis, the estimated posterior distribution for the human potency of plutonium was found to be approximately a gamma distribution with 1.9 cancers out of 4700 PY-rad, that is, $\Gamma(1.9, 4700)$. Now combining this with the observed human plutonium data yields an *updated* posterior distribution of $\Gamma(1.9, 5149)$ and $\Gamma(1.9, 5024)$ for the two isotopes, barely shifted at all from those estimated from the combined analysis of the human radium and animal radium and plutonium data. More importantly, however, this analysis provides a basis for estimating a posterior credibility interval on the estimated risk of 0.3 per 1,000 PY-rad (1.9/5149) ranging from 0.08 to 1.07, which takes account not just of the sampling variability in the data but the uncertainty about the validity of the multiplicative assumption used in the inter-species scaling.

Peters et al. (2005) describe related Bayesian methods for combining human and toxicologic evidence with an application to the association between low birthweight and trihalomethane exposures. Their framework is similar to the DuMouchel and Harris one, with discipline replacing species, allowing separate estimates of a relative risk coefficient by discipline, as well as an overall pooled estimate. In any such cross-species comparison, however, the problems of appropriate scaling of dose discussed earlier in this chapter remain. Peters et al. use a scaling by dose per body weight per day in a regression of the log(OR) on log(dose) for this purpose.

Probability of causation and compensation

In 1984, the U.S. Congress passed the Orphan Drug Act, which included a provision instructing the National Institutes of Health (NIH) to establish a set of "Radioepidemiologic Tables" that could be used to resolve various claims for compensation relating to exposure to ionizing radiation from government programs. These included the "atomic veterans" who had participated in nuclear weapons tests at the Nevada Test Site and the Pacific Proving Grounds in the Marshall Islands, the "downwinders" who were exposed to fallout from the Nevada Test Site, the uranium miners on the Colorado Plateau who provided the raw materials for the country's nuclear weapons program, and various nuclear workers. In due course, NIH established a committee that prepared such a report (NIH 1985), under the oversight of another committee of the National Academy of Sciences (NAS 1984). These two reports established the scientific principles that have guided the enactment of legislation setting up compensation programs for these various groups and served as a reference for numerous radiation-related litigations in the private and public sectors for individuals not covered by these programs. The Radioepidemiologic Tables were updated in 2000 in the form of an interactive computer program (the Interactive Radiation Epidemiology Program, IREP (Kocher et al. 2008), available on the Internet at https://www.niosh-irep.com/irep%5Fniosh/, again with oversight by a new committee of the National Academy of Sciences (NRC 2000), which commented in depth on the scientific principles involved.

Of course, epidemiologic evidence had been used in compensation claims and tort litigation long before the Radioepidemiologic Tables came into existence. Particularly influential were the enormous number of lawsuits relating to asbestos in the 1970s that threatened to bankrupt the entire industry and spawned a series of seminal papers about the relevance of epidemiologic evidence to individuals. Although cases of mesothelioma were known to be caused only by asbestos—so their cause could be readily established—the cause was not so clear cut for cases of lung cancer in smokers. In 1980, Philip Enterline (1980) published a paper on this question that defined the concept of Probability of Causation (PC) and established the basic scientific principles for apportioning this quantity between multiple causes. These concepts continue to be debated to this day (Chase et al. 1985; Grimson 1987; Wraith and Mengersen 2007).

Starting in the late 1980s, Sander Greenland and Jamie Robins published a series of theoretical papers (Greenland and Robins 1988; Robins and Greenland 1989; 1991) calling into question the statistical validity of the concept of probability of causation and arguing in favor of compensation schemes based instead on loss of life expectancy (LLE). Their arguments are complex and subtle, but central to the concepts developed in this chapter, so will be discussed following an exposition of the basic ideas. We begin with a discussion of what is meant by causation at the population and individual levels, the relation between the epidemiologic parameter population attributable risk and the individual PC, and how they can be apportioned between multiple causes.

Uncertainties arise at many levels of such debates: at the population level, whether there is compelling evidence of a causal connection between exposure and disease and the magnitude of the association; at the individual level, about the details of individual's history of exposure and other factors. In setting policy for resolving such claims, one must look beyond the point estimate of a probability of causation and consider how such uncertainties should be taken into account. Some agencies, like the U.S. Veterans Administration have tended to take a liberal stance, awarding compensation if some upper confidence limit attains a given threshold. We conclude this chapter with a discussion of some of the policy implications of alternative schemes and the ways such uncertainties can be taken into account.

Causation at the population and individual levels

The word "evidence" takes different meanings in epidemiologic, statistical, general scientific, legal, and policy-making contexts and is used in different ways to support inferences about association and causation or recommendations about compensation policy. We begin by exploring some general principles about the use of different kinds of evidence at the population level, before proceeding to a discussion of the concept of probability of causation as applied to individual compensation decisions.

Epidemiologists have long recognized a distinction between association and causation (Chapter 2). The term "association" refers to a general tendency for there to be a relationship between an exposure and risk of disease (or more generally, between any risk factor and the distribution of any outcome) across individuals within a population. Beyond the exposure being a real cause of disease, there are many ways such an association could arise artifactually, including lack of comparability of the individuals being compared in terms of other risk factors (confounding), any of many other sources of bias, or chance, as discussed in Chapter 2. Here, by "exposure having *caused* disease," we mean that, for an individual who was exposed and subsequently developed disease, that outcome would not have occurred had that person not been exposed ("but for exposure" in legal parlance). In this sense, one could define a population association as being causal if at least some of the cases in the population would not have occurred in the absence of exposure.

Association does not necessarily imply a causal relationship. The inference of causality is a matter of scientific judgment that involves consideration of factors beyond the statistical analysis of the data that provide evidence of association. For many years, epidemiologists have relied on a series of criteria first outlined by Sir Austin Bradford Hill (1965), including dose-response, temporal relationship, consistency across multiple studies, biological plausibility, lack of credible alternative explanations, and coherence of the totality of the evidence (human or experimental). These criteria have been used in one form or another by virtually all expert bodies charged with making judgments about the scientific evidence of human health hazards.

In recent years, a formal theory of causal inference has been developed by a number of authors (Greenland 1990; Greenland et al. 1999a; Dawid 2000; Pearl 2000; Parascandola and Weed 2001; Maldonado and Greenland 2002; Cox and Wermuth 2004). Much of this literature is based on a formalization of the "but for" concept mentioned above, known as "counterfactual inference." In essence, one postulates the unobservable outcomes an individual could have experienced in the hypothetical scenario of having had an exposure history different from the one that actually transpired. Individuals whose outcomes would have been the same, whether exposed or unexposed, can be thought of as "immune" or "doomed," while those whose outcomes would have differed depending on their exposure status can be considered to have been "caused" or "prevented" by exposure (Table 16.1). Because such comparisons are made *within* the same individual (under hypothetically different exposure situations), they are not subject to most of the potential biases like confounding that threaten the validity of real epidemiologic studies, which are necessarily based on comparisons *between* individuals with different histories. Unfortunately, the comparisons can only be "thought experiments," but they help clarify some of the foundational principles on which the inference of causality is based, and have suggested some novel statistical approaches in some special situations (Hernan et al. 2000; Robins et al. 2000).

It is important to distinguish the thought processes involved in making judgments about causality in the population and in individuals. In general, it is necessary to have demonstrated that there can be a causal connection between an exposure and a disease across a population before it would be meaningful to claim a causal connection for any exposed individual who

Table 16.1. Representation of counter-factual outcomes and their relationship to observable data

Hypothetical outcome if unexposed	Hypothetical outcome if exposed		Observed exposed individuals
	Unaffected	Affected	
Unaffected	Immune	Caused	Controls
Affected	Prevented	Doomed	Cases
Observed unexposed individuals	Controls	Cases	

has been affected. Of course, it is always possible that such a relationship could exist in a population, but not have been demonstrated in an epidemiologic study because of inadequate statistical power, confounding or other biases, or chance (see Chapter 2), or conceivably because the number of cases caused by exposure was approximately offset by the number prevented by exposure. But one would be on shaky ground to claim a causal connection for an individual seeking redress or compensation in the absence of a population association.

The principles of evaluating the evidence for association and causation at the population level have been discussed in earlier chapters. Chapters 2 and 15 have discussed the causal interpretation of epidemiologic associations and Chapter 3 outlined the basic statistical principles for judging the statistical significance of an observed association. In the Agent Orange Act of 1991 (Public Law 102-04 105 STAT. 11), the U.S. Congress established as a basic ground rule for setting presumptions of causation for veterans that "the credible evidence for an association is greater than the credible evidence against an association." The interpretation of this language has proven quite ambiguous, as discussed at length in a recent Institute of Medicine report (IOM 2007). Since one can always find an alternative hypothesis with a higher likelihood than the null hypothesis, this language cannot simply mean that the likelihood ratio is greater than 1. Conventionally, scientists require that the alternative hypothesis be "significantly" more likely, as judged by a formal hypothesis test, but the level of significance (e.g., $p < 0.05$) is completely arbitrary. A Bayesian framework arguably provides a more natural interpretation of this statement, namely that the posterior odds in favor of the alternative hypothesis is greater than one, but this would require a consensus about the prior odds. In any event, it boils down to a matter of scientific judgment, combining the results of all relevant epidemiologic data with a subjective evaluation of their validity and any available biological or experimental evidence.

Attributable risk and probability of causation

So far, we have focused on what might be called the *strength* of the evidence. More important for evaluating causality for individuals is the *magnitude* of the relative risk. These are different concepts: there can be strong evidence for the existence of a small excess risk, or conversely only weak evidence for a large risk. Such factors as the study sample size and the frequency of exposure and/or disease can cause such discrepancies.

A useful summary measure of public health impact is the population attributable risk (PAR), defined as the proportion of disease *in a population* that is attributable to (loosely, "caused by") exposure (see Chapter 3). This quantity is a function of both the RR and the population frequency or distribution of exposure. More relevant in setting compensation policy is the attributable risk among the exposed (AR_E), given by the simple formula (RR−1)/RR. The relationship between two quantities depends upon the population frequency of exposure (p_E):

$$\text{PAR} = \frac{p_E(\text{RR} - 1)}{p_E(\text{RR} + (1 - p_E))} = \frac{p_E AR_E}{p_E AR_E + 1/\text{RR}}$$

(Levin 1953). Here, the numerator is the excess risk in the portion of the population that is exposed and the denominator is the total risk in exposed and unexposed subpopulations combined.

The AR_E can be interpreted as the proportion of disease *among exposed individuals* that is attributable to that exposure. In particular, if the RR > 2, this implies that > 50% of all exposed cases are attributable to exposure. Hence, in a population of similar individuals with similar exposures, on average any case that develops among them is "more likely than not" to have been caused by exposure. This is the commonly accepted criterion in tort litigation for a plaintiff to prevail. For a specific individual from a homogeneous population for which this RR applies, this quantity has been called the probability of causation (PC) and is similarly calculated as (RR−1)/RR. Of course, individuals differ in many ways. To the extent that such differences can be quantified and a person-specific RR estimated by some appropriate stratified analysis or statistical model, then this would be the quantity that would be used in this calculation.

In a commentary on the original Radioepidemiologic Tables (NIH 1985), the NAS Oversight Committee (NAS 1984; Lakagos and Mosteller 1986) pointed out that the PC was not strictly speaking a probability in the usual sense, but rather an estimate of the *proportion* of cancers that were caused by exposure in a hypothetical group of similarly exposed cases, an estimate that was then *assigned* to all members of the group. Therefore, they recommended the use of the term "assigned shares" (AS) for this quantity.

Assigned shares for multiple causes

Chapter 12 discussed the analysis of multiple causes of disease and a range of risk models, such as multiplicative and additive, that might be fitted to epidemiologic data. We now consider the application of such models to the apportionment of causation in an individual case having multiple factors that could have been the cause, singly or jointly. Consider a group of smokers exposed to some environmental toxin. The attributable risk among the exposed can be decomposed as follows:

$$AR_0 = 1/RR_{joint}$$
$$AR_E = (RR_E - 1)/RR_{joint}$$
$$AR_S = (RR_S - 1)/RR_{joint}$$
$$AR_{Int} = (RR_{joint} - RR_E - RR_S + 1)/RR_{joint}$$
$$= RR_{Int(add)}/RR_{joint}$$

where AR_0 is the proportion of cases among exposed smokers that is attributable to background factors unrelated to both, AR_E the proportion attributable to the environmental toxin alone, AR_S the proportion attributable to smoking acting alone, and AR_{Int} is the *additional* proportion attributable to the two factors acting in combination.

Specifically, for the hypothetical data shown in Table 12.4, we would compute the various components of the population attributable risk as shown in Table 16.4. Note that in both cases, the total adds up to 100%. In the multiplicative model, one might be tempted to conclude that the

Table 16.2. Calculation of population attributable fractions under multiplicative and additive models

	Multiplicative model		Additive model	
	RR	AR	RR	AR
Background	1	$1/30 = 3.3\%$	1	$1/12 = 8.3\%$
Exposure	3	$(3-1)/30 = 6.7\%$	3	$(3-1)/12 = 16.7\%$
Smoking	10	$(10-1)/30 = 30\%$	10	$(10-1)/12 = 75\%$
Interaction	30	$(30-3-10+1)/30$ $= 60\%$	12	$(12-3-10+1)/12$ $= 0\%$

contribution of the environmental toxin was very small, but this would be a wrong: combined with its interaction with smoking, the total contribution of the environmental toxin is 66.7%. (Note that this is the same AR_E as for nonsmokers, $2/3 = 67\%$, revealing that under a multiplicative model, the total AR_E does not depend upon smoking.) Of course, the same calculation applied to smoking would yield an estimate of 90% for the total attributable to that factor. If one were naively to add these two figures, it would appear that we had accounted for more than 100%, but of course this is simply because we had counted the interaction contribution twice. A more appropriate way to think about these calculations is that 66.7% of cases in this group could have been prevented if the environmental hazard had not occurred (or 90% if they had not smoked, or 97% if both causes were eliminated). We will revisit this point below when we discuss their implications for compensation policy.

In nonsmokers, there is only a single factor to consider, so we would compute their AR_E as $(3 - 1)/3 = 67\%$, the same value as the combined attributable risks for exposure alone and acting jointly with smoking among smokers. This is not a coincidence, but a logical consequence of the multiplicative model: under that model, the RR for exposure is the same for smokers and nonsmokers, so the AR_E does not depend on smoking status. In other words, if a multiplicative model holds, an individual's smoking status should be irrelevant in determining eligibility for compensation.

Under a purely additive model, the AR_{Int} becomes zero. Thus, if RR_{joint} were 12 in our hypothetical example, we would obtain $AR_E = 2/12 = 16.7\%$ in smokers with no additional interactive component. Among nonsmokers, however, $AR_E = 2/3 = 67\%$, revealing that under an additive model (or any less-than-multiplicative model), the AR_E will be larger for nonsmokers than for smokers.

The AR_Es computed in this manner have the same interpretation as probabilities of causation for individuals who are otherwise indistinguishable from all cases in the group from which these parameters were estimated. Thus, the total PC attributable to exposure is estimated by $AR_E + AR_{Int} = (RR_{joint} - RR_S)/RR_{joint}$ or $(30 - 10)/30 = 66.7\%$ in our hypothetical illustration.

The legal tradition that supports this interpretation for determining compensation policy is based on the notion that liability for harms caused should be based on whether the disease would not have occurred "but for" the fact that exposure had occurred. As explained above, the elimination of exposure would have eliminated both the excess cases due to exposure acting alone and those due to its joint action with smoking. Hence, both components should be counted in a PC calculation for purposes of deciding whether compensation should be provided to exposed cases among smokers. Another legal tradition that appears to support this interpretation is that compensation is due whether exposure either caused *or* contributed to a disease; here, the interactive effect can be interpreted as "contributing to" or "aggravating" a condition that may have been initially caused by something else. Some have argued that if both entities are liable for harm caused by their actions, then this interactive contribution should be shared in some way. However, it is rare that both claims would be considered concurrently, so in considering a claim for compensation for an environmental toxin, the entire smoking interaction effect should be included and conversely when a liability claim for tobacco use is considered.

In establishing a presumption, one might consider whether the definition of the class who will be entitled to compensation should involve other factors, for example, whether lung cancer claims by exposed veterans should be limited to nonsmokers. Setting aside the practical difficulties of documenting an individual's smoking history, compensation policy has seldom made such restrictions (an exception is the Radiation Exposure Compensation Act of 1990, subsequently amended to eliminate this restriction). Absent evidence that exposure has *no* effect in smokers, the total attributable risk for exposure will still be greater than zero for smokers, even though it may be smaller than that for nonsmokers. Since the main reason for establishing a presumption for a class in the first place is the difficulty of computing a reliable PC estimate for individuals so a decision has been made to compensate the entire class, it seems to defeat the purpose to try to define separate classes based on weak epidemiologic evidence for major differences in AR_E.

Loss of life expectancy

The basic calculation of LLE was described in the previous chapter. To recap, an individual's life expectancy (LE) under a given exposure scenario Z is given by

$$\text{LE}(Z) = \int_0^\infty S(t|Z)\,dt = \int_0^\infty t\lambda(t|Z)S(t|Z)\,dt$$

where $S(t) = \exp(-[\int_0^t \lambda(u|Z) + \mu(u)]\,du)$ is the probability of surviving both the cause of interest $\lambda(t|Z)$ and all competing causes $\mu(t)$. Thus, an individual's LLE attributable to exposure is simply $\text{LLE} = \text{LE}(0) - \text{LE}(Z)$. For this purpose, one might substitute for $\lambda(t|Z)$ a fitted model derived from epidemiologic data for the predicted hazard function given the specifics of the individual's exposure history. As we shall see below, however, this entails an assumption of homogeneity of the population risks that could be questionable.

Estimability of PCs and LLEs

In an important series of papers (Greenland and Robins 1988; Robins and Greenland 1989a, b; 1991), Robins and Greenland have argued that the PC is not estimable for individuals without making unverifiable assumptions concerning biological mechanisms and the extent of heterogeneity between individuals. In particular, they have demonstrated scenarios in which the relative risk could be arbitrarily close to 1 (and hence the PC would be close to zero), yet all cases were affected by exposure in the sense that their death times had been advanced by some small amount. They have also shown that even the population mean of a heterogeneous distribution of PCs is not estimable and that compensation schemes that pay in proportion to the PC are neither "robust" to model misspecification nor "economically rational." By robust, they mean a payment scheme under which the amounts paid to those individuals *actually harmed by exposure* are the same whether the model is misspecified or not. By economically rational, they mean that the total amounts paid to *all* individuals are the same whether the model is misspecified or not. Thus, while a truly robust scheme may be unattainable, an economically rational one would at least pay out the same total amount under a misspecified model as would have been paid under the correct model, even though the payments to specific individuals might be inequitably distributed under the misspecified model.

Under somewhat weaker conditions, however, they showed that individuals' *average* LLE can be validly estimated from epidemiologic data, but not *individually*, conditional on their observed ages at death. Thus, a compensation scheme based on the average LLE is robust, equitable, and economically rational in the sense that it will yield the targeted level of compensation on average (even if the statistical risk model is incorrect), although it will err upwards or downwards for specific individuals.

While these points are well taken and mathematically correct, these theoretical problems of estimability need not necessarily invalidate the usefulness of the PC for decision making. Realistic degrees of heterogeneity may produce only moderate bias, and in particular, Thomas (2000) has argued that the PC may nevertheless provide a reasonable *ranking* of claims.

To develop these ideas more formally, we follow the notation of Robins and Greenland (1989b) and let $\lambda_i(t, Z)$ denote the hypothetical hazard rate that individual i would experience if he were exposed to Z. Then the individual's true probability of causation is

$$p_i = \frac{\lambda_i(t_i, Z_i) - \lambda_i(t, 0)}{\lambda_i(t_i, Z_i)}$$

Of course, these individual hazard functions are not directly observable, and hence neither is p_i. The observable population rate is given by

$$\lambda_*(t, Z) = E[\lambda_i(t, Z)] = \frac{\sum_i \lambda_i(t, Z_i) S_i(t, Z_i)}{\sum_i S_i(t, Z_i)}$$

where $S_i(t, Z)$ is the corresponding individual survival function. The "naive PC" (or "rate fraction" in the terminology of Greenland and Robins) is given by

$$\tilde{p}_i = \frac{\lambda_*(t_i, Z_i) - \lambda_*(t_i, 0)}{\lambda_*(t_i, Z_i)}$$

which in general will not equal $E(p_i)$ if there is heterogeneity between individuals in their baseline hazards. In fact, they show that $\tilde{p}_i < E(p_i)$.

Greenland and Robins also discussed models in which all individuals' death times are advanced by an amount that depends upon exposure. For example, suppose *every* case's date of death was advanced by exposure, but the amount that was so small that the excess relative risk (and hence the PC) was very small. They interpret this situation as implying that exposure "contributed to" disease in 100% of the cases—and hence some compensation would be deserved—yet no individual would qualify under a PC-based criterion that required at least a 50% probability.

This idea can be formalized in terms of the "accelerated failure time model", in which $S_i(t, Z) = S_{0i}(t - Z_i\beta)$, where $S_{0i}(t)$ is, for argument sake, the survival curve derived from a gamma frailty model. Under this model with frailty variance $\theta = 2$ and an acceleration of 2 years for all subjects, Thomas (2000) demonstrated a pattern of declining PCs with increasing age, the naive PC beginning to underestimate the true PC starting about age 50. Thus, under this model, the true PC is not 100%, even though the entire survival distribution has been shifted by a constant two years.

How much heterogeneity is it reasonable to expect might actually exist? Unfortunately, for a nonrecurrent event like death, survival times of independent individuals provide no information about the variability in individual hazard rates. However, such information *can* be obtained from study of related individuals, particularly monozygotic twins, who are perfectly matched on genotype and tend to have experienced similar environments (identical *in utero* and very similar childhood). Frailty models can be used to estimate the variance *between* twin pairs in the component of their baseline risks they share. This can be interpreted as an estimate of the variance in baseline risks between unrelated individuals. Data on Danish identical twins leads to an estimate of about a 7-fold range in multiplicative factors ("frailties") between 80% of individuals (Hougaard et al. 1992).

Despite the bias due to differential survival, the strongest determinants of an individual's PC are still dose and age at death, so it is reasonable to inquire whether the naive PC could still be used to *rank* individuals. Using the frailty variance estimate from the Danish study, Thomas (2000) estimated the correlation between $\tilde{p}_i$ and p_i as 0.995 under the accelerated failure time model and 0.969 under the conditional proportional hazards model. Even the correlation between the true PCs under the true accelerated failure time model and that under a misspecified constant RR model was still 0.971. Thus, it appears that the naive PC can indeed be used to *approximately* rank individual's claims.

Schemes for resolving compensation claims

The "balance of probabilities" principle in tort law has been widely interpreted by the courts as requiring that the point estimate of the PC (or AS) be at least 50%. A number of authors, including the 1984 Oversight Committee, have pointed out various inconsistencies that can arise in a compensation scheme that provides full payment to those with greater than 50% PC and nothing to those with PCs less than 50%. For one thing, the difference between the treatment of two individuals with very similar claims, one of whom had the good fortune to have an estimated PC of 51%, the other one with the bad luck to have a PC of 49% seems unfair, particularly view of the inherent uncertainty of these estimates. Furthermore, situations could arise where hazardous exposures were high enough to cause a substantial proportion of disease in exposed individuals, yet none of them would be entitled to compensation under such a scheme. This could provide an employer with no incentive to lower exposure levels beyond what would be needed to avoid compensation claims, while still allowing them to cause an unacceptable burden of disease.

Various alternative compensation schemes have been proposed, most based on some kind of sliding scale in which the amount of compensation awarded depends on the magnitude of the estimated PC. For example, the 1984 Oversight Committee suggested a scheme in which individuals with PCs greater than 50% would receive full compensation, those with less than 10% would receive none (to discourage "frivolous" claims), and the award would be linearly scaled by the PC in the range between 10% and 50%. Such a compensation scheme was implemented by British Nuclear Fuels Ltd and the U.K. Atomic Energy Agency even before the 1984 NIH report (Thomas et al. 1991).

It is worth pointing out, however, that in addition to the theoretical objections to PC-based compensation schemes raised by Greenland and Robins, such schemes do not distinguish between those whose lives might have been shortened only a little or a lot. In this sense, an LLE-based scheme is more attractive. Consider two populations with the same absolute excess risk of cancer due to radiation exposure but different baseline risks. In the high baseline risk population, the PC will be lower because it is more likely that the cancer was caused by factors other than radiation. On the other hand, the expected LLE due to radiation in the total population (cases and survivors combined) would be approximately the same in the two populations. Thus, the expected LLE due to radiation *amongst all cases* would also be higher in the low-risk population. A policy based on either the PC or the LLE would therefore correctly favor the low-risk population but would do so differently: a PC-based policy would reward *a higher proportion* of cases, but the size of the award would be the same for compensable cases in both populations; an LLE-based policy would reward *all* cases in both populations, but the average size of the award per case would be higher in the low-risk population. Which is a more equitable solution is an important policy decision.

Dealing with uncertainties in compensation

Estimates of either the PC or the LLE are inherently uncertain. Uncertainties include (i) statistical sampling errors in the estimated relative risk parameters, whose variance can be directly computed from the data, (ii) various potential biases and sources of uncertainty such as measurement error, dose-rate effects, and risk transport problems, where some data may exist to suggest a reasonable range of values, and (iii) other uncertainties for which it is impossible to justify any particular choice of value. Setting aside the last of these, one could in principle express the effects of chance and other quantifiable uncertainties in the form of a confidence limit on the PC, but this raises a number of important policy implications. Some agencies such as the Veterans Administration (VA) have used an upper confidence limit on the PC (the 99% percentile for the VA), at least as a screening criterion to weed out claims that had very little chance of being successful. In practice, however, few claims that attain the VA's screening criterion appear to have been eliminated in the subsequent adjudication process, presumably because there is little additional information that can be brought to bear on the question of causality that has not already been taken into account. Hence, the upper 99% percentile is effectively the *de facto* compensation criterion.

Under such a policy, are two individuals with the same point estimate of PC = 55%, but with confidence limits (52–57%) or (45–65%), equally deserving of compensation? What about an individual with PC = 75%

(95% CI 0–90%), reflecting a high probability, but based on a nonsignificant association? Or conversely, one with PC = 40% (95% CI 10–95% based on a significant association), whose best estimate fails to attain the 50% criterion, but, taking account of the various uncertainties, could have a value well over 50%?

A policy based on upper confidence limits has the merit of liberality, in that few claimants whose true PC was greater than 50% would be denied compensation as a result of uncertainties bringing their naïve PC below the threshold. But this should be weighed against the anticipated number of payments to individuals with low "true PCs" that would be provided at the expense of all taxpayers or at the expense of other groups who may be more deserving of government benefits. Arguably, the loss to these other groups is individually minuscule in comparison to the benefit to those who do merit compensation but might otherwise be denied it under a more restrictive policy. Even under a policy which aims at equity rather than resolving uncertainties in favor of the claimant, there should be a reasonable balance between "false positive" and "false negative" decisions. Using the full probability distribution produced by uncertainty propagation methods like that used in IREP is one way of accomplishing this.

Use of upper confidence limits also has the unsatisfactory feature of favoring claims for which the evidence of a causal association is the weakest. At the individual level, for example, a claimant whose PC was 45% with confidence limits 42–48% would lose, while one with a PC of 10% and confidence limits 0–90% would win, even though there is stronger evidence of a population association for the former than the latter. Schemes that pay in proportion to the expected value of the PC or the posterior probability that the PC is greater than 50% to some extent avoid this difficulty.

Such a policy also can favor *groups* with the weakest evidence of causation at the population level. Consider the following two scenarios:

1. A common cancer for which the relative risk is estimated to be 1.8 with 95% confidence limits (1.7–1.9), translating to a PC of 44% (CI 41–47%).
2. A rare cancer for which the RR was estimated at 1.1 with 95% confidence limits (0.22–5.5), translating to a PC of 9% (CI 0–82%).

Clearly, the first case provides much stronger evidence of a causal association in the population (highly statistically significant), even though the assigned share for the group as a whole is less than 50%. In the second case, it is not at all clear that there even is a causal connection in the population, as the association is far from statistically significant and the estimated assigned share is much lower. Nevertheless, under a policy that provides full compensation to those with upper 99% confidence limits on the assigned share of at least 50%, none of the members of the first group would obtain compensation while all of those in the latter would be successful. In short, such a policy would appear to reward ignorance.

This observation has important implications for the question of how to select and group cancer sites at the stage of developing the scientific evidence to support policy recommendations. Unnecessarily fine grouping would tend to produce upper confidence limits on associations that are unreasonably high in relation to what might be expected based on similar cancers and would tend to favor compensating cancers for which there is little or no evidence of a dose–response relationship (or even a negative relationship) over those for which there is much stronger evidence of a positive relationship. On the other hand, for some exposures like radiation, virtually all cancer sites may be radiosensitive, but to somewhat different extents. Failing to group rare sites, for which significant evidence of an association would be difficult to obtain, would unfairly penalize individuals with these cancers under a policy of requiring a significant association.

Questions remain about how to handle the remaining components of uncertainty that cannot be quantified, yet undoubtedly exist to some extent and about how such uncertainties should be addressed in resolving compensation claims. While it would be impossible to anticipate all possible sources of uncertainty, such a list for radiogenic cancers might include differences between populations (e.g., the United States and Japan) in the ERR coefficients that are being transported, misspecification of the form of the fitted models, uncontrolled confounding and other biases in the epidemiologic studies, and so on. As noted at the beginning of this chapter, the decision about potential causation needs to be addressed at the population level before considering the merits of an individual's claim. This includes consideration not just of the level of statistical significance, but also such subjective features as the reliability of the epidemiologic data base (freedom from bias, etc.), biological plausibility, and other criteria that are widely used to assess causality in epidemiologic associations.

17 Further challenges

Reproductive endpoints

Although most of this book has been concerned with chronic diseases, the developing fetus is particularly sensitive to environmental insults, as is male and female fertility. Studies of the reproductive effects of environmental exposures pose unique methodological challenges.

There are several distinct outcomes of pregnancy that are potentially of interest:

- Spontaneous abortion and stillbirths
- Congenital anomalies
- Prematurity
- Low birth weight (small for gestational age)
- Failure to thrive after birth

In addition, various maternal conditions, such as preeclampsia, can affect both the mother's health and her child's. These various endpoints can be interrelated in complex ways. Fetuses with anomalies are more likely to spontaneously abort, but all that may be recorded might be the spontaneous abortion event, not the presence of an anomaly. Indeed, a high proportion of unrecognized pregnancies end in early fetal loss, so neither the numerators nor the denominators may be completely ascertained. Gestational age can be an outcome in itself, a determinant of birth weight and subsequent growth, or the consequence of some anomaly. Statistical analysis of these interrelated endpoints requires particular care to avoid paradoxical results.

The best way to overcome the problem of early fetal losses is to use a cohort study design in which all women in some defined population (e.g., a health maintenance organization) who are *attempting to get pregnant* are invited to participate and enrolled in the cohort on the date of their first positive pregnancy test. Of course, even this is no guarantee that very early losses (within a month or so of last menstruation) will be detected, but at least a relatively complete record of first-trimester spontaneous abortions will be available. Such data are best treated as censored survival data, with the endpoint being spontaneous abortion or stillbirth and a live birth treated as a censoring event. In this analysis, gestational age is the time scale—women entering the cohort on the date of diagnosis of the pregnancy to avoid bias due to early fetal loss. Thus, at each gestational age, the comparison is between those known to be pregnant who aborted at that age and those whose fetuses survived beyond that point (irrespective of whether a spontaneous abortion or stillbirth occurred later or whether there was a congenital anomaly).

Congenital anomalies are systematically assessed only for live births, so must be treated as binary outcomes, with gestational age handled as a covariate. Because of the complexity of the relationship between anomalies and gestational age, however, one might wish to handle that part of the model flexibly, say with a generalized additive model, while the relationship to exposure is handled parametrically. Likewise, birth weight can be treated a continuous outcome, adjusted parametrically or nonparametrically for gestational age, or as a binary outcome using some percentile of the distribution of birth weight by gestational age. The presence of a congenital anomaly may be recorded for some spontaneous abortions or stillbirths, but unless this is done systematically and blindly with respect to exposure, there is considerable risk of ascertainment bias. For large cohort or case-control studies, systematic ascertainment of congenital anomalies in spontaneous abortions and stillbirths is generally only possible in subsamples. If available, however, such data could be used in a multilevel model treating the presence or absence of an anomaly as a latent variable for those who are not in the subsample.

Another common difficulty with the analysis of congenital anomaly data is the multitude of possible anomaly types, many quite rare. As their etiologies are likely to be heterogeneous, the question arises how they should be grouped to provide sufficient numbers of events, but also be sufficiently homogeneous. Broad categories based on organ systems and developmental stage are natural choice, but still one must decide how far down the hierarchy to draw the line. Hierarchical Bayes models are attractive, as one can allow an established hierarchy to determine the potential subdivisions, while letting the data determine whether groups are similar enough to be validly pooled and ensuring that the uncertainly about the degree of pooling will be taken into account in the final variance estimates. Having time-resolved exposure data would further inform such an analysis, as one could test whether exposure associations are strongest during the critical gestational ages for each endpoint.

These issues are illustrated in a study of the effects of aerial application of the pesticide Malathion to combat an infestation with the Medfly in the Bay Area of California in 1982–84 (Thomas et al. 1992c). A cohort of 7450 women who were registered with one of three Kaiser–Permanente facilities in the Bay Area and confirmed as pregnant during the spraying period was enrolled and followed using the organization's clinical records. While it is possible that some women opted for care outside the Kaiser system and their outcomes unknown, there were probably few of these as it was a prepaid comprehensive care organization. A case-cohort design was used to compare the 933 women with adverse outcomes with 1,000 randomly selected cohort members with normal outcomes. Exposure for this subsample was assessed by digitization of maps of the spray corridors and geocoded residence histories, using an algorithm to assign week-by-week exposure indices in relation to the overlap of each spray corridor with circular buffers around each residence. Cox regression with time-dependent covariates was used for the analysis of the spontaneous abortion data, logistic regression for congenital anomalies, and ordinary linear regression for birth weight and gestational age. No associations with Malathion spraying were found for spontaneous abortions, stillbirths, small for gestational age, or most categories of congenital anomalies. There was a 2.6-fold elevated risk of gastrointestinal anomalies with second-trimester exposures, but based on only 13 cases and not specific to any particular ICD code; thus, this could be simply a chance occurrence, considering the number of anomaly groups analyzed

Failure to conceive represents another quite different aspect of reproductive epidemiology, some of which could have an environmental etiology. Either or both of male or female infertility could play a role. Nelson and Bunge (1974) first reported reduced sperm densities for the period 1970–73 compared with those found two decades previously. In 1992, Carlsen et al. (1992) reported what they interpreted as a world-wide decline in sperm counts over the period 1938-90 based on a meta-analysis of 61 studies that had been published to date. This conclusion generated considerable controversy and two further meta-analyses (Becker and Berhane 1997; Swan et al. 2000), which confirmed the decline in North America, but not in developing countries, and differed in their conclusions about whether a decline was seen in Europe, depending upon their inclusion criteria. Hypotheses concerning the marked geographic variation and/or secular trends include environmental endocrine-disrupting chemicals, elevated temperature, and season, amongst other factors. Indeed, negative correlations between sperm density and organochlorine levels in seminal fluid have been reported (Dougherty et al. 1981), as well as with various other environmental exposures. More recently, Sokal et al. (2006) conducted a longitudinal analysis of specimens from donors to a sperm bank in Los Angeles in relation to air pollution levels in the subjects' area of residence during three critical periods of spermatogenesis and found a highly significant inverse associations in each period with ambient ozone concentrations, but not with any other air pollutant, suggesting a possible mechanism mediated by oxidative stress.

Disasters: natural and manmade

In a sense, it could be argued that most of environmental epidemiology is concerned with disasters of one kind or another. As an observational science, epidemiologists do not deliberately expose subjects to hazardous agents, so must content themselves with the aftermath of "natural experiments" in which individuals have been exposed for one reason or another for nonexperimental purposes. Of course, not all exposure scenarios that have been studied would rise to the level that would be called "disasters," but many of those described earlier—the atomic bombs, the London Fog, for example—clearly would. Kinston and Rosser (1974) define a disaster as a "situation of massive collective stress," and Logue et al. (1981) limit their review to "those events which affect whole communities or a sizeable segment of a community." The United Nations defines a disaster as "a disruption of the human ecology that exceeds the capacity of the community to function normally." The U.S. Disaster Relief Act of 1974 (PL 93-288) adopted as an operational definition any of several designated categories of catastrophic events "which causes damage of sufficient severity and magnitude to warrant major disaster assistance." In this section, we review several examples of epidemiologic investigations of monumental disasters and then try to identify some common methodological challenges such studies pose.

Some examples

Tropical cyclones and hurricanes

In November 1970, the severe tropical cyclone Bhola ripped through the East Bengal region of the Indian subcontinent now known as Bangladesh, killing half a million people. A rapid epidemiologic survey (Sommer and Mosley 1972) was mounted that is still widely regarded as a model for post-disaster needs assessment. Two surveys were conducted, the first mounted only two weeks after the cyclone visited 18 sites over five days, the second conducted two months later visited many more sites in greater depth over a three-week period. The first essentially confirmed the immediate adequacy of water supplies and the absence of unexpected morbidity or epidemic disease. (Contrary to popular impressions, outbreaks of infectious diseases are not commonly associated with cyclones, floods, and similar disasters (Shultz et al. 2005).) Using cluster sampling methods, the second survey estimated the age/sex-specific death rates, and calculated that there would have to have been at least 224,000 deaths (the official death toll is listed at 500,000 and the true number may have been even higher). They also estimated that 180,000 homes had been destroyed, 600,000 individuals were still homeless, and 1,000,000 dependent on outside food aid. Data on crop and livestock losses and needs to re-establish agricultural self-sufficiency were also provided. Arguably, lessons learned from this experience and the epidemiologic methods established were responsible for the dramatically lower mortality from the cyclone Gorky that hit the same region in 1991 (138,000 deaths) and cyclone Sidr in November 2007, the latter producing fewer than 5000 deaths despite an intensity rivaling the two 1970 and 1991 ones. In the 1980s, a network of shelters was established, but early warning systems proved inadequate to alert the population to reach them in time for the Gorky cyclone. The much lower death rates for the Sidr cyclone could be attributed in large part to improvements in early warning systems. Sadly, as this book was going to press, Cyclone Nargis hit Myanmar on May 2–3, 2008, causing a death toll that could exceed 100,000, illustrating that lessons about early warnings and speedy delivery of aid could still be learned.

The United States experienced the worst hurricane season on record in 2005, with 15 tropical storms becoming hurricanes, three of these reaching Category 5 (sustained winds over 155 mph), and four Category 3 or higher making landfall. One of these, Hurricane Wilma, was the most intense ever measured, but the damage from Hurricane Katrina was unprecedented, due to the breaching of the levees surrounding New Orleans resulting in flooding of 80% of the city. Deaths were estimated at 1000 in Louisiana and 220 in other states, the worst since the 1928 Florida hurricane (1828 deaths) and the 1900 Galveston Texas hurricane (more than 8000 deaths, the worst in U.S. history). With property damage in the billions of dollars, it was also the costliest on record. For details of the public health response, see the series of Morbidity and Mortality Weekly Reports published by the Centers for Disease Control (CDC 2006). Despite the potential utility of such reports for developing guidelines for future disaster preparedness and recovery, the CDC notes the difficulty of calculating rates because of lack of suitable denominators, variability in reporting sites over time, lack of specificity of numerators, and the subsequent dispersal of the population.

A review of epidemiologic methods for assessment of tropical cyclone effects (Shultz et al. 2005) concludes that there has been little advance in epidemiologic methods since the 1970 East Bengal studies. Most studies have been short-term assessments, postimpact needs and surveillance of mortality, injuries, and infectious disease. However, their cyclical nature raises the possibility of establishing networks for continuous surveillance in high-frequency impact zones to allow prospective, longitudinal evaluation. The authors also suggest greater reliance on case-control methods for analyzing the effects of specific storm exposure attributes and preparedness behaviors, perhaps combined with more objective assessment of physical forces, and more attention to vulnerable subgroups.

Asian tsunami

The magnitude 9.1–9.3 Great Sumatra-Andaman Earthquake off the coast of Indonesia on December 26, 2004 caused one of the worst tidal waves on record, affecting sites around the Indian Ocean as far away as Africa. Most heavily hit was the province of Aceh in Indonesia, already reeling from civil conflict, where the official toll was listed as 130,000 dead and 40,000 missing, with another half a million persons homeless. Such numbers were derived in part from a multistage cluster sample of nine affected districts of Aceh conducted in February, March, and August of 2005 (Doocy et al. 2007). Sampling proportional to estimates of the number of internally displaced persons was used, identifying clusters based on lists of known displaced-persons' locations (camps and host communities), selecting 20 clusters in each survey and 20 or 24 households with at least one survivor within each cluster. Estimate of crude death rates ranged from 5.3% to 23.6%, depending upon location, and were highest in the very young and elderly, and 44% higher in females than males. For contrasting views of the utility of death toll estimates (see Fleck 2005; Thieren 2005).

California and Indonesia fires

In October 2003, a series of devastating wildfires burned more than 3000 km^2 in Southern California, including several areas close to the communities participating in the Children's Health Study. To assess the health effects of these short-term, high-intensity exposures to the resulting particulate pollution, a special questionnaire was administered to over 6000 children in 16 communities (Kunzli et al. 2006). These outcomes were correlated with the daily PM_{10} data from the already-established central-site monitoring stations. Risks of all respiratory symptoms, including cough, bronchitis, wheezing, and asthma attacks, medication usage, and physician visits increased monotonically with the number of reported smoky days and with measured PM_{10} levels. Interestingly, associations were stronger for nonasthmatics, as asthmatics were more likely to wear masks or stay indoors during the fire.

Much more massive were the peat fires in Indonesia in 1997 that produced widespread haze. Source apportionment and chemical mass balance methods demonstrated that peat smoke could travel long distances (See et al. 2007), producing "very unhealthy" to "hazardous" levels of carbon monoxide and particulates, as well as greatly elevated levels of polycyclic aromatic hydrocarbons (Kunii et al. 2002). Several epidemiologic studies (Aditama 2000; Kunii et al. 2002; Sastry 2002; Frankenberg et al. 2005) found significant differences in respiratory health between haze and nonhaze areas as far away as Malaysia.

Bhopal

The release of about 30 tons of methyl isocyanate gas from the Union Carbide plant upon the sleeping city of Bhopal India on December 3, 1984 caused the worst industrial disaster in history. Despite early warnings before the accident, Union Carbide failed to implement adequate safety systems or to warn the surrounding community. The actual death toll will probably never be known, but is estimated to have been about 3800 instantly, perhaps 2–3 times that number in the next few days, and even more *excess* deaths in the following decades. The aftermath—the legal maneuvering by Union Carbide to avoid liability, the failure to establish a large-scale, long-term epidemiologic research program, the inadequacy of the compensation provided to the victims—is shameful and instructive. The history of the catastrophe and its aftermath is compelling as told by Broughton (2005). The causality of early deaths and injuries was never in serious dispute and these victims did eventually receive compensation from Union Carbide in a settlement of $470 million arranged by the Supreme Court of India (averaging $2200 to families of the dead, the rest distributed across half a million injury claims).

Various independent epidemiologic research organizations converged on the scene and conducted about a dozen cross-sectional surveys. For reviews of this early literature, see (Mehta et al. 1990; Dhara 1992; Dhara and Dhara 2002). A cohort of about 80,000 exposed survivors was enrolled for long-term follow-up. The Indian Council of Medical Research coordinated this activity, but its toxicology and human health effects reports were never released, although a few brief epidemiology papers were (Vijayan et al. 1995; Vijayan and Sankaran 1996). A completely separate International Medical Commission on Bhopal, a group of medical professionals, did a follow-up study ten years after the event, which was published in the open literature (Cullinan et al. 1997; Dhara et al. 2001; Dhara et al. 2002). With the confiscation of some Union Carbide assets, the Indian government established the Bhopal Memorial Hospital and Research Center (BMHRC) with a mandate to provide free care to exposed victims suffering from ailments on a list of designated effects (mainly respiratory, gastrointestinal, ocular, or neurological). The research activity, however, did not get established until 2004, originally with the aim of reactivating the epidemiologic follow-up of the survivor cohort, but soon shifted its focus to basic research. A new cohort of individuals who were *in utero* during the event has been established at BMHRC, however, and has shown significant elevations in a broad spectrum of immunologic markers (Mishra et al. 2007).

In addition to the health effects listed above, clear associations with chromosomal abnormalities have been demonstrated and the known genotoxic effect of the gas raises concern about a potential cancer risk. To date, only one earlier study has been published (Dikshit and Kanhere 1999), showing only moderate, nonsignificant excesses of lung and oropharnyx cancers through 1992, but it is obviously still too early to assess this risk. High rates of fetal loss have also been reported, concordant with experimental studies in rats (Varma 1987). As attention shifts to chronic effects of exposure, the question of causality becomes increasingly important, particularly considering the dearth of epidemiologic evidence from other contexts about the effects of short-term exposures to this or other toxic gasses. The difficulties of causal inference are exacerbated by the fact that this is a cohort of *survivors*, and hence potentially selected in favor of more resistant individuals. Such differential selection, if it exists, is likely to be stronger at higher exposures, thereby diluting exposure–response relationships relative to what would have been seen in the entire exposed population, possibly even producing apparently protective effects of exposure! Compounding this problem is that these are also cohorts of nonmigrants, it being likely that out-migration rates could also be related to both exposure and health status.

Most epidemiologic publications have relied on simple exposed versus unexposed comparisons (often with control groups of questionable comparability) or simple gradients with distance from the plant. An atmospheric dispersion model has been constructed (Singh and Ghosh 1987), but has not been used to date in any epidemiologic analyses. In lieu of this, Dhara et al. (2002) compared a variety of individual exposure indices combining distance, time, activity levels, and protective measures (such as whether the windows were open or they ran outside) in various ways. Exposure surrogates such as the number of deaths in the household were also collected on all cohort members at the time of enrollment and have been used in some other analyses.

World Trade Center attack

The September 11, 2001 attack on the New York World Trade Center (WTC) produced, amongst other effects, an environmental disaster with broad health consequences for both the exposed survivors and the rescue workers. Exposure assessments by the U.S. Environmental Protection Agency (EPA) (Lorber et al. 2007) and independent investigators (Landrigan et al. 2004), based on sampling of ambient air and outdoor and indoor settled dust combined with satellite imaging and modeling of the atmospheric plume, found extremely high concentrations of particulates, cement dust, lead, PAHs, PCBs, asbestos, synthetic vitreous fibers, and dioxins immediately after the collapse. Levels remained significantly elevated for several days after, although only limited data were available during this period. The EPA assessment concluded that most members of the general public would be unlikely to have experienced adverse health effects from exposures beyond this period, although concerns about underestimation of asbestos contamination persist. Various epidemiologic investigations found significant associations with various respiratory symptoms and asthma among both clean-up workers and the general population, and low birth weight offspring of women exposed in pregnancy (Landrigan et al. 2004; Mauer et al. 2007; Tao et al. 2007; Wheeler et al. 2007). Experimental instillation of WTC dust samples into mice produced bronchial hyper-reactivity but little inflammation, consistent with its high alkalinity. See Samet et al. (2007) for a review and editorial comment.

Iraq civilian death toll

Following the invasion of Iraq in 2003, a team from Johns Hopkins Bloomberg School of Public Health conducted two surveys of civilian deaths. The first (Roberts et al. 2004) surveyed 33 clusters of 30 households each and found all-cause mortality rates during the year and a half after the invasion to be 2.5-fold higher than in the previous year. They estimated the total excess deaths at about 100,000, even excluding the one outlier cluster from Fallujah that had been randomly selected, and far more if it were included. The second survey (Burnham et al. 2006) included 50 clusters from 16 Governorates, each comprising 40 households and estimated even more excess deaths—about 650,000 (95% CI 393,000-943,000), the vast majority due to violence. These amount to 2.5% of the total population of the study area.

Various methodological criticisms have been suggested (Bohannon 2006; Hicks 2007; von Schreeb et al. 2007), notably a concern that smaller streets may have been underrepresented. The authors have vigorously rebutted these criticisms (Burnham and Roberts 2006; Roberts and Burnham 2007). Nevertheless, these issues highlight the practical difficulties posed by research in a war zone, where, for example, protection of the confidentiality of the respondents and the safety of the interviewers becomes paramount.

In addition to the civilian death toll, the Gulf and Iraq wars produced serious environmental hazards, such as pollution from the Kuwait oil well fires (Lange et al. 2002) and depleted uranium from munitions (Pearce and Cardis 2001).

Refugees

Although perhaps not an "environmental" disaster *per se*, refugees from either massive natural disasters or armed conflicts can pose a humanitarian crisis with serious public health consequences (Murray et al. 2002). Notable are several instances of genocide in the last few decades—Rwanda, Bosnia, and Darfur, for example. In the latter instance, 1.6 million people had been displaced and 30,000 killed in the immediate attacks, countless more due to starvation, communicable disease, and other problems in the subsequent migration and encampment. Here, the tools of descriptive epidemiology go beyond simply the description of the numbers of persons displaced and their morbidity and mortality to guide the provision of humanitarian relief, but can also address the legal question of whether the situation rises to the level of "genocide." Leaning (2004) describes how the 1946 Convention on the Prevention and Punishment of the Crime of Genocide requires two elements—intent and acts—to be present and, if met, obliges signatories to the Convention to intervene to stop the attacks. (It is this latter requirement that makes governments so cautious about declaring a crisis to be genocide.) Standard survey sampling methods (Brown et al. 2001; Grais et al. 2006) can be applied to describe the social, economic, and demographic context, estimate mortality and nutritional status, and infer causal connections. For example, Depoortere et al. (2004) describe cluster sampling surveys in Darfur that found death rates before arrival at refugee camps ranging from 5.9 to 9.5 per 10,000 per day (the majority due to violence), somewhat lower in the camps (up to 5.6 per 10,000 per day), but still above the widely used "emergency" benchmark of 1/10,000. The authors note that when death rates are high, entire families may disappear and hence not be reflected in numerators or denominators, biasing estimated rates downwards. (Recall a similar problem in the tsunami study above.) In principle, this problem could be overcome by likelihood-based estimation based on the conditional probability of the numbers of deaths in each family given that there is at least one survivor, but this would require additional assumptions about the distribution of family sizes, dependency of deaths within families, and representativeness of the sample. Furthermore, recall bias could bias estimates upward or downward. Although clear operational guidelines are lacking, epidemiologic methods can be used to make inferences about intent to build the case for designation of a crisis as genocide. Epidemiologists and other public health workers can make valuable contributions to addressing such situations.

Chernobyl

April 26, 1986, saw the world's worst nuclear accident, when two explosions at the Chernobyl plant in Ukraine released more than six tons of radioisotopes, including iodine, cesium, and strontium. Naturally, the heaviest exposures occurred to the workers at the plant itself, the emergency and clean-up workers ("liquidators"), and immediately surrounding population, 116,000 of whom were evacuated immediately, another 220,000 after 1986. Another five million persons continued to live in contaminated areas of Belarus, Ukraine, and the Russian Federation. The contamination was also widely dispersed across Europe (Figure 17.1). Average thyroid doses to children under 1 year of age ranged from 0.01 mGy in Portugal to 750 mGy in parts of Belarus, while average doses from external radiation plus ingestion of from long-lived cesium reached 10 mSv in parts of Belarus and Russia (Drozdovitch et al. 2007).

While the total burden of cancer will not be known for many years, marked excesses of thyroid cancer have already been shown [reviewed by

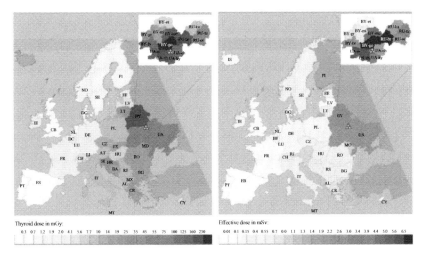

Thyroid dose in mGy:

0.3 0.7 1.2 1.9 2.0 4.1 5.6 7.7 10 14 19 25 35 45 55 75 100 125 160 230

Effective dose in mSv:

0.01 0.1 0.15 0.4 0.55 0.7 0.0 1.1 1.3 1.4 1.9 2.2 2.6 3.0 3.4 3.9 4.4 5.0 5.6 6.5

Figure 17.1. Estimated doses from radioiodine to the thyroid (left panel) and effective whole-body doses 1986–2002 from the Chernobyl accident. (Reproduced with permission from Drozdovitch et al. 2007.)

Cardis et al. (2006a)]. Using a case-control design, Cardis et al. (2005) studied thyroid cancer in the Ukraine, Belarus, and the four most heavily exposed regions of the Russian Federation, and found a strong linear dose–response relationship [RR at 1 Gy 5.5–8.4 depending on the risk model, compatible with an estimate of 7.7 derived from a pooled analysis of seven other studies (Ron et al. 1995)]. The excess risk was three-times higher in iodine deficient areas, but was reduced three-fold amongst those who were given potassium iodide. Results for leukemia and other cancers have so far been inconclusive.

A 2006 report by the United Nations estimated that no more than 4000 deaths would occur as a result, but the uncertainties in this number have been subject to much criticism. For example, this number related only to the 600,000 most heavily exposed people, with a similar number of deaths projected among the 7 million others living further away. Applying standard dose–response models to the estimates of the dose distribution for the European population, Cardis et al. (2006a,b) estimated that about 1000 thyroid cancers and 4000 other cancers attributable to Chernobyl (or 0.01% of all cancer deaths) have occurred within the first 20 years after the accident, and project an ultimate toll for the population of Europe (570 million) at 16,000 cases of thyroid cancer (95% uncertainty interval 3,400–72,000) and 25,000 cases of other cancers (11,000–59,000), against a background of several hundred millions. Such an increase would be virtually impossible to detect by routine surveillance of cancer rates, emphasizing the need for long-term follow-up of the most heavily-exposed cohorts of liquidators, evacuees, and nearby residents who were not evacuated. See Sali et al. (1996); Moysich et al. (2002); Hatch et al. (2005); Cardis (2007) for further reviews of the Chernobyl experience.

In contrast, releases from the Three Mile Island accident in Pennsylvania in March 1979 were relatively low and no convincing associations with cancer have been found (Hatch et al. 1990). In general, most studies have been negative, although a few marginally significant trends have been found (Wing et al. 1997; Talbott et al. 2003), but see (Hatch et al. 1997) for a critique of Wing et al.'s reanalysis of their 1990 data. These associations are potentially confounded, however, by the high levels of domestic radon in the area (Field 2005). The event clearly generated considerable psychological stress (Prince-Embury and Rooney 1988), and interestingly, a weak association of cancer rates in 1982–83 with proximity to the plant was found—weaker, but still elevated after adjustment for pre-accident rates—but this association could not be explained by estimated radiation doses (Hatch et al. 1991).

Similar to nuclear weapons manufacturing and testing facilities in the United States (e.g., the Hanford (Davis et al. 2004) and Nevada Test Site (Stevens et al. 1990; Kerber et al. 1993) studies described in earlier chapters), the Soviet nuclear program left behind a legacy of contamination, but on a much larger scale. The Techa River cohort study (Krestinina et al. 2007) of the population near the Mayak nuclear weapons facility at Chelyabinsk in the southern Urals found a strong association of individual dose estimates with solid cancers (ERR/Gy = 1.0 (95% CI 0.3-1.9) in a linear dose–response model) and leukemia (ERR/Gy = 6.5 (1.8–24) excluding chronic lymphatic leukemia), accounting for about 3% of solid cancers and 68% of leukemias. Excesses of thyroid nodules ($RR = 1.4$ comparing exposed and unexposed groups) have also been found (Mushkacheva et al. 2006). Mayak workers themselves also suffered excesses of lung, bone, and liver cancers (Gilbert et al. 2000; Koshurnikova et al. 2000; Gilbert et al. 2004). Studies around the Semipalatinsk nuclear weapons test site are on-going (Gilbert et al. 2002).

Global warming

The "Mother of All Disasters" could be the threat of global warming. The United Nations Intergovernmental Panel on Climate Change (Houghton et al. 2001) predicts an increase of 1.8–5.6°C and a rise in sea levels of 9-88 cm over the next century. Evidence for the existence of the phenomenon and the contribution of human activities (greenhouse gas emissions and loss of carbon sinks like tropical rain forests) is compelling (Keller 2003; 2007), despite persistent skepticism from some critics. Potential human health consequences include heat-related illnesses and deaths (like the thousands of deaths from the 2003 European heat wave), allergic and infectious diseases, malnutrition, effects of extreme weather events and air pollution, drowning from coastal flooding and consequences of population displacement (Haines and Patz 2004). Controversy remains, however, about some of the human health consequences such as the predicted changes in infectious disease patterns (Taubes 1997). Time-series analyses over several El Niño/La Niña cycles have documented strong associations with malaria and cholera incidence in Latin America and Southeast Asia, weaker for other mosquito- and rodent-borne illnesses (Kovats et al. 2003), with potential global consequences for populations affected by drought and other natural disasters (Bouma et al. 1997). See Patz et al. (2005); McMichael et al. (2006) for other reviews of the evidence for the impacts of recent regional climate change events on human health. Methodological challenges include the long time scale over which human activities affect climate and consequent health outcomes, the absence of suitable control groups, the numerous potential health outcomes, and the many nonclimate confounding variables (Patz et al. 2008).

Whether mankind can reach a consensus and commitment to deal with the problem will be crucial to the determining the outcome for future generations. The strategy of "Contraction and Convergence" (Stott 2007) is one such proposal for reducing carbon emissions, involving capping global emissions at a level at or below the globe's carrying capacity and equal allocation of tradable carbon credits to every person in the world, while also preserving carbon sinks like tropical rain forests. Patz et al. (2008) discuss risk assessment strategies for predicting the effects of such intervention on measures of health as disability-adjusted life years (DALYs; Pruss et al. 2001; Zhang et al. 2007, see Chapter 15). Patz et al. point out, for example, that prediction of effects of interventions on health mediated through climate change often do not account for concurrent mediating effects of such nonclimate variables as economic development and demographic changes. The "comparative risk assessment" approach advocated by the World Health Organization (Ezzati et al. 2002; Murray et al. 2003) involves identifying climate-sensitive health outcomes, determining current exposure–response relationships, selecting future climate change scenarios, and estimating the resulting attributable risk. Estimates of the current contribution of climate change to human health number about 5 million DALYs per year (WHO 2004), small in comparison to tobacco (55 million) and unsafe water and sanitation (54 million) (Ezzati et al. 2002), but of course, this is expected to rise in the future.

Methodological issues

In addition to the reviews of specific disasters or categories of disasters mentioned above, several textbooks (Seaman 1984; Noji 1997) and general reviews (Logue et al. 1981; Lechat 1990; Bromet and Dew 1995;

Noji 2005) of disaster epidemiology are worth studying. One review by Dominici et al. (2005a) is notable for its focus on methodological challenges. In the following section, we briefly summarize some of these.

Unpredictability

Disasters like earthquakes, tsunamis, or industrial explosions, occur with little or no advance warning, making careful planning of an epidemiologic investigation and collection of baseline data impossibility. At the same time, the resources needed for conducting epidemiologic research may be diverted to more urgent tasks of rescue and recovery. Needless to say, the safety of survey personnel and their subjects and the practical constraints of working in a disaster area should take precedence over research needs.

Nevertheless, some types of events like tropical cyclones and wildfires occur frequently enough in the same general area to consider setting up surveillance systems that would provide a basis of comparison with the health effects of a disaster when it does occur. Although established for research purposes, not in anticipation of a disaster, the study of the effects of the Southern California wildfires was strengthened by the availability of established cohorts in several communities near the fires (the Children's Health Study described in earlier chapters), not only permitting rapid deployment of a special-purpose field survey, but also estimates of baseline levels of the various health endpoints. This is particularly important as these communities had been selected specifically because of differences in usual ambient air pollution levels and the resulting differences in health endpoints had already been documented.

Scientific objectives

Clarity about the purposes of an epidemiologic investigation is of course essential to select appropriate study designs, measurement tools, and statistical analyses. Much of disaster epidemiology is concerned with collecting data that will be useful for public health response—estimation of mortality and prevalence of adverse outcomes, identifying needs for treatment, predicting future disasters to develop preparedness strategies and surveillance mechanisms. In addition, however, one might wish to characterize the population at risk, determine relevant exposure factors and risk modifiers (e.g., building characteristics, behaviors, support systems, etc.), or predict long-term consequences.

Defining the sampling frame and sampling scheme

It is possible to do good epidemiology under difficult circumstances, as illustrated by several of the case studies above. A key feature is the clear definition of a sampling frame and sampling scheme to avoid the biases that otherwise would be inevitable if subjects selected themselves for study (which would likely over-represent the most heavily exposed and severely affected individuals). This requires unbiased assessment of both numerators and denominators, a task which is often much more difficult to accomplish in developing countries. Where systematic assessment is infeasible for the entire exposed population (let alone a suitable control population), cluster sampling methods can offer an appealing balance between the need for a comprehensive (many clusters spread over a broad area) and in-depth (many individuals in each cluster) assessment, perhaps stratified by known risk factors like age.

Defining the endpoints

Objective measures of outcome and exposure are also critical, as self-reports of each are likely to be biased, particularly if there are hopes of compensation. Thus, objective measures like lung function testing may be less subject to reporting bias than symptoms like cough. For some endpoints, it may be helpful to have multiple, quasi-independent, measures which can be used in latent variable or other methods for allowing for outcome measurement error. Although the majority of studies in disaster epidemiology are concerned with prevalence, ideally one would want to establish a cohort (of exposed and unexposed individuals) and follow them longitudinally to assess incidence of disease and changes in continuous measures of health over time.

Assessing exposure–response relationships

The majority of epidemiologic studies of disasters have relied on relatively simple classifications of exposure, such as exposed or not. Finding a suitable control group that would be expected to have had comparable outcomes—but for the disaster experience—can be challenging. More attention needs to be paid to finer subgrouping within a broadly defined exposed group by degree of exposure, so that exposure–response gradients can be estimated. For this to be possible, it is essential that some exposure assessment be conducted on spatial and temporal scales to support such comparisons, perhaps using exposure modeling techniques to fill in missing observations. In some circumstances, biomarkers of exposure on samples of individuals may also be helpful. While it is likely that exposure–response relationships will be nonlinear, large sample sizes are generally needed to effectively test for nonlinearity and estimate threshold, saturation, or other parameters of the curve. An example discussed in earlier chapters is the assessment of the nonlinear relationship between mortality and temperature that could be useful in predicting the consequences of heat waves.

Global burden of environmental diseases

As part of the World Health Organization's Global Burden of Disease project (Ezzati et al. 2002; 2003; Murray et al. 2003; Rodgers et al. 2004), an environmental working group (Cohen et al. 2005) estimated the relative contributions of various environmental factors to human disease worldwide. They estimated that the two leading environmental causes of death are unsafe water (1,730,000 excess deaths per year) and indoor air pollution (1,619,000/year) followed by urban outdoor air pollution (799,000/year) and lead (234,000/year). This figure for outdoor air pollution translates to about 1.2% of all deaths or 6.4 million person years of life lost (0.5% of the total). For comparison, these "environmental" factors are outranked by underweight (9.5% of all DALYs), unsafe sex, blood pressure, tobacco, and alcohol, with large differences in ranking between developed and developing countries. Overall, the 19 avoidable risk factors considered in the most recent assessment (Lopez et al. 2006) accounted for 45% of global mortality and 36% of global morbidity, suggesting that considerable progress could still be made by appropriate interventions.

Future challenges

It is obviously difficult to forecast what new environmental challenges with potential impact on human health the world will face, beyond those described earlier in this chapter. Bioterrorist attacks have yet to have occurred on a scale meriting epidemiologic study, but there is obvious concern about the risk and need for preparedness (Bossi et al. 2006). Modern methods of toxicogenomics (proteomics, metabolomics, siRNAs, etc.) are being explored as part of the Environmental Genome Project and are likely to play an increasingly important role in such investigations (Bower and Shi 2005). Other more traditional environmental agents may continue to merit attention, likely at lower and lower levels of exposure, pushing the limit of what can be accomplished by epidemiologic methods (Davey Smith and Ebrahim 2001). Investigation of susceptible subgroups is also needed, hopefully inspiring the development of new methods for gene–environment interactions exploiting novel molecular technologies. Finally, globalization has importance in environmental health consequences: as the environment is cleaned up in developing countries, the burden of dirty work is often shifted to developing countries with weaker environmental and occupational standards.

These new epidemiological problems will doubtless continue to challenge statisticians with novel problems that are difficult to anticipate.

Whereas the basic methods of multivariate, exposure–time–response, longitudinal, and time-series analysis are now well established and likely to see only incremental refinements, methods for spatial, aggregate, and measurement error analysis continue to be fertile areas of methodologic research. Non- and semiparametric models are becoming more widely used and are an exciting area of statistical methods development. Novel study designs, such as multi-phase studies and hybrid individual/aggregate data studies, will likely be increasingly used and their statistical properties and optimal design merit further study. Other long-standing problems, like multiple hypothesis testing (Thomas et al. 1985), is becoming ever more important as it becomes feasible to test orders of magnitude more associations at once, such as gene–environment interactions in a genomewide context or spatial clustering of disease. Investigators must remain skeptical about *post-hoc* explanations for "interesting" but unanticipated findings in their data and avoid publication bias by clearly distinguishing a priori hypotheses and those that arise from hypothesis generation studies. Finally, as an observational science, the problem of residual confounding and selection bias will always remain, but novel methods (e.g., Wakefield 2003; Greenland 2005 and references therein) may provide some realistic bounds on the magnitude of such biases.

References

Abbey D.E., Nishino N., McDonnell W.F., Burchette R.J., Knutsen S.F., Lawrence Beeson W., Yang J.X. (1999). Long-term inhalable particles and other air pollutants related to mortality in nonsmokers. *American Journal of Respiratory and Critical Care Medicine* 159: 373–82.

Abrahamowicz M., Schopflocher T., Leffondré K., du Berger R., Krewski D. (2003). Flexible modeling of exposure–response relationship between long-term average levels of particulate air pollution and mortality in the American Cancer Society study. *Journal of Toxicology and Environmental Health, Part A* 66: 1625–54.

Academie des Sciences (1997). Problems associated with the effects of low doses of ionizing radiation. Rapport de L'Academie des Sciences, No. 38. Paris, Technique et Documentation.

Aditama T.Y. (2000). Impact of haze from forest fire to respiratory health: Indonesian experience. *Respirology* 5: 169–74.

Akaike H. (1973). Information theory and an extension of the maximum likelihood principle. In: *2nd International Symposium on Information Theory*, Petrov BN, Csaki F (eds.). Budapest, Akademia Kiado, pp. 267–81.

Albert P.S., Ratnasinghe D., Tangrea J., Wacholder S. (2001). Limitations of the case-only design for identifying gene–environment interactions. *American Journal of Epidemiology* 154: 687–93.

Allen B.C., Crump K.S., Shipp A.M. (1988). Correlation between carcinogenic potency of chemicals in animals and humans. *Risk Analysis* 8: 531–44.

Andersen P.K., Borgan O., Gill R.D., Keiding N. (1993). *Statistical Models Based on Counting Processes*. New York, Springer Verlag.

Andersen P.K., Gill R.D. (1982). Cox's regression model for counting processes: a large sample study. *Annals of Statistics* 10: 1100–20.

Andrieu N., Goldstein A.M., Thomas D.C., Langholz B. (2001). Counter-matching in studies of gene–environment interaction: efficiency and feasibility. *American Journal of Epidemiology* 153: 265–74.

Antenucci J.C., Brown K., Crosswell P.L., Kevany M.J. (1991). *Geographical Information Systems: A Guide to the Technology*. New York, Van Nostrand Reinhold.

Armitage P., Doll R. (1954). The age distribution of cancer and a multistage theory of carcinogenesis. *British Journal of Cancer* 8: 1–12.

Armstrong B.G. (1998). Effect of measurement error on epidemiological studies of environmental and occupational exposures. *Occupational and Environmental Medicine* 55: 651–6.

Armstrong B.G., Whittemore A.S., Howe G.R. (1989). Analysis of case-control data with covariate measurement error: application to diet and colon cancer. *Statistics in Medicine* 8: 1151–63; discussion 1165–6.

Armstrong B.K., White E., Saracci R. (1992). *Principles of exposure measurement in epidemiology*. Oxford, Oxford University Press.

Arratia R., Goldstein L., Langholz B. (2005). Local central limit theorems, the high order correlations of rejective sampling, and applications to conditional logistic likelihood asymptotics. *Annals of Statistics* 33: 871–914.

Arrighi H.M., Hertz-Picciotto I. (1994). The evolving concept of the healthy worker survivor effect. *Epidemiology* 5: 189–96.

Atkinson R.W., Anderson H.R., Sunyer J., Ayres J., Baccini M., Vonk J.M., Boumghar A., Forastiere F., Forsberg B., Touloumi G., Schwartz J., Katsouyanni K. (2001). Acute effects of particulate air pollution on respiratory admissions: results from APHEA 2 project. Air Pollution and Health: a European Approach. *American Journal of Respiratory and Critical Care Medicine* 164: 1860–6.

Austin H., Flanders W.D., Rothman K.J. (1989). Bias arising in case-control studies from selection of controls from overlapping groups. *International Journal of Epidemiology* 18: 713–16.

Avol E.L., Gauderman W.J., Tan S.M., London S.J., Peters J.M. (2001). Respiratory effects of relocating to areas of differing air pollution levels. *American Journal of Respiratory and Critical Care Medicine* 164: 2067–72.

Bagnardi V., Zambon A., Quatto P., Corrao G. (2004). Flexible meta-regression functions for modeling aggregate dose–response data, with an application to alcohol and mortality. *American Journal of Epidemiology* 159: 1077–86.

Bailer A.J., Hoel D.G. (1989). Metabolite-based internal doses used in a risk assessment of benzene. *Environmental Health Perspectives* 82: 177–84.

Baker G.S., Hoel D.G. (2003). Corrections in the atomic bomb data to examine low dose risk. *Health Physics* 85: 709–20.

Bardies M., Pihet P. (2000). Dosimetry and microdosimetry of targeted radiotherapy. *Current Pharmaceutical Design* 6: 1469–502.

Barlow R.E., Bartholomew D.J., Bremmer J.M., Brunk H.D. (1972). *Statistical inference under order restrictions*. New York, Wiley.

Bates D.V. (1992). Health indices of the adverse effects of air pollution: the question of coherence. *Environmental Research* 59: 336–49.

Bayer-Oglesby L., Grize L., Gassner M., Takken-Sahli K., Sennhauser F.H., Neu U., Schindler C., Braun-Fahrlander C. (2005). Decline of ambient air pollution levels and improved respiratory health in Swiss children. *Environmental Health Perspectives* 113: 1632–7.

Becker S., Berhane K. (1997). A meta-analysis of 61 sperm count studies revisited. *Fertility and Sterility* 67: 1103–8.

Begg C.B., Berlin J.A. (1988). Publication bias: A problem in interpreting medical data. *Journal of the Royal Statistical Society, Series A* 151: 419–63.

Bell C.M., Coleman D.A. (1987). Models of the healthy worker effect in industrial cohorts. *Statistics in Medicine* 6: 901–9.

Bell M.L., Davis D.L. (2001). Reassessment of the lethal London fog of 1952: novel indicators of acute and chronic consequences of acute exposure to air pollution. *Environmental Health Perspectives* 109(Suppl. 3): 389–94.

Bell M.L., McDermott A., Zeger S.L., Samet J.M., Dominici F. (2004a). Ozone and short-term mortality in 95 US urban communities, 1987–2000. *Journal of the American Medical Association* 292: 2372–8.

Bell M.L., Samet J.M., Dominici F. (2004b). Time-series studies of particulate matter. *Annual Review of Public Health* 25: 247–80.

Bellander T., Berglind N., Gustavsson P., Jonson T., Nyberg F., Pershagen G., Jarup L. (2001). Using geographic information systems to assess individual historical exposure to air pollution from traffic and house heating in Stockholm. *Environmental Health Perspectives* 109: 633–9.

Bender M.A., Awa A.A., Brooks A.L., Evans H.J., Groer P.G., Littlefield L.G., Pereira C., Preston R.J., Wachholz B.W. (1988). Current status of cytogenetic procedures to detect and quantify previous exposures to radiation. *Mutation Research* 196: 103–59.

Bennett J., Little M.P., Richardson S. (2004). Flexible dose–response models for Japanese atomic bomb survivor data: Bayesian estimation and prediction of cancer risk. *Radiation and Environmental Biophysics* 43: 233–45.

Benson P. (1989). CALINE4—A dispersion model for predicting air pollution concentration near roadways. Sacramento, C.A., Office of Transportation Laboratory, California Department of Transportation, 205.

Bergman R.N., Zaccaro D.J., Watanabe R.M., Haffner S.M., Saad M.F., Norris J.M., Wagenknecht L.E., Hokanson J.E., Rotter J.I., Rich S.S. (2003). Minimal model-based insulin sensitivity has greater heritability and a different genetic basis than homeostasis model assessment or fasting insulin. *Diabetes* 52: 2168–74.

Berhane K., Gauderman W., Stram D., Thomas D. (2004). Statistical issues in studies of the long term effects of air pollution: The Southern California Children Health Study. *Statistical Science* 19: 414–40.

Berhane K., Molitor N.-T. (2007). A Bayesian approach to functional based multi level modeling of longitudinal data: With applications to environmental epidemiology. *Biostatistics* (in press).

Berhane K., Thomas D.C. (2002). A two-stage model for multiple time series data of counts. *Biostatistics* 3: 21–32.

Berlin J.A., Longnecker M.P., Greenland S. (1993). Meta-analysis of epidemiologic dose–response data. *Epidemiology* 4: 218–28.

Bernstein J.L., Langholz B., Haile R.W., Bernstein L., Thomas D.C., Stovall M., Malone K.E., Lynch C.F., Olsen J.H., Anton-Culver H., et al. (2004). Study design: evaluating gene–environment interactions in the etiology of breast cancer—the WECARE study. *Breast Cancer Research* 6: R199–214.

Berrington A., Darby S.C., Weiss H.A., Doll R. (2001). 100 years of observation on British radiologists: mortality from cancer and other causes 1897–1997. *British Journal of Radiology* 74: 507–19.

Besag J. (1974). Spatial interaction and the statistical analysis of lattice systems. *Journal of the Royal Statistical Society* 36: 192–236.

Besag J., York J., Mollie A. (1991). Bayesian image restoration with two applications in spatial statistics (with discussion). *Annals of the Institute of Statistical Mathematics* 43: 1–59.

Beyea J. (1999). Geographic exposure modeling: a valuable extension of geographic information systems for use in environmental epidemiology. *Environmental Health Perspectives* 107(Suppl. 1): 181–90.

Bijwaard H., Brugmans M.J., Leenhouts H.P. (2001). A consistent two-mutation model of lung cancer for different data sets of radon-exposed rats. *Radiation and Environmental Biophysics* 40: 269–77.

Bijwaard H., Brugmans M.J., Leenhouts H.P. (2002). A consistent two-mutation model of bone cancer for two data sets of radium-injected beagles. *Journal of Radiological Protection* 22: A67–70.

Bijwaard H., Brugmans M.J., Leenhouts H.P. (2004). Two-mutation models for bone cancer due to radium, strontium and plutonium. *Radiation Research* 162: 171–84.

Bijwaard H., Dekkers F. (2007). Bone cancer risk o.f. (239)Pu in humans derived from animal models. *Radiation Research* 168: 582–92.

Bithell J.F., Stone R.A. (1989). On statistical methods for analysing the geographical distribution of cancer cases near nuclear installations. *Journal of Epidemiology and Community Health* 43: 79–85.

Blair A., Burg J., Foran J., Gibb H., Greenland S., Morris R., Raabe G., Savitz D., Teta J., Wartenberg D., et al. (1995). Guidelines for application of meta-analysis in environmental epidemiology. ISLI Risk Science Institute. *Regulatory Toxicology and Pharmacology* 22: 189–97.

Bohannon J. (2006). Epidemiology. Iraqi death estimates called too high; methods faulted. *Science* 314: 396–7.

Boice J.D., Jr, Blettner M., Kleinerman R.A., Stovall M., Moloney W.C., Engholm G., Austin D.F., Bosch A., Cookfair D.L., Krementz E.T. (1987). Radiation dose and leukemia risk in patients treated for cancer of the cervix. *Journal of the National Cancer Institute* 79: 1295–311.

Boice J.D., Jr, Harvey E.B., Blettner M., Stovall M., Flannery J.T. (1992). Cancer in the contralateral breast after radiotherapy for breast cancer. *New England Journal of Medicine* 326: 781–5.

Borgan O., Langholz B. (1997). Estimation of excess risk from case-control data using Aalen's linear regression model. *Biometrics* 53: 690–7.

Borgan O., Langholz B., Samuelsen S.O., Goldstein L., Pogoda J. (2000). Exposure stratified case-cohort designs. *Lifetime Data Analysis* 6: 39–58.

Bossi P., Garin D., Guihot A., Gay F., Crance J.M., Debord T., Autran B., Bricaire F. (2006). Bioterrorism: management of major biological agents. *Cell and Molecular Life Sciences* 63: 2196–212.

Bouma M.J., Kovats R.S., Goubet S.A., Cox J.S., Haines A. (1997). Global assessment of El Nino's disaster burden. *Lancet* 350: 1435–8.

Bower J.J., Shi X. (2005). Environmental health research in the post-genome era: new fields, new challenges, and new opportunities. *Journal of Toxicology and Environmental Health, Part B: Critical Reviews* 8: 71–94.

Braga A.L., Zanobetti A., Schwartz J. (2001). The lag structure between particulate air pollution and respiratory and cardiovascular deaths in 10 US cities. *Journal of Occupational and Environmental Medicine* 43: 927–33.

Brauer M., Brumm J., Vedal S., Petkau A.J. (2002). Exposure misclassification and threshold concentrations in time series analyses of air pollution health effects. *Risk Analysis* 22: 1183–93.

Brauer M., Hoek G., van Vliet P., Meliefste K., Fischer P., Gehring U., Heinrich J., Cyrys J., Bellander T., Lewn M., Brunekreef B. (2003). Estimating long-term average particulate air pollution concentrations: application of traffic indicators and geographic information systems. *Epidemiology* 14: 228–239.

Brennan P. (2002). Gene–environment interaction and aetiology of cancer: what does it mean and how can we measure it? *Carcinogenesis* 23: 381–7.

Brenner D.J., Little J.B., Sachs R.K. (2001). The bystander effect in radiation oncogenesis: II. A quantitative model. *Radiation Research* 155: 402–8.

Brenner D.J., Sachs R.K. (2002). Do low dose-rate bystander effects influence domestic radon risks? *International Journal of Radiation Biology* 78: 593–604.

Brenner D.J., Sachs R.K. (2003). Domestic radon risks may be dominated by bystander effects—but the risks are unlikely to be greater than we thought. *Health Physics* 85: 103–8.

Brenner H., Greenland S., Savitz D.A. (1992a). The effects of nondifferential confounder misclassification in ecologic studies. *Epidemiology* 3: 456–9.

Brenner H., Savitz D.A., Gefeller O. (1993). The effects of joint misclassification of exposure and disease on epidemiologic measures of association. *Journal of Clinical Epidemiology* 46: 1195–202.

Brenner H., Savitz D.A., Jockel K.H., Greenland S. (1992b). Effects of nondifferential exposure misclassification in ecologic studies. *American Journal of Epidemiology* 135: 85–95.

Breslow N. (1981). Odds ratio estimators when the data are sparse. *Biometrika* 68: 73–84.

Breslow N., Cain K. (1988). Logistic regression for two-stage case-control data. *Biometrika* 75: 11–20.

Breslow N., Powers W. (1978). Are there two logistic regressions for retrospective studies? *Biometrics* 34: 100–5.

Breslow N.E. (1984). Extra-Poisson variation in log-linear models. *Applied Statistics* 33: 38–44.

Breslow N.E., Chatterjee N. (1999). Design and analysis of two-phase studies with binary outcome applied to Wilms tumor prognosis. *Applied Statistics* 48: 457–68.

Breslow N.E., Day N.E. (1980). Statistical methods in cancer research: I. The analysis of case-control studies. Lyon, IARC Scientific publications, No. 32.

Breslow N.E., Day N.E. (1987). Statistical methods in cancer research. II. The design and analysis of cohort studies. Lyon, IARC Scientific Publications, No. 82.

Breslow N.E., Holubkov R. (1997a). Maximum likelihood estimation of logistic regression parameters under two-phase, outcome-dependent sampling. *Journal of the Royal Statistical Society: Series B* 59: 447–61.

Breslow N.E., Holubkov R. (1997b). Weighted likelihood, pseudo-likelihood and maximum likelihood methods for logistic regression analysis of two-stage data. *Statistics in Medicine* 16: 103–16.

Breslow N.E., Lubin J.H., Marek P., Langholz B. (1983). Multiplicative models and cohort analysis. *Journal of the American Statistical Association* 78: 1–12.

Breslow N.E., Storer B.E. (1985). General relative risk functions for case-control studies. *American Journal of Epidemiology* 122: 149–62.

Breslow N.E., Zhao L.P. (1988). Logistic regression for stratified case-control studies. *Biometrics* 44: 891–9.

Briggs D. (2005). The role of GIS: coping with space (and time) in air pollution exposure assessment. *Journal of Toxicology and Environmental Health, Part A* 68: 1243–61.

Briggs D.J., de Hoogh C., Gulliver J., Wills J., Elliott P., Kingham S., Smallbone K. (2000). A regression-based method for mapping traffic-related air pollution: application and testing in four contrasting urban environments. *Science of the Total Environment* 253: 151–67.

Brody J.G., Vorhees D.J., Melly S.J., Swedis S.R., Drivas P.J., Rudel R.A. (2002). Using GIS and historical records to reconstruct residential exposure to large-scale pesticide application. *Journal of Exposure Analysis Environmental Epidemiology* 12: 64–80.

Bromet E., Dew M.A. (1995). Review of psychiatric epidemiologic research on disasters. *Epidemiologic Reviews* 17: 113–19.

Broughton E. (2005). The Bhopal disaster and its aftermath: a review. *Environmental Health* 4: 6.

Brown C.C., Chu K.C. (1983a). Implications of the multistage theory of carcinogenesis applied to occupational arsenic exposure. *Journal of the National Cancer Institute* 70: 455–63.

Brown C.C., Chu K.C. (1983b). A new method for the analysis of cohort studies: implications of the multistage theory of carcinogenesis applied to occupational arsenic exposure. *Environmental Health Perspectives* 50: 293–308.

Brown C.C., Chu K.C. (1987). Use of multistage models to infer stage affected by carcinogenic exposure: example of lung cancer and cigarette smoking. *Journal Chronic Diseases* 40(Suppl. 2): 171S–179S.

Brown P.J., Le N.D., Zidek J.V. (1994). Multivariate spatial interpolation and exposure to air pollutants. *Canadian Journal of Statistics* 22: 489–509.

Brown V., Jacquier G., Coulombier D., Balandine S., Belanger F., Legros D. (2001). Rapid assessment of population size by area sampling in disaster situations. *Disasters* 25: 164–71.

Brunekreef B., Hoek G. (2000). Beyond the body count: air pollution and death. *American Journal of Epidemiology* 151: 449–51.

Bucher J.R., Portier C. (2004). Human carcinogenic risk evaluation, Part V: The national toxicology program vision for assessing the human carcinogenic hazard of chemicals. *Toxicological Sciences* 82: 363–6.

Burnett R., Ma R., Jerrett M., Goldberg M., Cakmak S., Pope C., Krewski D. (2001). The spatial association between community air pollution and mortality: a new method of analyzing correlated geographic cohort data. *Environmental Health Perspectives* 109(Suppl.): 375–80.

Burnett R.T., Dewanji A., Dominici F., Goldberg M.S., Cohen A., Krewski D. (2003). On the relationship between time-series studies, dynamic population studies, and estimating loss of life due to short-term exposure to environmental risks. *Environmental Health Perspectives* 111: 1170–4.

Burnham G., Lafta R., Doocy S., Roberts L. (2006). Mortality after the 2003 invasion of Iraq: a cross-sectional cluster sample survey. *Lancet* 368: 1421–8.

Burnham G., Roberts L. (2006). A debate over Iraqi death estimates. *Science* 314: 1241; author reply 1241.

Burrough P.A., McDonnell R.A. (1998). *Principles of geographical information systems.* Oxford, Oxford University Press.

Cain K., Breslow N. (1988). Logistic regression analysis and efficient design for two-stage studies. *American Journal of Epidemiology* 128: 1198–206.

Campbell D.T., Stanley J.C. (1963). *Experimental and quasi-experimental designs for research on teaching.* Boston, Houghton-Mifflin.

Cardis E. (2007). Current status and epidemiological research needs for achieving a better understanding of the consequences of the Chernobyl accident. *Health Physics* 93: 542–6.

Cardis E., Howe G., Ron E., Bebeshko V., Bogdanova T., Bouville A., Carr Z., Chumak V., Davis S., Demidchik Y., et al. (2006a). Cancer consequences of the Chernobyl accident: 20 years on. *Journal of Radiological Protection* 26: 127–40.

Cardis E., Kesminiene A., Ivanov V., Malakhova I., Shibata Y., Khrouch V., Drozdovitch V., Maceika E., Zvonova I., Vlassov O., et al. (2005). Risk of thyroid cancer after exposure to 131I in childhood. *Journal of the National Cancer Institute* 97: 724–32.

Cardis E., Krewski D., Boniol M., Drozdovitch V., Darby S.C., Gilbert E.S., Akiba S., Benichou J., Ferlay J., Gandini S., et al. (2006b). Estimates of the cancer burden in Europe from radioactive fallout from the Chernobyl accident. *International Journal of Cancer* 119: 1224–35.

Cardis E., Richardson L., Deltour I., Armstrong B., Feychting M., Johansen C., Kilkenny M., McKinney P., Modan B., Sadetzki S., et al. (2007). The INTERPHONE study: design, epidemiological methods, and description of the study population. *European Journal of Epidemiology*.

Carlsen E., Giwercman A., Keiding N., Skakkebaek N.E. (1992). Evidence for decreasing quality of semen during past 50 years. *British Medical Journal* 305: 609–13.

Carroll R.J. (1997). Surprising effects of measurement error on an aggregate data estimator. *Biometrika* 84: 231–4.

Carroll R.J., Gail M.H., Lubin J.H. (1993). Case-control studies with errors in covariates. *Journal of the American Statistical Association* 88: 185–99.

Carroll R.J., Ruppert D., Stefanski L.A. (1995). *Measurement error in nonlinear models.* London, Chapman & Hall.

Carroll R.J., Ruppert D., Stefanski L.A., Crainiceanu C.M. (2006). *Measurement error in nonlinear models: a modern perspective* (2nd edn.). London, Chapman &Hall, CRC Press.

Carroll R.J., Spiegelman C.H., Lan KKG, Bailey K.T., Abbott R.D. (1984). On errors-in-variables for binary regression models. *Biometrika* 71: 19–25.

Carroll R.J., Stefanski L.A. (1990). Approximate quasi-likelihood estimation in models with surrogate predictors. *Journal of the American Statistical Association* 85: 652–63.

Carroll R.J., Wand M.P. (1991). Semiparametric estimation in logistic measurement error models. *Journal of the Royal Statistical Society, Series B* 53: 573–85.

CDC (2006). Public health response to Hurricanes Katrina and Rita—United States, 2005. *MMWR Morbidity and Mortality Weekly Report* 55: 229–31.

Chakraborty S., Ghosh M., Maiti T., Tewari A. (2005). Bayesian neural networks for bivariate binary data: an application to prostate cancer study. *Statistics in Medicine* 24: 3645–62.

Chambers R.L., Steel D.G. (2001). Simple methods for ecological inference in 2×2 tables. *Journal of the Royal Statistical Society, Series A* 164: 175–92.

Chase G.R., Kotin P., Crump K., Mitchell R.S. (1985). Evaluation for compensation of asbestos-exposed individuals. II. Apportionment of risk for lung cancer and mesothelioma. *Journal of Occupational Medicine* 27: 189–98.

Chatterjee N., Wacholder S. (2002). Validation studies: bias, efficiency, and exposure assessment. *Epidemiology* 13: 503–6.

Checkoway H., Pearce N.E., Crawford-Brown D.J. (1989). *Research methods in occupational epidemiology*. Oxford, Oxford University Press.

Chen C.L., Hsu L.I., Chiou H.Y., Hsueh Y.M., Chen S.Y., Wu M.M., Chen C.J. (2004). Ingested arsenic, cigarette smoking, and lung cancer risk: a follow-up study in arseniasis-endemic areas in Taiwan. *Journal of the American Medical Association* 292: 2984–90.

Chen T.T. (1989). A review of methods for misclassified categorical data in epidemiology. *Statistics in Medicine* 8: 1095–106; discussion 1107–8.

Cheney W., Kincaid D. (1999). *Numerical mathematics and computing.* Pacific Grove, C.A., Brooks/Cole.

Chetwynd A.G., Diggle P.J., Marshall A., Parslow R. (2001). Investigation of spatial clustering from individually matched case-control studies. *Biostatistics* 2: 277–93.

Chib S., Greenberg E. (1995). Understanding the Metropolis–Hastings algorithm. *American Statistician* 46: 167–74.

Chipman H. (1996). Bayesian variable selection with related predictors. *Canadian Journal of Statistics* 24: 17–36.

Chrisman N.R. (2002). *Exploring geographic information systems* (2nd edn.). New York, Wiley.

Christakos G., Li X. (1998). Bayesian maximum entropy analysis and mapping: a farewell to kriging estimators. *Mathematical Geology* 30: 435–62.

Ciocco A., Thompson D.J. (1961). A follow-up of Donora ten years after: methodology and findings. *American Journal of Public Health* 51: 155–64.

Clancy L., Goodman P., Sinclair H., Dockery D.W. (2002). Effect of air-pollution control on death rates in Dublin, Ireland: an intervention study. *Lancet* 360: 1210–4.

Clayton D.G. Models for the analysis of cohort and case-control studies with inaccurately measured exposures. In: Dwyer J.H., Lippert P., Feinleib M. et al. (eds.). *Statistical methods for longitudinal studies of health.* Oxford, Oxford University Press. 1991: 301–31.

Clayton D., Kaldor J. (1987). Empirical Bayes estimates of age-standardized relative risks for use in disease mapping. *Biometrics* 43: 671–81.

Clayton D.G., Bernardinelli L., Montomoli C. (1993). Spatial correlation in ecological analysis. *International Journal of Epidemiology* 22: 1193–202.

Cleave N., Brown P.J., Payne C.D. (1995). Evaluation of Methods for Ecological Inference. *Journal of the Royal Statistical Society, Series A* 158: 55–72.

Cleveland W.S., Devlin S.J. (1988). Locally weighted regression: an approach to regression by local fitting. *Journal of the American Statistical Association* 83: 596–610.

Clifford P., Richardson S., Hemon D. (1989). Assessing the significance of the correlation between two spatial processes. *Biometrics* 45: 123–34.

Clyde M. (2000). Model uncertainty and health effect studies for particulate matter. *Environmetrics* 11: 745–63.

Cohen A.J., Ross Anderson H., Ostro B., Pandey K.D., Krzyzanowski M., Kunzli N., Gutschmidt K., Pope A., Romieu I., Samet J.M., Smith K. (2005). The global burden of disease due to outdoor air pollution. *Journal of Toxicology and Environmental Health, Part A* 68: 1301–7.

Cohen B.L. (1986). A national survey of 222Rn in U.S. homes and correlating factors. *Health Physics* 51: 175–83.

Cohen B.L. (1987). Tests of the linear, no-threshold dose-response relationship for high-LET radiation. *Health Physics* 52: 629–36.

Cohen B.L. (1990a). Ecological versus case-control studies for testing a linear-no threshold dose–response relationship. *International Journal of Epidemiology* 19: 680–4.

Cohen B.L. (1990b). A test of the linear-no threshold theory of radiation carcinogenesis. *Environmental Research* 53: 193–220.

Cohen B.L. (1995). Test of the linear-no threshold theory of radiation carcinogenesis for inhaled radon decay products. *Health Physics* 68: 157–74.

Cohen B.L. (2001). Updates and extensions to tests of the linear no-threshold theory. *Technology* 7: 657–62.

Cohen B.L., Colditz G.A. (1994). Tests of the linear-no threshold theory for lung cancer induced by exposure to radon. *Environmental Research* 64: 65–89.

Cohen B.L., Shah R.S. (1991). Radon levels in United States homes by states and counties. *Health Physics* 60: 243–59.

Cohen B.L., Stone C.A., Schilken C.A. (1994). Indoor radon maps of the United States. *Health Physics* 66: 201–5.

Cologne J., Langholz B. (2003). Selecting controls for assessing interaction in nested case-control studies. *Journal of Epidemiology* 13: 193–202.

Cologne J.B. (1997). Counterintuitive matching. *Epidemiology* 8: 227–9.

Cologne J.B., Pawel D.J., Preston D.L. (1998). Statistical issues in biological radiation dosimetry for risk assessment using stable chromosome aberrations. *Health Physics* 75: 518–29.

Cologne J.B., Preston D.L. (2001). Impact of comparison group on cohort dose response regression: an example using risk estimation in atomic-bomb survivors. *Health Physics* 80: 491–6.

Cologne J.B., Sharp G.B., Neriishi K., Verkasalo P.K., Land C.E., Nakachi K. (2004). Improving the efficiency of nested case-control studies of interaction by selecting controls using counter matching on exposure. *International Journal of Epidemiology* 33: 485–92.

Cologne J.B., Shibata Y. (1995). Optimal case-control matching in practice. *Epidemiology* 6: 271–5.

Conti D.V., Cortessis V., Molitor J., Thomas D.C. (2003). Bayesian modeling of complex metabolic pathways. *Human Heredity* 56: 83–93.

Conti D.V., Lewinger J.P., Swan G., Tyndale R., Benowitz N., Thomas P. (2007). Using ontologies in hierarchical modeling of genes and exposures in biologic pathways. In: *Phenotypes, Endophenotypes, and Genetic Studies of Nicotine Dependence*, NCI Monograph. 22 (in press).

Cook D.G., Pocock S.J. (1983). Multiple regression in geographical mortality studies, with allowance for spatially correlated errors. *Biometrics* 39: 361–71.

Cook J.R., Stefanski L.A. (1994). Simulation-extrapolation estimation in parametric measurement error models. *Journal of the American Statistical Association* 89: 1314–1328.

Cook N.R., Zee R.Y., Ridker P.M. (2004). Tree and spline based association analysis of gene–gene interaction models for ischemic stroke. *Statistics in Medicine* 23: 1439–53.

Cortessis V., Thomas D.C. (2003). Toxicokinetic genetics: An approach to gene-environment and gene–gene interactions in complex metabolic pathways. In: *Mechanistic considerations in the molecular epidemiology of cancer*. Bird P., Boffetta P., Buffler P., Rice J. (eds.). Lyon, France, IARC Scientific Publications, 157: pp. 127–50.

Court Brown W.M., Doll R. (1958). Expectation of life and mortality from cancer among British radiologists. *British Medical Journal* 2: 181–7.

Cox D. (1972). Regression models and life tables (with discussion). *Journal of the Royal Statistical Society, Series B* 34: 187–220.

Cox D., Hinkley D. (1974). *Theoretical statistics*. Longon, Chapman & Hall.

Cox D.R. (1975). Partial likelihood. *Biometrika* 62: 269–76.

Cox D.R., Wermuth N. (2004). Causality: a statistical view. *International Statistical Review* 72: 285–305.

Crainiceanu C.M., Dominici F., Parmigiani G. (2007). Adjustment uncertainty in effect estimation. *Biometrika*: under review.

Cressie NAC (1993). *Statistics for spatial data*. New York, Wiley & Sons, Inc.

Crowther J.A. (1924). Some considerations relative to the action of x-rays on tissue cells [in German]. *Proceedings of the Royal Society (London), Series B* 96: 207.

Crump K., Allen B., Shipp A. (1989). Choice of dose measure for extrapolating carcinogenic risk from animals to humans: an empirical investigation of 23 chemicals. *Health Physics* 57(Suppl. 1): 387–93.

Crump K.S. (1984). An improved procedure for low-dose carcinogenic risk assessment from animal data. *Journal of Environmental Pathology, Toxicology and Oncology* 5: 339–48.

Crump K.S. (1996). The linearized multistage model and the future of quantitative risk assessment. *Human and Experimental Toxicology* 15: 787–98.

Crump K.S., Allen B.C., Howe R.B., Crockett P.W. (1987). Time-related factors in quantitative risk assessment. *Journal Chronic Diseases* 40(Suppl. 2): 101S–11S.

Crump K.S., Hoel D.G., Langley C.H., Peto R. (1976). Fundamental carcinogenic processes and their implications for low dose risk assessment. *Cancer Research* 36: 2973–9.

Cullinan P., Acquilla S., Dhara V.R. Respiratory morbidity 10 years after the union carbide gas leak at bhopal: A cross sectional survey. *BMJ*, 1997; 314:338–342.

Cullings H.M., Fujita S., Funamoto S., Grant E.J., Kerr G.D., Preston D.L. (2006). Dose estimation for atomic bomb survivor studies: its evolution and present status. *Radiation Research* 166: 219–54.

Curtis S.B., Luebeck E.G., Hazelton W.D., Moolgavkar S.H. (2001). The role of promotion in carcinogenesis from protracted high-LET exposure. *Physica Medica* 17(Suppl. 1): 157–60.

Cuzick J., Edwards R. (1990). Spatial clustering for inhomogeneous populations. *Journal of the Royal Statistical Society, Series B* 52: 73–104.

Daniels M., Dominici F., Samet J., Zeger S. (2000). Estimating particulate matter-mortality dose–response curves and threshold levels: an analysis of daily time-series for the 20 largest US cities. *American Journal of Epidemiology* 152: 397–406.

Darby S., Deo H., Doll R., Whitley E. (2001). A parallel analysis of individual and ecological data on residential radon and lung cancer in south-west England. *Journal of the Royal Statistical Society, Series A* 164: 193–203.

Darby S., Hill D., Auvinen A., Barros-Dios J.M., Baysson H., Bochicchio F., Deo H., Falk R., Forastiere F., Hakama M., et al. (2005). Radon in homes and risk of lung cancer: collaborative analysis of individual data from 13 European case-control studies. *British Medical Journal* 330: 223.

Davey Smith G., Ebrahim S. (2001). Epidemiology—is it time to call it a day? *International Journal of Epidemiology* 30: 1–11.

Davis S., Kopecky K.J., Hamilton T.E., Onstad L., and the Hanford Thyroid Disease Study Team (2004). Thyroid Neoplasia, Autoimmune Thyroiditis, and Hypothyroidism in Persons Exposed to Iodine 131 From the Hanford Nuclear Site. *Journal of the American Medical Association* 292: 2600–13.

Dawid A.P. (2000). Causal inference without counterfactuals (with discussion). *Journal of the American Statistical Association* 95: 407–48.

Day N.E., Boice J.D.J. (1983). *Second cancer in relation to radiation treatment for cervical cancer*. Lyon, IARC Scientific Publication, No. 52.

Day N.E., Brown C.C. (1980). Multistage models and primary prevention of cancer. *Journal of the National Cancer Institute* 64: 977–89.

Day N.E., Byar D.P., Green S.B. (1980). Overadjustment in case-control studies. *American Journal of Epidemiology* 112: 696–706.

de Boor C. (1978). *A practical guide to splines*. New York, Springer-Verlag.

Deltour I., Richardson S., Thomas D. (1999). A Bayesian approach to measurement error in dose response analysis: application to the atomic bomb survivors cohort. In: *Uncertainties in radiation dosimetry and their impact on dose-response analyses (DHHS Publication no. 99-4541)*. Ron E., Hoffman F.O. (eds.). Bethesda, M.D., National Cancer Institute: pp. 100–9.

DeMers M.N. (2000). *Fundamentals of Geographic Information Systems* (2nd ed.). New York, Wiley.

Demirtas H. (2005). Multiple imputation under Bayesianly smoothed pattern-mixture models for non-ignorable drop-out. *Statistics in Medicine* 24: 2345–63.

Denison D.G., Holmes C.C. (2001). Bayesian partitioning for estimating disease risk. *Biometrics* 57: 143–9.

Depoortere E., Checchi F., Broillet F., Gerstl S., Minetti A., Gayraud O., Briet V., Pahl J., Defourny I., Tatay M., Brown V. (2004). Violence and mortality in West Darfur, Sudan (2003–04): epidemiological evidence from four surveys. *Lancet* 364: 1315–20.

Dewanji A., Moolgavkar S.H. (2000). A Poisson process approach for recurrent event data with environmental covariates. *Environmetrics* 11: 665–73.

Dewanji A., Moolgavkar S.H. (2002). Choice of stratification in Poisson process analysis of recurrent event data with environmental covariates. *Statistics in Medicine* 21: 3383–93.

Dhara R. (1992). Health effects of the Bhopal gas leak: a review. *Epidemiologia e prevenzione* 52: 22–31.

Dhara R., Acquilla S., Cullinan P. Has the world forgotten bhopal? *Lancet*, 2001; 357: 809–10.

Dhara V.R., Dhara R. (2002). The Union Carbide disaster in Bhopal: a review of health effects. *Archives of Environmental Health* 57: 391–404.

Dhara V.R., Dhara R., Acquilla S.D., Cullinan P. (2002). Personal exposure and long-term health effects in survivors of the union carbide disaster at Bhopal. *Environmental Health Perspectives* 110: 487–500.

Dickersin K., Berlin J.A. (1992). Meta-analysis: state-of-the-science. *Epidemiologic Reviews* 14: 154–76.

Dietrich T., Hoffmann K. (2004). A comprehensive index for the modeling of smoking history in periodontal research. *Journal of Dental Research* 83: 859–63.

Diggle P., Lophaven S. (2005). Bayesian geostatistical design. *Scandinavian Journal of Statistics* 33: 53–64.

Diggle P., Morris S., Morton-Jones T. (1999). Case-control isotonic regression for investigation of elevation in risk around a point source. *Statistics in Medicine* 18: 1605–13.

Diggle P.J., Gomez-Rubio V., Brown P.E., Chetwynd A.G., Gooding S. (2007). Second-order analysis of inhomogeneous spatial point processes using case-control data. *Biometrics* 63: 550–7.

Diggle P.J., Morris S.E., Wakefield J.C. (2000). Point-source modelling using matched case-control data. *Biostatistics* 1: 89–105.

Diggle P.J., Tawn J.A., Moyeed R.A. (1998). Model-based geostatistics. *Applied Statistics* 47: 299–350.

Dikshit R.P., Kanhere S. (1999). Cancer patterns of lung, oropharynx and oral cavity cancer in relation to gas exposure at Bhopal. *Cancer Causes and Control* 10: 627–36.

DiMajo V., Coppola M., Rebessi S., Covelli V. (1990). Age-related susceptibility of mouse liver to incuction of tumors by neutrons. *Radiation Research* 124: 227–234.

Dobson A.J. (1988). Proportional hazards models for average data for groups. *Statistics in Medicine* 7: 613–18.

Dockery D., Pope C., Xu X., Spengler J., Ware J., Fay M., Ferris B., Speizer F. (1993). An association between air pollution and mortality in six U.S. cities. *New England Journal of Medicine* 329: 1753–9.

Doll R. (1998). Effects of small doses of ionising radiation. *Journal of Radiological Protection* 18: 163–74.

Doll R., Peto R. (1978). Cigarette smoking and bronchial carcinoma: dose and time relationships among regular smokers and lifelong non-smokers. *Journal of Epidemiology and Community Health* 32: 303–13.

Dominici F., Daniels M., Zeger S.L., Samet J.M. (2002a). Air pollution and mortality: estimating regional and national dose-response relationships. *Journal of the American Statistical Association* 97: 100–11.

Dominici F., Levy J.I., Louis T.A. (2005a). Methodological challenges and contributions in disaster epidemiology. *Epidemiologic Reviews* 27: 9–12.

Dominici F., McDermott A., Daniels M., Zeger S.L., Samet J.M. (2005b). Revised analyses of the National Morbidity, Mortality, and Air Pollution Study: mortality among residents of 90 cities. *Journal of Toxicology and Environmental Health, Part A* 68: 1071–92.

Dominici F., McDermott A., Zeger S.L., Samet J.M. (2002b). On the use of generalized additive models in time-series studies of air pollution and health. *American Journal of Epidemiology* 156: 193–203.

Dominici F., McDermott A., Zeger S.L., Samet J.M. (2003a). Airborne particulate matter and mortality: timescale effects in four US cities. *American Journal of Epidemiology* 157: 1055–65.

Dominici F., Peng R.D., Bell M.L., Pham L., McDermott A., Zeger S.L., Samet J.M. (2006). Fine particulate air pollution and hospital admission for cardiovascular and respiratory diseases. *Journal of the American Medical Association* 295: 1127–34.

Dominici F., Samet J., Zeger S.G., Xu J. (2000). Combining evidence on air pollution and daily mortality from the largest 20 US cities: A hierarchical modeling strategy. *Journal of the Royal Statistical Society, Series A* 163: 263–302.

Dominici F., Sheppard L., Clyde M. (2003b). Health effects of air pollution: a statistical review. *International Statistical Review* 71: 243–76.

Dominici F., Zanobetti A., Zeger S.L., Schwartz J., Samet J.M. (2004). Hierarchical bivariate time series models: a combined analysis of the effects of particulate matter on morbidity and mortality. *Biostatistics* 5: 341–60.

Dong M.H., Redmond C.K., Mazumdar S., Costantino J.P. (1988). A multistage approach to the cohort analysis of lifetime lung cancer risk among steelworkers exposed to coke oven emissions. *American Journal of Epidemiology* 128: 860–73.

Doocy S., Rofi A., Moodie C., Spring E., Bradley S., Burnham G., Robinson C. (2007). Tsunami mortality in Aceh Province, Indonesia. *Bulletin of the World Health Organization* 85: 273–8.

Dosemeci M., Wacholder S., Lubin J.H. (1990). Does nondifferential misclassification of exposure always bias a true effect toward the null value? *American Journal of Epidemiology* 132: 746–8.

Dougherty R.C., Whitaker M.J., Tank S-Y, Bottcher R., Keller M., Kuehl D.W. (1981). Sperm density and toxic substances: a potential key to environmental health hazards. In: *Environmental Health Chemistry: The Chemistry of Environmental Agents as Potential Human Hazards*. McKinney J.D. (eds.). Ann Arbor, M.I., Ann Arbor Science Publication: 263–78.

Downs S.H., Schindler C., Liu L.J., Keidel D., Bayer-Oglesby L., Brutsche M.H., Gerbase M.W., Keller R., Kunzli N., Leuenberger P., et al. (2007). Reduced exposure to PM10 and attenuated age-related decline in lung function. *New England Journal of Medicine* 357: 2338–47.

Drozdovitch V., Bouville A., Chobanova N., Filistovic V., Ilus T., Kovacic M., Malatova I., Moser M., Nedveckaite T., Volkle H., Cardis E. (2007). Radiation exposure to the population of Europe following the Chernobyl accident. *Radiation Protection and Dosimetry* 123: 515–28.

DuMouchel W.H., Harris J.E. (1983). Bayes methods for combining the results of cancer studies in humans and other species. *Journal of the American Statistical Association* 78: 293–308.

Durkheim E. (1951). *Suicide* (translated by Spaulding JA and Simpson G). Glencoe, I.L., Free Press.

Eatough J.P., Henshaw D.L. (1993). Radon and monocytic leukaemia in England. *Journal of Epidemiology and Community Health* 47: 506–7.

Ebi K.L., Zaffanella L.E., Greenland S. (1999). Application of the case-specular method to two studies of wire codes and childhood cancers. *Epidemiology* 10: 398–404.

Elgethun K., Fenske R.A., Yost M.G., Palcisko G.J. (2003). Time-location analysis for exposure assessment studies of children using a novel global positioning system instrument. *Environmental Health Perspectives* 111: 115–22.

Elgethun K., Yost M.G., Fitzpatrick C.T., Nyerges T.L., Fenske R.A. (2007). Comparison of global positioning system (GPS) tracking and parent-report diaries to characterize children's time-location patterns. *Journal of Exposure Science and Environmental Epidemiology* 17: 196–206.

Elliott P., Cuzick J., English D., Stern R., Eds. (1992). *Geographical and environmental epidemiology methods for small-area studies*. Oxford, Oxford University Press.

Elliott P., Wakefield J. (2001). Disease Clusters: Should They Be Investigated, and, If S.o., When and How? *Journal of the Royal Statistical Society. Series A. (Statistics in Society)* 164: 3–12.

Elliott P., Wakefield J., Best N., Briggs D. (2001). *Spatial epidemiology: methods and applications*. Oxford, Oxford University Press.

Elliott P., Wartenberg D. (2004). Spatial epidemiology: current approaches and future challenges. *Environmental Health Perspectives* 112: 998–1006.

Elton R.A., Duffy S.W. (1983). Correcting for the effect of misclassification bias in a case-control study using data from two different questionnaires. *Biometrics* 39: 659–63.

Enterline P.E. (1980). Asbestos and lung cancer. Attributability in the face of uncertainty. *Chest* 78: 377–9.

EPA (1997). The benefits and costs of the Clean Air Act, 1970 to 1990. Washington, D.C., Environmental Protection Agency, Office of Radiation and Air, 58 p.p., plus appendices and front matter.

EPA (1999). Report to Congress on the Benefits and Costs of the Clean Air Act, 1990 to 2010 (EPA 410-R-97-002). Washington, D.C., U.S. Environmental Protection Agency, Office of Air and Radiation.

Evrard A.S., Hemon D., Billon S., Laurier D., Jougla E., Tirmarche M., Clavel J. (2005). Ecological association between indoor radon concentration and childhood leukaemia incidence in France, 1990–1998. *Eururopean Journal of Cancer Prevention* 14: 147–57.

Evrard A.S., Hemon D., Billon S., Laurier D., Jougla E., Tirmarche M., Clavel J. (2006). Childhood leukemia incidence and exposure to indoor radon, terrestrial and cosmic gamma radiation. *Health Physics* 90: 569–79.

Ezzati M., Hoorn S.V., Rodgers A., Lopez A.D., Mathers C.D., Murray C.J. (2003). Estimates of global and regional potential health gains from reducing multiple major risk factors. *Lancet* 362: 271–80.

Ezzati M., Lopez A.D., Rodgers A., Vander Hoorn S., Murray C.J. (2002). Selected major risk factors and global and regional burden of disease. *Lancet* 360: 1347–60.

Ezzati T.M., Massey J.T., Waksberg J., Chu A., Maurer K.R. (1992). Sample design: Third National Health and Nutrition Examination Survey. *Vital Health Statistics, Series 2* 113: 1–35.

Faes C., Aerts M., Geys H., Molenberghs G. (2007). Model averaging using fractional polynomials to estimate a safe level of exposure. *Risk Analysis* 27: 111–23.

Faucett C.L., Schenker N., Elashoff R.M. (1998). Analysis of censored survival data with intermittently observed time-dependent binary covariates. *Journal of the American Statistical Association* 93: 427–37.

Faucett C.L., Thomas D.C. (1996). Simultaneously modelling censored survival data and repeatedly measured covariates: a Gibbs sampling approach. *Statistics in Medicine* 15: 1663–85.

Fearn T., Hill D.C., Darby S.C. (2008). Measurement error in the explanatory variable of a binary regression: Regression calibration and integrated conditional likelihood in studies of residential radon and lung cancer. *Statistics in Medicine* 27: 2159–76.

Fears T., Brown C. (1986). Logistic regression methods for retrospective case-control studies using complex sampling procedures. *Biometrics* 42: 955–60.

Fears T.R., Gail M.H. (2000). Analysis of a two-stage case-control study with cluster sampling of controls: application to nonmelanoma skin cancer. *Biometrics* 56: 190–8.

Ferlay J., Bray F., Pisani P., Parkin D.M. (2001). "GLOBOCAN 2002: Cancer Incidence, Mortality and Prevalence Worldwide. IARC CancerBase No. 5." from http://www-dep.iarc.fr/.

Field R.W. (2005). Three Mile Island epidemiologic radiation dose assessment revisited: 25 years after the accident. *Radiation Protection and Dosimetry* 113: 214–17.

Field R.W., Smith B.J., Lynch C.F. (1999). Cohen's paradox. *Health Physics* 77: 328–9.

Firebaugh G. (1978). A rule for inferring individual-level relationships from aggregate data. *American Sociological Review* 43: 557–72.

Firket J. (1931). The cause of the symptoms found in the Meuse Valley during the fog of December, 1930. *Bulletin de l'Acadamie Royale Medicale de Belge* 11: 683–741.

Fisher R.A. (1935). *The design of experiments*. Edinburgh, Oliver and Boyd.

Fisher R.A. (1958). Cancer and smoking. *Nature* 182: 596.

Fitzmaurice G.M., Laird N.M., Lipsitz S.R. (1994). Analysing incomplete longitudinal binary responses: a likelihood-based approach. *Biometrics* 50: 601–12.

Flanders W.D., DerSimonian R., Rhodes P. (1990). Estimation of risk ratios in case-base studies with competing risks. *Statistics in Medicine* 9: 423–35.

Flanders W.D., Greenland S. (1991). Analytic methods for two-stage case-control studies and other stratified designs. *Statistics in Medicine* 10: 739–47.

Flanders W.D., Harland A. (1986). Possibility of selection bias in matched case-control studies using friend controls. *American Journal of Epidemiology* 124: 150–3.

Flanders W.D., Lally C.A., Zhu B.P., Henley S.J., Thun M.J. (2003). Lung cancer mortality in relation to age, duration of smoking, and daily cigarette consumption: results from Cancer Prevention Study II. *Cancer Research* 63: 6556–62.

Fleck F. (2005). Tsunami body count is not a ghoulish numbers game. *Bulletin of the World Health Organization* 83: 88–9.

Flegal K.M., Keyl P.M., Nieto F.J. (1991). Differential misclassification arising from non-differential errors in exposure measurement. *American Journal of Epidemiology* 134: 1233–44.

Fleming T.R., Harrington D.P. (1991). *Counting Processes and Survival Analysis*. New York, Wiley.

Foster D.P., George E.I. (1994). The risk inflation criterion for multiple regression. *Annals of Statistics* 22: 1947–75.

Fotheringham A.S., Wong DWS (1991). The modifiable area unit problem in multivariate analysis. *Environ Plan A* 23: 1025–44.

Frankenberg E., McKee D., Thomas D. (2005). Health consequences of forest fires in Indonesia. *Demography* 42: 109–29.

Fraser G.E., Stram D.O. (2001). Regression calibration in studies with correlated variables measured with error. *American Journal of Epidemiology* 154: 836–44.

Fraumeni J.F., Jr (1987). Keynote lecture. Etiologic insights from cancer mapping. *Princess Takamatsu Symp* 18: 13–25.

Freedman D.A. (2006). On the so-called "Huber sandwich estimator" and "robust standard errors". *American Statistician* 60: 299–302.

Freedman D.A., Navidi W.C. (1989). Multistage models for carcinogenesis. *Environmental Health Perspectives* 81: 169–88.

Freedman D.A., Navidi W.C. (1990). Ex-smokerivess and the multistage model for lung cancer. *Epidemiology* 1: 21–9.

Friedman M.S., Powell K.E., Hutwagner L., Graham L.M., Teague W.G. (2001). Impact of changes in transportation and commuting behaviors during the 1996 Summer Olympic Games in Atlanta on air quality and childhood asthma. *Journal of the American Medical Association* 285: 897–905.

Frye C., Hoelscher B., Cyrys J., Wjst M., Wichmann H.E., Heinrich J. (2003). Association of lung function with declining ambient air pollution. *Environmental Health Perspectives* 111: 383–7.

Fuller W. (1987). *Measurement error models*. New York, Wiley.

Fung K., Krewski D., Burnett R., Dominici F. (2005a). Testing the harvesting hypothesis by time-domain regression analysis. I: baseline analysis. *Journal of Toxicology and Environmental Health, Part A* 68: 1137–54.

Fung K., Krewski D., Burnett R., Ramsay T., Chen Y. (2005b). Testing the harvesting hypothesis by time-domain regression analysis. II: covariate effects. *Journal of Toxicology and Environmental Health, Part A* 68: 1155–65.

Fung K.Y., Howe G.R. (1984). Methodological issues in case-control studies. III–The effect of joint misclassification of risk factors and confounding factors upon estimation and power. *International Journal of Epidemiology* 13: 366–70.

Gail M.H., Mark S.D., Carroll R.J., Green S.B., Pee D. (1996). On design considerations and randomization-based inference for community intervention trials. *Statistics in Medicine* 15: 1069–92.

Gail M.H., Tan W.Y., Piantadosi S. (1988). Tests for no treatment effect in randomized clinical trials. *Biometrika* 75: 57–64.

Gant T.W., Zhang S.D. (2005). In pursuit of effective toxicogenomics. *Mutation Research* 575: 4–16.

Gauderman W.J., McConnell R., Gilliland F., London S., Thomas D., Avol E., Vora H., Berhane K., Rappaport E.B., Lurmann F., Margolis H.G., Peters J. Association between air pollution and lung function growth in southern california children. *American Journal of Respiratory & Critical Care Medicine*, 2000; 162: 1383–90.

Gauderman W.J., Avol E., Gilliland F., Vora H., Thomas D., Berhane K., McConnell R., Kuenzli N., Lurmann F., Rappaport E., Margolis H., Bates D., Peters J. (2004). The effect of air pollution on lung development from 10 to 18 years of age. *New England Journal of Medicine* 351: 1057–67.

Gauderman W.J., Avol E., Lurmann F., Kuenzli N., Gilliland F., Peters J., McConnell R. (2005). Childhood asthma and exposure to traffic and nitrogen dioxide. *Epidemiology* 16: 737–43.

Gelman A., Park D.K., Ansolabehere S., Price P.N., Minnite L.C. (2001). Models, assumptions and model checking in ecological regressions. *Journal of the Royal Statistical Society, Series A* 164: 101–18.

George E.I., Foster D.P. (2000). Calibration and empirical Bayes variable selection. *Biometrika* 87: 731–47.

George E.I., McCulloch R.E. (1993). Variable selection via Gibbs sampling. *Journal of the American Statistical Association* 88: 881–89.

Gilbert E.S. (1984). Some effects of random dose measurement errors on analyses of atomic bomb survivor data. *Radiation Research* 98: 591–605.

Gilbert E.S., Koshurnikova N.A., Sokolnikov M., Khokhryakov V.F., Miller S., Preston D.L., Romanov S.A., Shilnikova N.S., Suslova K.G., Vostrotin V.V. (2000). Liver cancers in Mayak workers. *Radiation Research* 154: 246–52.

Gilbert E.S., Koshurnikova N.A., Sokolnikov M.E., Shilnikova N.S., Preston D.L., Ron E, Okatenko P.V., Khokhryakov V.F., Vasilenko E.K., Miller S., Eckerman K., Romanov S.A. (2004). Lung cancer in Mayak workers. *Radiation Research* 162: 505–16.

Gilbert E.S., Land C.E., Simon S.L. (2002). Health effects from fallout. *Health Physics* 82: 726–35.

Gilbert E.S., Tarone R., Bouville A., Ron E. (1998). Thyroid cancer rates and 131I doses from Nevada atmospheric nuclear bomb tests. *Journal of the National Cancer Institute* 90: 1654–60.

Gilks W., Richardson S., Spiegelhalter D., Eds. (1996). *Markov Chain Monte Carlo in practice*. London, Chapman & Hall.

Gilks W., Wilde P. (1992). Adaptive rejection sampling for Gibbs sampling. *Applied Statistics* 41: 337–48.

Gilliland F.D., Berhane K., Rappaport E.B., Thomas D.C., Avol E., Gauderman W.J., London S.J., Margolis H.G., McConnell R., Islam K.T., Peters J.M. (2001). The effects of ambient air pollution on school absenteeism due to respiratory illnesses. *Epidemiology* 12: 43–54.

Gilliland F.D., Berhane K.T., Li Y.F., Gauderman W.J., McConnell R., Peters J. (2003). Children's lung function and antioxidant vitamin, fruit, juice, and vegetable intake. *American Journal of Epidemiology* 158: 576–84.

Gilliland F.D., Gauderman W.J., Vora H., Rappaport E., Dubeau L. (2002a). Effects of glutathione-S-transferase M1, T1, and P1 on childhood lung function growth. *American Journal of Respiratory and Critical Care Medicine* 166: 710–16.

Gilliland F.D., Li Y.F., Dubeau L., Berhane K., Avol E., McConnell R., Gauderman W.J., Peters J.M. (2002b). Effects of glutathione S-transferase M1, maternal smoking during pregnancy, and environmental tobacco smoke on asthma and wheezing in children. *American Journal of Respiratory and Critical Care Medicine* 166: 457–63.

Gilliland F.D., Li Y.F., Saxon A., Diaz-Sanchez D. (2004). Effect of glutathione-S-transferase M1 and P1 genotypes on xenobiotic enhancement of allergic responses: randomised, placebo-controlled crossover study. *Lancet* 363: 119–25.

Gilliland F.D., McConnell R., Peters J., Gong H., Jr (1999). A theoretical basis for investigating ambient air pollution and children's respiratory health. *Environmental Health Perspectives* 107: 403–7.

Gilliland F.D., Rappaport E.B., Berhane K., Islam T., Dubeau L., Gauderman W.J., McConnell R. (2002c). Effects of glutathione S-transferase P1, M1, and T1 on acute respiratory illness in school children. *American Journal of Respiratory and Critical Care Medicine* 166: 346–51.

Gold M.R., Stevenson D., Fryback D.G. (2002). HALYS and QALYS and DALYS, Oh My: similarities and differences in summary measures of population Health. *Annual Review of Public Health* 23: 115–34.

Goldberg M.S., Burnett R.T. (2005). A new longitudinal design for identifying subgroups of the population who are susceptible to the short-term effects of ambient air pollution. *Journal of Toxicology and Environmental Health, Part A* 68: 1111–25.

Goldsmith J.R. (1999). The residential radon-lung cancer association in U.S. counties: a commentary. *Health Physics* 76: 553–7.

Goldstein L., Langholz B. (1992). Asymptotic theory for nested case-control sampling in the Cox regression model. *Annals of Statistics* 20: 1903–28.

Goodhead D.T. (2006). Energy deposition stochastics and track structure: what about the target? *Radiation Protection and Dosimetry* 122: 3–15.

Gordis L. (1982). Should dead cases be matched to dead controls? *American Journal of Epidemiology* 115: 1–5.

Gotway C.A., Wolfinger R.D. (2003). Spatial prediction of counts and rates. *Statistics in Medicine* 22: 1415–32.

Grais R.F., Coulombier D., Ampuero J., Lucas M.E., Barretto A.T., Jacquier G., Diaz F., Balandine S., Mahoudeau C., Brown V. (2006). Are rapid population estimates accurate? A field trial of two different assessment methods. *Disasters* 30: 364–76.

Gray R., LaFuma J., Parish S.E., Peto R. (1986). Lung tumors and radon inhalation in over 2000 rats: Approximate linearity across a wide range of doses and potentiation by tobacco smoke. In: *Life-Span Radiation Effects Studies in Animals: What Can They Tell Us? (Publ No DE9301860 (CONF-830951))*. Thompson RC, Mahaffey JA (eds.). Washington, D.C., U.S. Department of Energy.

Green P., Richardson S. (2001). Hidden Markov models and disease mapping. *Journal of the American Statistical Association* 97: 1055–70.

Green P.J., Silverman B.W. (1994). *Nonparametric regression and generalized linear models: a roughness penalty approach*. London, Chapman & Hall.

Greenland S. (1980). The effect of misclassification in the presence of covariates. *American Journal of Epidemiology* 112: 564–9.

Greenland S. (1988a). Statistical uncertainty due to misclassification: implications for validation substudies. *Journal of Clinical Epidemiology* 41: 1167–74.

Greenland S. (1988b). Variance estimation for epidemiologic effect estimates under misclassification. *Statistics in Medicine* 7: 745–57.

Greenland S. (1990). Randomization, statistics, and causal inference. *Epidemiology* 1: 421–9.

Greenland S. (1992). Divergent biases in ecologic and individual-level studies. *Statistics in Medicine* 11: 1209–23.

Greenland S. (1993). Methods for epidemiologic analyses of multiple exposures: a review and comparative study of maximum–likelihood, preliminary-testing, and empirical-Bayes regression. *Statistics in Medicine* 12: 717–36.

Greenland S. (1994a). Hierarchical regression for epidemiologic analyses of multiple exposures. *Environmental Health Perspectives* 102(Suppl. 8): 33–9.

Greenland S. (1994b). Invited commentary: a critical look at some popular meta-analytic methods. *American Journal of Epidemiology* 140: 290–6.

Greenland S. (1995). Dose-response and trend analysis in epidemiology: alternatives to categorical analysis. *Epidemiology* 6: 356–65.

Greenland S. (1997). Second-stage least squares versus penalized quasi-likelihood for fitting hierarchical models in epidemiologic analyses. *Statistics in Medicine* 16: 515–26.

Greenland S. (1999a). Multilevel modeling and model averaging. *Scandinavian Journal of Work, Environment and Health* 25(Suppl. 4): 43–8.

Greenland S. (1999b). A unified approach to the analysis of case-distribution (case-only) studies. *Statistics in Medicine* 18: 1–15.

Greenland S. (2000). Principles of multilevel modelling. *International Journal of Epidemiology* 29: 158–67.

Greenland S. (2001). Ecologic versus individual-level sources of bias in ecologic estimates of contextual health effects. *International Journal of Epidemiology* 30: 1343–50.

Greenland S. (2002). A review of multilevel theory for ecologic analyses. *Statistics in Medicine* 21: 389–95.

Greenland S. (2005). Multiple-bias modelling for analysis of observational data. *Journal of the Royal Statistical Society, Series A* 168: 267–91.

Greenland S. (2007). Bayesian perspectives for epidemiological research. II. Regression analysis. *International Journal of Epidemiology* 36: 195–202.

Greenland S., Brenner H. (1993). Correcting for non-differential misclassification in ecological analyses. *Applied Statistics* 42: 117–126.

Greenland S., Finkle W.D. (1995). A critical look at methods for handling missing covariates in epidemiologic regression analyses. *American Journal of Epidemiology* 142: 1255–64.

Greenland S., Kleinbaum D.G. (1983). Correcting for misclassification in two-way tables and matched-pair studies. *International Journal of Epidemiology* 12: 93–7.

Greenland S., Morgenstern H. (1989). Ecological bias, confounding, and effect modification. *International Journal of Epidemiology* 18: 269–74.

Greenland S., O'Rourke K. (2001). On the bias produced by quality scores in meta-analysis, and a hierarchical view of proposed solutions. *Biostatistics* 2: 463–71.

Greenland S., Pearl J., Robins J.M. (1999a). Causal diagrams for epidemiologic research. *Epidemiology* 10: 37–48.

Greenland S., Robins J. (1988). Conceptual problems in the definition and interpretation of attributable fractions. *American Journal of Epidemiology* 128: 1185–97.

Greenland S., Robins J. (1994). Ecologic studies—Biases, misconceptions, and counterexamples. *American Journal of Epidemiology* 139: 747–60.

Greenland S., Robins J.M., Pearl J. (1999b). Confounding and Collapsibility in Causal Inference. *Statistical Science* 14: 29–46.

Greenland S., Thomas D. (1982). On the need for the rare disease assumption in case-control studies. *American Journal of Epidemiology* 116: 547–53.

Greenland S., Thomas D.C., Morgenstern H. (1986). The rare-disease assumption revisited. A critique of "estimators of relative risk for case-control studies." *American Journal of Epidemiology* 124: 869–83.

Grimson R.C. (1987). Apportionment of risk among environmental exposures: application to asbestos exposure and cigarette smoking. *Journal of Occupational Medicine* 29: 253–5.

Grosche B., Kreuzer M., Kreisheimer M., Schnelzer M., Tschense A. (2006). Lung cancer risk among German male uranium miners: a cohort study, 1946–1998. *British Journal of Cancer* 95: 1280–7.

Grufferman S. (1977). Clustering and aggregation of exposures in Hodgkin's disease. *Cancer* 39: 1829–33.

Gryparis A., Coull B.A., Schwartz J., Suh H.H. (2007). Semiparametric latent variable regression models for spatiotemporal modelling of mobile source particles in the greater Boston area. *Journal of the Royal Statistical Society: Series C. (Applied Statistics)* 56: 183–209.

Guerro V.M., Johnson R.A. (1982). Use of the Box–Cox transformation with binary response models. *Biometrika* 69: 309–14.

Gulliver J., Briggs D.J. (2005). Time-space modeling of journey-time exposure to traffic-related air pollution using GIS. *Environmental Research* 97: 10–25.

Guthrie K.A., Sheppard L. (2001). Overcoming biases and misconceptions in ecological studies. *Journal of the Royal Statistical Society, Series A* 164: 141–54.

Guthrie K.A., Sheppard L., Wakefield J. (2002). A hierarchical aggregate data model with spatially correlated disease rates. *Biometrics* 58: 898–905.

Haines A., Patz J.A. (2004). Health effects of climate change. *Journal of the American Medical Association* 291: 99–103.

Hakoda M., Akiyama M., Hirai Y., Kyoizumi S., Awa A.A. (1988). In vivo mutant T cell frequency in atomic bomb survivors carrying outlying values of chromosome aberration frequencies. *Mutation Research* 202: 203–8.

Hamada N., Matsumoto H., Hara T., Kobayashi Y. (2007). Intercellular and intracellular signaling pathways mediating ionizing radiation-induced bystander effects. *Journal of Radiation Research (Tokyo)* 48: 87–95.

Hamilton T.E., Davis S., Onstad L., Kopecky K.J. (2005). Hyperparathyroidism in persons exposed to iodine-131 from the Hanford Nuclear Site. *Journal of Clinical Endocrinology and Metabolism* 90: 6545–8.

Hammitt J.K., Graham J.D. (1999). Willingness to pay for health protection: inadequate sensitivity to probability? *Journal of Risk and Uncertainty* 18.

Hammond E.C. (1966). Smoking in relation to the death rates of one million men and women. *National Cancer Institute Monographs* 19: 127–204.

Harel O., Zhou X.H. (2007). Multiple imputation: review of theory, implementation and software. *Statistics in Medicine* 26: 3057–77.

Hastie T., Tibshirani R. (1990). *Generalized additive models*. London, Chapman and Hall.

Hatch M., Ron E., Bouville A., Zablotska L., Howe G. (2005). The Chernobyl disaster: cancer following the accident at the Chernobyl nuclear power plant. *Epidemiologic Reviews* 27: 56–66.

Hatch M., Susser M., Beyea J. (1997). Comments on "A reevaluation of cancer incidence near the Three Mile Island nuclear plant". *Environmental Health Perspectives* 105: 12.

Hatch M.C., Beyea J., Nieves J.W., Susser M. (1990). Cancer near the Three Mile Island nuclear plant: radiation emissions. *American Journal of Epidemiology* 132: 397–412; discussion 413–17.

Hatch M.C., Wallenstein S., Beyea J., Nieves J.W., Susser M. (1991). Cancer rates after the Three Mile Island nuclear accident and proximity of residence to the plant. *American Journal of Public Health* 81: 719–24.

Hauptmann M., Berhane K., Langholz B., Lubin J. (2001). Using splines to analyse latency in the Colorado Plateau uranium miners cohort. *Journal of Epidemiology and Biostatistics* 6: 417–24.

Haynes R.M. (1988). The distribution of domestic radon concentrations and lung cancer mortality in England and Wales. *Radiation Protection and Dosimetry* 25: 93–6.

Hazelton W.D., Clements M.S., Moolgavkar S.H. (2005). Multistage carcinogenesis and lung cancer mortality in three cohorts. *Cancer Epidemiology, Biomarkers, and Prevention* 14: 1171–81.

Hazelton W.D., Luebeck E.G., Heidenreich W.F., Moolgavkar S.H. (2001). Analysis of a historical cohort of Chinese tin miners with arsenic, radon, cigarette smoke, and pipe smoke exposures using the biologically based two-stage clonal expansion model. *Radiation Research* 156: 78–94.

Heath C.W., Jr, Bond P.D., Hoel D.G., Meinhold C.B. (2004). Residential radon exposure and lung cancer risk: commentary on Cohen's county-based study. *Health Physics* 87: 647–55; discussion 656–8.

Hedley A.J., Wong C.M., Thach T.Q., Ma S., Lam T.H., Anderson H.R. (2002). Cardiorespiratory and all-cause mortality after restrictions on sulphur content of fuel in Hong Kong: an intervention study. *Lancet* 360: 1646–52.

HEI (2004). Measuring the health impact of actions taken to improve air quality (RFA 04-4; available at http://www.healtheffects.org/RFA/RFA2004Fall.pdf). Boston, MA, Health Effects Institute.

HEI Accountability Working Group (2003). Assessing health impact of air quality regulations: concepts and methods for accountability research (Communication No. 11). Boston, Health Effects Institute: 113.

Heidenreich W.F., Brugmans M.J., Little M.P., Leenhouts H.P., Paretzke H.G., Morin M., Lafuma J. (2000). Analysis of lung tumour risk in radon-exposed rats: an intercomparison of multi-step modelling. *Radiation and Environmental Biophysics* 39: 253–64.

Heidenreich W.F., Jacob P., Paretzke H.G. (1997). Exact solutions of the clonal expansion model and their application to the incidence of solid tumors of atomic bomb survivors. *Radiation and Environmental Biophysics* 36: 45–58.

Heidenreich W.F., Luebeck E.G., Hazelton W.D., Paretzke H.G., Moolgavkar S.H. (2002a). Multistage models and the incidence of cancer in the cohort of atomic bomb survivors. *Radiation Research* 158: 607–14.

Heidenreich W.F., Luebeck E.G., Moolgavkar S.H. (2004a). Effects of exposure uncertainties in the TSCE model and application to the Colorado miners data. *Radiation Research* 161: 72–81.

Heidenreich W.F., Paretzke H.G. (2001). The two-stage clonal expansion model as an example of a biologically based model of radiation-induced cancer. *Radiation Research* 156: 678–81.

Heidenreich W.F., Tomasek L., Rogel A., Laurier D., Tirmarche M. (2004b). Studies of radon-exposed miner cohorts using a biologically based model: comparison of current Czech and French data with historic data from China and Colorado. *Radiation and Environmental Biophysics* 43: 247–56.

Heidenreich W.F., Wellmann J., Jacob P., Wichmann H.E. (2002b). Mechanistic modelling in large case-control studies of lung cancer risk from smoking. *Statistics in Medicine* 21: 3055–70.

Heinrich J., Hoelscher B., Frye C., Meyer I., Pitz M., Cyrys J., Wjst M., Neas L., Wichmann H.E. (2002). Improved air quality in reunified Germany and decreases in respiratory symptoms. *Epidemiology* 13: 394–401.

Heinrich J., Hoelscher B., Wichmann H.E. (2000). Decline of ambient air pollution and respiratory symptoms in children. *American Journal of Respiratory and Critical Care Medicine* 161: 1930–6.

Henderson R., Diggle P., Dobson A. (2000). Joint modelling of longitudinal measurements and event time data. *Biostatistics* 1: 465–80.

Hernan M.A., Brumback B., Robins J.M. (2000). Marginal structural models to estimate the causal effect of zidovudine on the survival of HIV-positive men. *Epidemiology* 11: 561–70.

Hicks M.H. (2007). Mortality in Iraq. *Lancet* 369: 101–2; author reply 103–4.

Hill A.B. (1965). The environment and disease: association or causation? *Journal of the Royal Society of Medicine* 58: 295–300.

Hoeffding (1948). A class of statistics with asymptotically normal distributions. *Annals of Mathematical Statistical* 19: 293–325.

Hoel D.G. (1979). Animal experimentation and its relevance to man. *Environmental Health Perspectives* 32: 25–30.

Hoel D.G., Haseman J.K., Hogan M.D., Huff J., McConnell E.E. (1988). The impact of toxicity on carcinogenicity studies: implications for risk assessment. *Carcinogenesis* 9: 2045–52.

Hoel D.G., Kaplan N.L., Anderson M.W. (1983). Implication of nonlinear kinetics on risk estimation in carcinogenesis. *Science* 219: 1032–7.

Hoel D.G., Li P. (1998). Threshold models in radiation carcinogenesis. *Health Physics* 75: 241–50.

Hoeting J.A., Madigan D., Raftery A.E., Volinsky C.T. (1999). Bayesian model averaging: a tutorial. *Statistical Science* 14: 382–417.

Hoffman F.O., Ruttenber J., Apostoaei A.I., Carroll R.J., Greenland S. (2006). The Hanford Thyroid Disease Study: An alternative view of the findings. *Health Physics*: under review.

Hogan J.W., Laird N.M. (1997). Mixture models for the joint distribution of repeated measures and event times. *Statistics in Medicine* 16: 239–57.

Hogan J.W., Laird N.M. (1998). Increasing efficiency from censored survival data by using random effects to model longitudinal covariates. *Statistical Methods in Medical Research* 7: 28–48.

Hogue C.J., Gaylor D.W., Schulz K.F. (1983). Estimators of relative risk for case-control studies. *American Journal of Epidemiology* 118: 396–407.

Hoh J., Ott J. (2003). Mathematical multi-locus approaches to localizing complex human trait genes. *Nature Review Genetics* 4: 701–9.

Holcroft C.A., Spiegelman D. (1999). Design of validation studies for estimating the odds ratio of exposure–disease relationships when exposure is misclassified. *Biometrics* 55: 1193–201.

Holford T.R. (1976). Life tables with concomitant information. *Biometrics* 32: 587–97.

Holford T.R. (2006). Approaches to fitting age–period–cohort models with unequal intervals. *Statistics in Medicine* 25: 977–93.

Holford T.R., Stack C. (1995). Study design for epidemiologic studies with measurement error. *Stat Methods Med Res* 4: 339–58.

Hornung R.W., Deddens J.A., Roscoe R.J. (1998). Modifiers of lung cancer risk in uranium miners from the Colorado Plateau. *Health Physics* 74: 12–21.

Horvitz D., Thompson D. (1952). A generalization of sampling without replacement from a finite population. *Journal of the American Statistical Association* 47: 663–85.

Hougaard P., Harvald B., Holm N. (1992). Measuring similarities between the lifetimes of adult Danish twin born between 1881-1930. *Journal of the American Statistical Association* 87: 17–24.

Houghton J., Ding Y., Griggs M., Noguer M., van der Linden P. (2001). *Climate Change 2001: The Scientific Basis (Report of the U.N. Intergovernmental Panel on Climate Change).* Cambridge, Cambridge University Press.

Huber P.J. (1967). The behavior of maximum likelihood estimates under nonstandard conditions. *Proceedings of the Fifth Berkeley Symposium on Mathematical Statistics and Probability.*

Huberman M., Langholz B. (1999). Application of the missing-indicator method in matched case-control studies with incomplete data. *American Journal of Epidemiology* 150: 1340–5.

Hunter D.J. (2005). Gene–environment interactions in human diseases. *Nature Reviews Genetics* 6: 287–98.

IOM (2007). *Evaluation of the presumptive disability decision-making process for veterans.* Washington, DC, National Academy Press.

Isager H., Larsen S. (1980). Pre-morbid factors in Hodgkin's disease III. School contact between patients. *Scandinavian Journal of Haematology* 25: 158–64.

Jablon S. (1971). Atomic bomb radiation dose estimation at ABCC (TR 23-71). Hiroshima, Radiation Effects Research Foundation.

Jackson C., Best a.N., Richardson S. (2008). Hierarchical related regression for combining aggregate and individual data in studies of socio-economic disease risk factors. *Journal of the Royal Statistical Society, Series A. (Statistics in Society)* 0: in press.

Jackson C., Best N., Richardson S. (2006). Improving ecological inference using individual-level data. *Statistics in Medicine* 25: 2136–159.

Jacob P., Meckbach R., Sokolnikov M., Khokhryakov V.V., Vasilenko E. (2007). Lung cancer risk of Mayak workers: modelling of carcinogenesis and bystander effect. *Radiation and Environmental Biophysics* 46: 383–94.

Jacob V., Jacob P. (2004). Modelling of carcinogenesis and low-dose hypersensitivity: an application to lung cancer incidence among atomic bomb survivors. *Radiation and Environmental Biophysics* 42: 265–73.

Jacob V., Jacob P., Meckbach R., Romanov S.A., Vasilenko E.K. (2005). Lung cancer in Mayak workers: interaction of smoking and plutonium exposure. *Radiation and Environmental Biophysics* 44: 119–29.

Janes H., Dominici F., Zeger S.L. (2007). Trends in air pollution and mortality: an approach to the assessment of unmeasured confounding. *Epidemiology* 18: 416–23.

Janes H., Sheppard L., Lumley T. (2005). Overlap bias in the case-crossover design, with application to air pollution exposures. *Statistics in Medicine* 24: 285–300.

Jansen R.C. (2003). Studying complex biological systems using multifactorial perturbation. *Nature Reviews Genetics* 4: 145–51.

Jarup L. (2004). Health and environment information systems for exposure and disease mapping, and risk assessment. *Environmental Health Perspectives* 112: 995–7.

Jarup L., Best N. (2003). Editorial comment on Geographical differences in cancer incidence in the Belgian Province of Limburg by Bruntinx and colleagues. *European Journal of Cancer* 39: 1973–5.

Jarup L., Best N., Toledano M.B., Wakefield J., Elliott P. (2002). Geographical epidemiology of prostate cancer in Great Britain. *International Journal of Cancer* 97: 695–9.

Jerrett M., Burnett R., Willis A., Krewski D., Goldberg M., DeLuca P., Finkelstein N. (2003). Spatial analysis of the air pollution–mortality relationship in the context of ecologic confounders. *Journal of Toxicology and Environmental Health, Part A* 66: 1735–78.

Jerrett M., Burnett R.T., Ma R., Pope C.A., 3rd, Krewski D., Newbold K.B., Thurston G., Shi Y., Finkelstein N., Calle E.E., Thun M.J. (2005). Spatial analysis of air pollution and mortality in Los Angeles. *Epidemiology* 16: 727–36.

Jerrett M., Shankardass K., Berhane K., Gauderman W.J., Kunzli N., Avol E., Gilliland F., Lurmann F., Molitor J.N., Molitor J.T., Thomas D.C., Peters J., McConnell R. (2008). Traffic-related air pollution and asthma onset in children: a prospective cohort study with individual exposure measurement. *Environmental Health Perspectives* 116: 1433–8.

Johnston R.J., Hay A.M. (1982). On the parameters of uniform swing in single-member constituency electoral systems. *Environment and Planning A* 14: 61–74.

Kai M., Luebeck E.G., Moolgavkar S.H. (1997). Analysis of the incidence of solid cancer among atomic bomb survivors using a two-stage model of carcinogenesis. *Radiation Research* 148: 348–58.

Kaiser J. (2003). How much are human lives and health worth? *Science* 299: 36–7.

Kalbfleisch J., Prentice R. (1980). *The statistical analysis of failure time data*, Wiley-Interscience.

Kanaroglou P.S., Jerrett M., Morrison J., Beckerman B., Arain M.A., Gilbert N.L., Brook J.R. (2005). Establishing an air pollution monitoring network for intra-urban population exposure assessment: A location-allocation approach. *Atmospheric Environment* 39: 2399–409.

Kaplan S., Novikov I., Modan B. (1998). A methodological note on the selection of friends as controls. *International Journal of Epidemiology* 27: 727–9.

Kass R., Raftery A. (1995). Bayes factors. *Journal of the American Statistical Association* 90: 773–95.

Katsouyanni K., Touloumi G., Spix C., Schwartz J., Balducci F., Medina S., Rossi G., Wojtyniak B., Sunyer J., Bacharova L., Schouten J.P., Ponka A., Anderson H.R. (1997). Short-term effects of ambient sulphur dioxide and particulate matter on mortality in 12 European cities: results from time series data from the APHEA project. Air Pollution and Health: a European Approach. *British Medical Journal* 314: 1658–63.

Katsouyanni K., Zmirou D., Spix C., Sunyer J., Schouten J.P., Ponka A., Anderson H.R., Le Moullec Y., Wojtyniak B., Vigotti M.A., et al. (1995). Short-term effects of air pollution on health: a European approach using epidemiological time-series data. The APHEA project: background, objectives, design. *European Respiratory Journal* 8: 1030–8.

Keller C.F. (2003). Global warming: the balance of evidence and its policy implications. A review of the current state-of-the-controversy. *Scientific World Journal* 3: 357–411.

Keller C.F. (2007). Global warming 2007. An update to global warming: the balance of evidence and its policy implications. *Scientific World Journal* 7: 381–99.

Kellerer A.M. (1996). Radiobiological challenges posed by microdosimetry. *Health Physics* 70: 832–6.

Kellerer A.M., Rossi H.D. (1972). Theory of dual radiation action. *Current Topics in Radiation Research Quarterly* 8: 85–158.

Kellerer A.M., Rossi H.H. (1978). A generalized formulation of dual radiation action. *Radiation Research* 75: 471–88.

Kelsall J.E., Diggle P.J. (1998). Spatial variation in risk of disease: a nonparametric binary regression approach. *Applied Statistics* 47: 559–73.

Kelsall J.E., Zeger S.L., Samet J.M. (1999). Frequency Domain Log-linear Models; Air Pollution and Mortality. *Journal of the Royal Statistical Society: Series C. (Applied Statistics)* 48: 331–44.

Kempf A.M., Remington P.L. (2007). New challenges for telephone survey research in the twenty-first century. *Annual Review of Public Health* 28: 113–26.

Kerber R.A., Till J.E., Simon S.L., Lyon J.L., Thomas D.C., Preston-Martin S., Rallison M.L., Lloyd R.D., Stevens W. (1993). A cohort study of thyroid disease in relation to fallout from nuclear weapons testing. *Journal of the American Medical Association* 270: 2076–82.

Kiefer J., Wolfowitz J. (1956). Consistency of the Maximum Likelihood Estimator in the Presence of Infinitely Many Incidental Parameters. *The Annals of Mathematical Statistics* 27: 887–906.

Kinston W., Rosser R. (1974). Disaster: effects on mental and physical state. *Journal of Psychosomatic Research* 18: 437–56.

Kleinbaum D.G., Kupper L.L., Morgentern H. (1982). *Epidemiologic research: Principles and quantitative methods*. Belmont, CA, Lifetime Learning Publications.

Knoke J.D., Shanks T.G., Vaughn J.W., Thun M.J., Burns D.M. (2004). Lung cancer mortality is related to age in addition to duration and intensity of cigarette smoking: an analysis of CPS-I data. *Cancer Epidemiology, Biomarkers, and Prevention* 13: 949–57.

Knorr-Held L., Rasser G. (2000). Bayesian detection of clusters and discontinuities in disease maps. *Biometrics* 56: 13–21.

Knox E.G., Bartlett M.S. (1964). The detection of space–time interactions. *Applied Statistics* 13: 25–30.

Kocher D.C., Apostoaei A.I., Henshaw R.W., Hoffman F.O., Schubauer-Berigan M.K., Stancescu D.O., Thomas B.A., Trabalka J.R., Gilbert E.S., Land C.E. Interactive radioepidemiological program (IREP): A web-based tool for estimating probability of causation/assigned share of radiogenic cancers. *Health Phys*, 2008; 95: 119–47.

Kohli S., Noorlind Brage H., Lofman O. (2000). Childhood leukaemia in areas with different radon levels: a spatial and temporal analysis using GIS. *Journal of Epidemiology and Community Health* 54: 822–6.

Koop L., Tole L. (2004). Measuring the health effects of air pollution: to what extent can we really say that people are dying from bad air? *Journal of Environmental Economics and Management* 47: 30–54.

Kooperberg C., Ruczinski I. (2005). Identifying interacting SNPs using Monte Carlo logic regression. *Genetic Epidemiology* 28: 157–70.

Koopman J.S. (1977). Causal models and sources of interaction. *American Journal of Epidemiology* 106: 439–44.

Kopecky K.J., Davis S., Hamilton T.E., Saporito M.S., Onstad L.E. (2004). Estimation of thyroid radiation doses for the Hanford thyroid disease study: results and implications for statistical power of the epidemiological analyses. *Health Physics* 87: 15–32.

Kopecky K.J., Onstad L., Hamilton T.E., Davis S. (2005). Thyroid ultrasound abnormalities in persons exposed during childhood to 131I from the Hanford nuclear site. *Thyroid* 15: 604–13.

Korn E.L., Whittemore A.S. (1979). Methods for analyzing panel studies of acute health effects of air pollution. *Biometrics* 35: 795–802.

Koshurnikova N.A., Gilbert E.S., Sokolnikov M., Khokhryakov V.F., Miller S., Preston D.L., Romanov S.A., Shilnikova N.S., Suslova K.G., Vostrotin V.V. (2000). Bone cancers in Mayak workers. *Radiation Research* 154: 237–45.

Kovats R.S., Bouma M.J., Hajat S., Worrall E., Haines A. (2003). El Nino and health. *Lancet* 362: 1481–9.

Kraft P., Hunter D. (2005). Integrating epidemiology and genetic association: the challenge of gene–environment interaction. *Philosophical Transactions of the Royal Society London, Series B* 360: 1609–16.

Kramer U., Behrendt H., Dolgner R., Ranft U., Ring J., Willer H., Schlipkoter H.W. (1999). Airway diseases and allergies in East and West German children during the first 5 years after reunification: time trends and the impact of sulphur dioxide and total suspended particles. *International Journal of Epidemiology* 28: 865–73.

Kreft I., de Leeuw J. (1998). *Introducing multilevel modeling*. London, Sage.

Krestinina L.Y., Davis F., Ostroumova E., Epifanova S., Degteva M., Preston D., Akleyev A. (2007). Solid cancer incidence and low-dose-rate radiation exposures in the Techa River cohort: 1956–2002. *International Journal of Epidemiology* 36: 1038–46.

Krewski D., Burnett R., Goldberg M., Hoover B.K., Siemiatycki J., Jerrett M., Abrahamowicz M., White W. (2003). Overview of the reanalysis of the Harvard Six Cities Study and American Cancer Society study of particulate air pollution and mortality. *Journal of Toxicology and Environmental Health, Part A* 66: 1507–52.

Krewski D., Burnett R., Jerrett M., Pope C.A., Rainham D., Calle E., Thurston G., Thun M. (2005a). Mortality and Long-Term Exposure to Ambient Air Pollution: Ongoing Analyses Based on the American Cancer Society Cohort. *Journal of Toxicology and Environmental Health, Part A* 68: 1093–1109.

Krewski D., Lubin J.H., Zielinski J.M., Alavanja M., Catalan V.S., Field R.W., Klotz J.B., eacute, tourneau E.G., Lynch C.F., Lyon J.L., Sandler D.P., Schoenberg J.B., Steck D.J.,

Stolwijk J.A., Weinberg C., Wilcox H.B. (2006). A Combined Analysis of North American Case-Control Studies of Residential Radon and Lung Cancer. *Journal of Toxicology and Environmental Health, Part A* 69: 533–97.

Krewski D., Lubin J.H., Zielinski J.M., Alavanja M., Catalan V.S., Field R.W., Klotz J.B., Letourneau E.G., Lynch C.F., Lyon J.I., et al. (2005b). Residential radon and risk of lung cancer: a combined analysis of 7 North American case-control studies. *Epidemiology* 16: 137–45.

Krieger N. (2003). Place, space, and health: GIS and epidemiology. *Epidemiology* 14: 384–5.

Krupnick A. (2002). The value of reducing risk of death: a policy perspective. *Journal of Policy Analysis and Management* 21: 275–82.

Krupnick A., Morgenstern R. (2002). The future of benefit-cost analyses of the Clean Air Act. *Annual Review of Public Health* 23: 427–48.

Krzyzanowski M., Lebret E., Younes M. (1995). Methodology for assessment of exposure to environmental factors in application to epidemiological studies: preface. *Science of The Total Environment* 168: 91–2.

Kuha J. (1994). Corrections for exposure measurement error in logistic regression models with an application to nutritional data. *Statistics in Medicine* 13: 1135–48.

Kulldorff M. (2001). Prospective time periodic geographical disease surveillance using a scan statistic. *Journal of the Royal Statistical Society. Series A* 164: 61–72.

Kunii O., Kanagawa S., Yajima I., Hisamatsu Y., Yamamura S., Amagai T., Ismail I.T. (2002). The 1997 haze disaster in Indonesia: its air quality and health effects. *Arch Environ Health* 57: 16–22.

Kunzli N., Avol E., Wu J., Gauderman W.J., Rappaport E., Millstein J., Bennion J., McConnell R., Gilliland F.D., Berhane K., Lurmann F., Winer A., Peters J.M. (2006). Health effects of the 2003 Southern California wildfires on children. *American Journal of Respiratory and Critical Care Medicine* 174: 1221–8.

Kunzli N., Kaiser R., Medina S., Studnicka M., Chanel O., Filliger P., Herry M., Horak F., Jr, Puybonnieux-Texier V., Quenel P., et al. (2000). Public-health impact of outdoor and traffic-related air pollution: a European assessment. *Lancet* 356: 795–801.

Kunzli N., McConnell R., Bates D., Bastain T., Hricko A., Lurmann F., Avol E., Gilliland F., Peters J. (2003). Breathless in Los Angeles: the exhausting search for clean air. *American Journal of Public Health* 93: 1494–9.

Kunzli N., Medina S., Kaiser R., Quenel P., Horak F., Jr, Studnicka M. (2001). Assessment of deaths attributable to air pollution: should we use risk estimates based on time series or on cohort studies? *American Journal of Epidemiology* 153: 1050–5.

Kunzli N., Perez L., Lurmann F., Hricko A., Penfold B., McConnell R. (2008). An attributable risk model for exposures assumed to cause both chronic disease and its exacerbations. *Epidemiology* 19: 179–85.

Kupper L.L., McMichael A.J., Spirtas R. (1975). A hybrid epidemiologic study design useful in estimating relative risk. *Journal of the American Statistical Association* 70: 524–8.

Lagarde F., Pershagen G. (1999). Parallel analyses of individual and ecologic data on residential radon, cofactors, and lung cancer in Sweden. *American Journal of Epidemiology* 149: 268–74.

Laird N. (1978). Nonparametric maximum likelihood estimation of a mixing distribution. *Journal of the American Statistical Association* 73: 805–10.

Lakagos S., Mosteller F. (1986). Assigned shares in compensation for radiation-related cancers (with discussion). *Risk Analysis* 6: 345–80.

Landrigan P.J., Lioy P.J., Thurston G., Berkowitz G., Chen L.C., Chillrud S.N., Gavett S.H., Georgopoulos P.G., Geyh A.S., Levin S., Perera F., Rappaport S.M., Small C. (2004). Health and environmental consequences of the world trade center disaster. *Environmental Health Perspectives* 112: 731–9.

Lange J.L., Schwartz D.A., Doebbeling B.N., Heller J.M., Thorne P.S. (2002). Exposures to the Kuwait oil fires and their association with asthma and bronchitis among gulf war veterans. *Environmental Health Perspectives* 110: 1141–6.

Langholz B. (2003). Counter-matching. In: *Encyclopedia of Biostatistics* (2nd edn). Armitage P, Colton T. (eds.). New York, John Wiley & Sons.

Langholz B., Borgan O. (1995). Counter-matching: a stratified nested case-control sampling method. *Biometrika* 82: 69–79.

Langholz B., Goldstein L. (1996). Risk set sampling in epidemiologic cohort studies. *Statistical Science* 11: 35–53.

Langholz B., Goldstein L. (2001). Conditional logistic analysis of case-control studies with complex sampling. *Biostatistics* 2: 63–84.

Langholz B., Thomas D., Xiang A., Stram D. (1999). Latency analysis in epidemiologic studies of occupational exposures: application to the Colorado Plateau uranium miners cohort. *American Journal of Industrial Medicine* 35: 246–56.

Langholz B., Thomas D.C. (1990). Nested case-control and case-cohort sampling: a critical comparison. *American Journal of Epidemiology* 131: 169–76.

Langholz B., Thomas D.C., Stovall M., Smith S., Boice J.D., Shore R.E., WECARE Study Group, Bernstein J. (2007). Analysis of radiation dose-response with tumor location and location-specific dose in the WECARE study of second breast cancer. *Radiation Research* 167: 358–9.

Lasserre V., Guihenneuc-Jouyaux C., Richardson S. (2000). Biases in ecological studies: utility of including within-area distribution of confounders. *Statistics in Medicine* 19: 45–59.

Laurier D., Valenty M., Tirmarche M. (2001). Radon exposure and the risk of leukemia: a review of epidemiological studies. *Health Physics* 81: 272–88.

Lavori P.W., Dawson R., Shera D. (1995). A multiple imputation strategy for clinical trials with truncation of patient data. *Statistics in Medicine* 14: 1913–25.

Lawson A. (2006). *Statistical Methods in Spatial Epidemiology* (2nd edn.). New York, Wiley.

Lawson A., Biggeri A., Dean C., Waller L. (2000a). Disease mapping with a focus on evaluation: preface. *Statistics in Medicine* 19: 2201–2.

Lawson A.B. (1993). On the analysis of mortality events associated with a prespecified fixed point. *Journal of the Royal Statistical Society, Series A* 156: 363–77.

Lawson A.B., Biggeri A., Bohning D., Lassaffre E., Viel J.F., Bertollini R., Eds. (1999). *Disease mapping and risk assessment for public health*. New York, Wiley.

Lawson A.B., Biggeri A.B., Boehning D., Lesaffre E., Viel J.F., Clark A., Schlattmann P., Divino F. (2000b). Disease mapping models: an empirical evaluation. Disease Mapping Collaborative Group. *Statistics in Medicine* **19**: 2217–41.

Le N.D., Zidek J.V. (1992). Interpolation with uncertain spatial covariances: a Bayesian alternative to kriging. *Journal of Multivariate Analysis* **43**: 351–74.

Leamer E.E. (1978). *Specification searches: ad hoc inference with nonexperimental data.* New York, Wiley.

Leaning J. (2004). Diagnosing genocide—the case of Darfur. *New England Journal of Medicine* **351**: 735–8.

Lechat M.F. (1990). The epidemiology of health effects of disasters. *Epidemiologic Reviews* **12**: 192–8.

Leffondre K., Abrahamowicz M., Siemiatycki J., Rachet B. (2002). Modeling smoking history: a comparison of different approaches. *American Journal of Epidemiology* **156**: 813–23.

Leffondre K., Abrahamowicz M., Xiao Y., Siemiatycki J. (2006). Modelling smoking history using a comprehensive smoking index: application to lung cancer. *Statistics in Medicine* **25**: 4132–46.

Levin M.L. (1953). The occurrence of lung cancer in man. *Acta Unio Int Contra Cancrum* **19**: 531–41.

Li X., Song X., Gray R.H. (2004). Comparison of the missing-indicator method and conditional logistic regression in 1:m matched case-control studies with missing exposure values. *American Journal of Epidemiology* **159**: 603–10.

Li Y.F., Gauderman W.J., Avol E., Dubeau L., Gilliland F.D. (2006). Associations of Tumor Necrosis Factor G-308A with Childhood Asthma and Wheezing. *American Journal of Respiratory and Critical Care Medicine.*

Li Y.F., Langholz B., Salam M.T., Gilliland F.D. (2005a). Maternal and grandmaternal smoking patterns are associated with early childhood asthma. *Chest* **127**: 1232–41.

Li Y.F., Tsao Y.H., Gauderman W.J., Conti D.V., Avol E., Dubeau L., Gilliland F.D. (2005b). Intercellular adhesion molecule-1 and childhood asthma. *Human Genetics* **117**: 476–84.

Liang K.Y., Zeger S.L. (1986). Longitudinal data analysis using generalized linear models. *Biometrika* **73**: 13–22.

Liddell FDK, McDonald J.C., C T.D., V C.S. (1977). Methods of cohort analysis: appraisal by application to asbestos mining. *Journal of the Royal Statistical Society, Series A* **140**: 469–91.

Link M.W., Kresnow M.J. (2006). The future of random-digit-dial surveys for injury prevention and violence research. *American Journal of Preventive Medicine* **31**: 444–50.

Lipsitz S.R., Parzen M., Ewell M. (1998). Inference Using Conditional Logistic Regression with Missing Covariates. *Biometrics* **54**: 295–303.

Littel R.C., Milliken G.A., Stroup W.W., Wolfinger R.D. (1996). *SAS system for mixed models.* Cary, NC, SAS Institute Inc.

Little M.P. (1995a). Are two mutations sufficient to cause cancer? Some generalizations of the two-mutation model of carcinogenesis of Moolgavkar, Venzon, and Knudson, and of the multistage model of Armitage and Doll. *Biometrics* **51**: 1278–91.

Little M.P. (1996). Generalisations of the two-mutation and classical multi-stage models of carcinogenesis fitted to the Japanese atomic bomb survivor data. *Journal of Radiological Protection* **16**: 7–24.

Little M.P. (1997). Variations with time and age in the relative risks of solid cancer incidence after radiation exposure. *Journal of Radiological Protection* **17**: 159–77.

Little M.P. (2000). A comparison of the degree of curvature in the cancer incidence dose–response in Japanese atomic bomb survivors with that in chromosome aberrations measured in vitro. *International Journal of Radiation Biology* **76**: 1365–75.

Little M.P. (2002). Absence of evidence for differences in the dose-response for cancer and non-cancer endpoints by acute injury status in the Japanese atomic-bomb survivors. *International Journal of Radiation Biology* **78**: 1001–10.

Little M.P. (2004a). The bystander effect model of Brenner and Sachs fitted to lung cancer data in 11 cohorts of underground miners, and equivalence of fit of a linear relative risk model with adjustment for attained age and age at exposure. *Journal of Radiological Protection* **24**: 243–55.

Little M.P. (2004b). Threshold and other departures from linear-quadratic curvature in the non-cancer mortality dose–response curve in the Japanese atomic bomb survivors. *Radiation and Environmental Biophysics* **43**: 67–75.

Little M.P., Deltour I., Richardson S. (2000). Projection of cancer risks from the Japanese atomic bomb survivors to the England and Wales population taking into account uncertainty in risk parameters. *Radiation and Environmental Biophysics* **39**: 241–52.

Little M.P., Filipe J.A., Prise K.M., Folkard M., Belyakov O.V. (2005). A model for radiation-induced bystander effects, with allowance for spatial position and the effects of cell turnover. *Journal of Theoretical Biology* **232**: 329–38.

Little M.P., Hawkins M.M., Charles M.W., Hildreth N.G. (1992). Fitting the Armitage–Doll model to radiation-exposed cohorts and implications for population cancer risks. *Radiation Research* **132**: 207–21.

Little M.P., Haylock R.G., Muirhead C.R. (2002). Modelling lung tumour risk in radon-exposed uranium miners using generalizations of the two-mutation model of Moolgavkar, Venzon and Knudson. *International Journal of Radiation Biology* **78**: 49–68.

Little M.P., Li G. (2007). Stochastic modelling of colon cancer: is there a role for genomic instability? *Carcinogenesis* **28**: 479–87.

Little M.P., Muirhead C.R. (1996). Evidence for curvilinearity in the cancer incidence dose-response in the Japanese atomic bomb survivors. *International Journal of Radiation Biology* **70**: 83–94.

Little M.P., Muirhead C.R. (1998). Curvature in the cancer mortality dose response in Japanese atomic bomb survivors: absence of evidence of threshold. *International Journal of Radiation Biology* **74**: 471–80.

Little M.P., Muirhead C.R., Charles M.W. (1999). Describing time and age variations in the risk of radiation-induced solid tumour incidence in the Japanese atomic bomb survivors using generalized relative and absolute risk models. *Statistics in Medicine* **18**: 17–33.

Little M.P., Wakeford R. (2001). The bystander effect in C3H 10T cells and radon-induced lung cancer. *Radiation Research* 156: 695–9.

Little M.P., Wright E.G. (2003). A stochastic carcinogenesis model incorporating genomic instability fitted to colon cancer data. *Mathematical Biosciences* 183: 111–34.

Little R. (1992). Regression with missing X's: A review. *Journal of the American Statistical Association* 87: 1227–37.

Little R., Rubin D. (1989a). *Statistical analysis with missing data*. New York, Wiley.

Little RJA (1993). Pattern-mixture models for multivariate incomplete data. *Journal of the American Statistical Association* 88: 125–34.

Little RJA (1994). A class of pattern-mixture models for normal incomplete data. *Biometrika* 81: 471–83.

Little RJA (1995b). Modeling the drop-out mechanism in repeated-measures studies. *Journal of the American Statistical Association* 90: 1112–21.

Little RJA, Rubin D.B. (1989b). The analysis of social science data with missing values. *Sociological Methods Research* 18: 292–326.

Little RJA, Wang Y. (1996). Pattern-mixture models for multivariate incomplete data with covariates. *Biometrics* 52: 98–111.

Logan W.P. (1953). Mortality in the London fog incident, 1952. *Lancet* 1: 336–8.

Logue J.N., Melick M.E., Hansen H. (1981). Research issues and directions in the epidemiology of health effects of disasters. *Epidemiologic Reviews* 3: 140–62.

Longley P.A., Goodchild M.F., Maguire D.J., Rhind D.W. (2005). *Geographic Information Systems and Science*. New York, Wiley.

Lopez A.D., Mathers C.D., Ezzati M., Jamison D.T., Murray CJ (2006). Global and regional burden of disease and risk factors, 2001: systematic analysis of population health data. *Lancet* 367: 1747–57.

Lorber M., Gibb H., Grant L., Pinto J., Pleil J., Cleverly D (2007). Assessment of Inhalation Exposures and Potential Health Risks to the General Population that Resulted from the Collapse of the World Trade Center Towers. *Risk Analysis* 27: 1203–21.

Lubin J.H. (1998). On the discrepancy between epidemiologic studies in individuals of lung cancer and residential radon and Cohen's ecologic regression. *Health Physics* 75: 4–10.

Lubin J.H. (2002). The potential for bias in Cohen's ecological analysis of lung cancer and residential radon. *Journal of Radiological Protection* 22: 141–8.

Lubin J.H., Alavanja M.C., Caporaso N., Brown L.M., Brownson RC, Field R.W., Garcia-Closas M, Hartge P., Hauptmann M., Hayes R.B., et al. (2007a). Cigarette smoking and cancer risk: modeling total exposure and intensity. *American Journal of Epidemiology* 166: 479–89.

Lubin J.H., Boice J.D., Jr. (1997). Lung cancer risk from residential radon: meta-analysis of eight epidemiologic studies. *Journal of the National Cancer Institute* 89: 49–57.

Lubin J.H., Boice J.D., Jr, Edling C., Hornung R.W., Howe GR, Kunz E., Kusiak R.A., Morrison HI, Radford E.P., Samet J.M., et al. (1995). Lung cancer in radon-exposed miners and estimation of risk from indoor exposure. *Journal of the National Cancer Institute* 87: 817–27.

Lubin J.H., Boice JDJ, Edling C. (1994). Radon and lung cancer risk: a joint analysis of 11 underground miners studies. Bethesda, MD, US DHHS. NIH publ 94-3644.

Lubin J.H., Caporaso N., Wichmann H.E., Schaffrath-Rosario A, Alavanja M.C. (2007b). Cigarette smoking and lung cancer: modeling effect modification of total exposure and intensity. *Epidemiology* 18: 639–48.

Lubin J.H., Caporaso N.E. (2006). Cigarette smoking and lung cancer: modeling total exposure and intensity. *Cancer Epidemiology, Biomarkers, and Prevention* 15: 517–23.

Lubin J.H., Gaffey W. (1988). Relative risk models for assessing the joint effects of multiple factors. *American Journal of Industrial Medicine* 13: 149–67.

Lubin J.H., Gail M.H. (1984). Biased selection of controls for case-control analysis of cohort studies. *Biometrics* 40: 63–75.

Luebeck E.G., Heidenreich W.F., Hazelton W.D., Paretzke HG, Moolgavkar S.H. (1999). Biologically based analysis of the data for the Colorado uranium miners cohort: age, dose and dose-rate effects. *Radiation Research* 152: 339–51.

Luebeck E.G., Moolgavkar S.H. (2002). Multistage carcinogenesis and the incidence of colorectal cancer. *Proceedings of the National Academy of Sciences* 99: 15095–100.

Lumley T., Heagerty P. (1999). Weighted empirical adaptive variance estimators for correlated data regression. *Journal of the Royal Statistical Society: Series B (Statistical Methodology)* 61: 459–77.

Lundin F., Wagoner J., Archer V. (1971). Radon Daughter Exposure and Respiratory Cancer: Quantitative and Temporal Aspects. Report from the Epidemiological Study of United States Uranium Miners. Bethesda, National Cancer Institute.

Ma R., Krewski D., Burnett R.T. (2003). Random effects Cox models: A Poisson modelling approach. *Biometrika* 90: 157–69.

Ma X., Buffler P.A., Layefsky M., Does M.B., Reynolds P. (2004). Control selection strategies in case-control studies of childhood diseases. *American Journal of Epidemiology* 159: 915–21.

Mack T.M. (2004). *Cancer in the urban environment.* New York, Academic.

MacLehose R.F., Dunson D.B., Herring A.H., Hoppin J.A. (2007). Bayesian methods for highly correlated exposure data. *Epidemiology* 18: 199–207.

Maclure M. (1991). The case-crossover design: a method for studying transient effects on the risk of acute events. *American Journal of Epidemiology* 133: 144–53.

MacMahon B., Pugh T.F. (1970). *Epidemiology: principles and methods.* Boston, Little, Brown and Co.

Madigan D., Raftery A.E. (1994). Model Selection and Accounting for Model Uncertainty in Graphical Models Using Occam's Window. *Journal of the American Statistical Association* 89: 1535–46.

Maheswaran R., Craglia M. (eds.) (2004). *GIS in public health practice.* New York, CRC.

Maldonado G., Greenland S. (2002). Estimating causal effects. *International Journal of Epidemiology* 31: 422–9.

Mallick B., Hoffman F.O., Carroll R.J. (2002). Semiparametric regression modeling with mixtures of Berkson and classical error, with application to fallout from the Nevada Test Site. *Biometrics* 58: 13–20.

Mallows C.L. (1973). Some comments on C_p. *Technometrics* 15: 661–76.

Mantel N. (1967). The detection of disease clustering and a generalized regression approach. *Cancer Research* 27: 209–20.

Mantel N. (1973). Synthetic retrospective studies and related topics. *Biometrics* 29: 479–86.

Mantel N., Haenszel W. (1959). Statistical aspects of the analysis of data from retrospective studies of disease. *Journal of the National Cancer Institute* 22: 719–48.

Marshall J.R. (1989). The use of dual or multiple reports in epidemiologic studies. *Statistics in Medicine* 8: 1041–9; discussion 1071–3.

Marshall R.J. (1990). Validation study methods for estimating exposure proportions and odds ratios with misclassified data. *Journal of Clinical Epidemiology* 43: 941–7.

Martin A.E. (1964). Mortality and Morbidity Statistics and Air Pollution. *Proceedings of the Royal Society of Medicine* 57(Suppl.): 969–75.

Martinez F.D. (2007a). CD14, endotoxin, and asthma risk: actions and interactions. *Proceedings of the American Thoracic Society* 4: 221–5.

Martinez F.D. (2007b). Gene–environment interactions in asthma: with apologies to William of Ockham. *Proceedings of the American Thoracic Society* 4: 26–31.

Matsumoto H., Hamada N., Takahashi A., Kobayashi Y., Ohnishi T (2007). Vanguards of paradigm shift in radiation biology: radiation-induced adaptive and bystander responses. *Journal of Radiation Research (Tokyo)* 48: 97–106.

Mauer M.P., Cummings K.R., Carlson G.A. (2007). Health effects in New York State personnel who responded to the World Trade Center disaster. *Journal of Occupational and Environmental Medicine* 49: 1197–205.

McConnell R., Berhane K., Gilliland F., Islam T., Gauderman WJ, London S.J., Avol E., Rappaport E.B., Margolis H.G., Peters JM (2002a). Indoor risk factors for asthma in a prospective study of adolescents. *Epidemiology* 13: 288–95.

McConnell R., Berhane K., Gilliland F., London S.J., Islam T, Gauderman W.J., Avol E., Margolis H.G., Peters J.M. (2002b). Asthma in exercising children exposed to ozone: a cohort study. *Lancet* 359: 386–91.

McConnell R., Berhane K., Gilliland F., London S.J., Vora H, Avol E., Gauderman W.J., Margolis H.G., Lurman F., Thomas D.C., Peters JM (1999). Air pollution and bronchitic symptoms in Southern California children with asthma. *Environmental Health Perspectives* 107: 757–60.

McConnell R., Berhane K., Gilliland F., Molitor J., Thomas D, Lurmann F., Avol E., Gauderman W.J., Peters J.M. (2003). Prospective study of air pollution and bronchitic symptoms in children with asthma. *American Journal of Respiratory and Critical Care Medicine* 168: 790–97.

McConnell R., Berhane K., Yao L., Jerrett M., Lurmann F, Gilliland F., Kunzli N., Gauderman J., Avol E., Thomas D., Peters J (2006). Traffic, susceptibility, and childhood asthma. *Environmental Health Perspectives* 114: 766–72.

McKnight B., Cook L.S., Weiss N.S. (1999). Logistic regression analysis for more than one characteristic of exposure. *American Journal of Epidemiology* 149: 984–92.

McLaughlin J.K., Blot W.J., Mehl E.S., al. e. (1985). Problems in the use of dead controls in case-control studies: I. General results. *American Journal of Epidemiology* 121: 131–9.

McMichael A.J., Woodruff R.E., Hales S. (2006). Climate change and human health: present and future risks. *Lancet* 367: 859–69.

Meade M., Florin J., Gesler W. (1988). *Medical geography*. New York, Guilford.

Mehta P.S., Mehta A.S., Mehta S.J., Makhijani A.B. (1990). Bhopal tragedy's health effects. A review of methyl isocyanate toxicity. *Journal of the American Medical Association* 264: 2781–7.

Miettinen O.S. (1969). Individual matching with multiple controls in the case of all-or-none responses. *Biometrics* 25: 339–55.

Miettinen O.S. (1974). Proportion of disease caused or prevented by a given exposure, trait or intervention. *American Journal of Epidemiology* 99: 325–32.

Miettinen O.S. (1982a). Causal and preventive interdependence. Elementary principles. *Scandinavian Journal of Work, Environment and Health* 8: 159–68.

Miettinen O.S. (1982b). Design options in epidemiology research: an update. *Scandinavian Journal of Work, Environment and Health* 8(Suppl. 1), 1295–311.

Miettinen O.S. (1985). *Theoretical epidemiology: Principles of occurrence research in medicine*. New York, John Wiley & Sons.

Millstein J., Conti D.V., Gilliland F.D., Gauderman W.J. (2006). A testing framework for identifying susceptibility genes in the presence of epistasis. *American Journal of Human Genetics* 78: 15–27.

Mishra P.K., Dabadghao S., Modi G., Desikan P., Jain A., Mittra I (2007). Immune status in individuals exposed in-utero to methyl isocyanate during the Bhopal gas tragedy, a study after two decades. 13th International Congress of Immunology, Rio De Janeiro, Brazil.

Molitor J., Jerrett M., Chang C.C., Molitor N.T., Gauderman J, Berhane K., McConnell R., Lurmann F., Wu J., Winer A., Thomas D (2007). Assessing uncertainty in spatial exposure models for air pollution health effects assessment. *Environmental Health Perspectives* 115: 1147–53.

Molitor J., Molitor N.T., Jerrett M., McConnell R., Gauderman J, Berhane K., Thomas D. (2006). Bayesian modeling of air pollution health effects with missing exposure data. *American Journal of Epidemiology*

Molitor N-T, Jackson C., Richardson S., Best N. (2008). Using Bayesian graphical models to model biases in observational studies and to combine multiple data sources: Application to low birth-weight and water disinfection by-products. *Journal of Royal Statistical Society, Series A*: under review.

Moolgavkar S., Knudson A. (1981). Mutation and cancer: a model for human carcinogenesis. *Journal of the National Cancer Institute* 66: 1037–52.

Moolgavkar S., Kresweki D., Zeise L., Cardis E., Moller H (1999a). *Quantitative estimation and prediction of human cancer risks*. Lyon, IARC Scientific Publications.

Moolgavkar S., Krewski D., Schwarz M. (1999b). Mechanisms of carcinogenesis and biologically based models for estimation and prediction of risk. In: *Quantitative Estimation and Prediction of Human Cancer Risks*. Moolgavkar S., Kresweki D., Zeise L, Cardis E., Moller H. (eds.). Lyon, IARC Scientific Publications, 131: pp. 179–237.

Moolgavkar S., Venzon D.J. (1987). General relative risk regression models for epidemiologic studies. *American Journal of Epidemiology* 126: 949–61.

Moolgavkar S.H. (1978). The multistage theory of carcinogenesis and the age distribution of cancer in man. *Journal of the National Cancer Institute* 61: 49–52.

Moolgavkar S.H. (1986). Carcinogenesis modelling: from molecular biology to epidemiology. *Annual Review of Public Health* 7: 151–169.

Moolgavkar S.H. (2005). A review and critique of the EPA's rationale for a fine particle standard. *Regulatory Toxicology and Pharmacology* 42: 123–44.

Moolgavkar S.H., Cross F.T., Luebeck G., Dagle G.E. (1990). A two-mutation model for radon-induced lung tumors in rats. *Radiation Research* 121: 28–37.

Moolgavkar S.H., Day N.E., Stevens R.G. (1980). Two-stage model for carcinogenesis: Epidemiology of breast cancer in females. *Journal of the National Cancer Institute* 65: 559–69.

Moolgavkar S.H., Dewanji A., Luebeck G. (1989). Cigarette smoking and lung cancer: reanalysis of the British doctors' data. *Journal of the National Cancer Institute* 81: 415–20.

Moolgavkar S.H., Dewanji A., Venzon D.J. (1988). A stochastic two-stage model for cancer risk assessment. I. The hazard function and the probability of tumor. *Risk Analysis* 8: 383–92.

Moolgavkar S.H., Luebeck E.G. (1992). Multistage carcinogenesis: population-based model for colon cancer. *Journal of the National Cancer Institute* 84: 610–18.

Moolgavkar S.H., Luebeck E.G., Krewski D., Zielinski JM (1993). Radon, cigarette smoke, and lung cancer: a re-analysis of the Colorado Plateau uranium miners' data. *Epidemiology* 4: 204–17.

Moolgavkar S.H., Venzon D.J. (1979). Two-event models for carcinogenesis: incidence curves for childhood and adult tumors. *Mathematical Biosciences* 47: 55–77.

Moore J.H. (2003). The ubiquitous nature of epistasis in determining susceptibility in common human diseases. *Human Heredity* 56: 73–82.

Morgan W.F., Sowa M.B. (2007). Non-targeted bystander effects induced by ionizing radiation. *Mutation Research* 616: 159–64.

Morgenstern H. (1982). Uses of ecologic analysis in epidemiologic research. *American Journal of Public Health* 72: 1336–44.

Morgenstern H. (1995). Ecologic studies in epidemiology: concepts, principles, and methods. *Annual Review of Public Health* 16: 61–81.

Morrissey M.J., Spiegelman D. (1999). Matrix methods for estimating odds ratios with misclassified exposure data: extensions and comparisons. *Biometrics* 55: 338–44.

Morton-Jones T., Diggle P., Elliott P. (1999). Investigation of excess environmental risk around putative sources: Stone's test with covariate adjustment. *Statistics in Medicine* 18: 189–97.

Mothersill C., Seymour C. (2003). Radiation-induced bystander effects, carcinogenesis and models. *Oncogene* 22: 7028–33.

Mothersill C., Seymour C. (2004). Radiation-induced bystander effects and adaptive responses–the Yin and Yang of low dose radiobiology? *Mutation Research* 568: 121–8.

Moysich K.B., Menezes R.J., Michalek A.M. (2002). Chernobyl-related ionising radiation exposure and cancer risk: an epidemiological review. *Lancet Oncology* 3: 269–79.

Muller P., Roeder K. (1997). A Bayesian semiparametric model for case-control studies with errors in variables. *Biometrika* 84: 523–37.

Murray C.J., Ezzati M., Lopez A.D., Rodgers A., Vander Hoorn S (2003). Comparative quantification of health risks conceptual framework and methodological issues. *Population Health Metrics* 1: 1.

Murray C.J., King G., Lopez A.D., Tomijima N., Krug E.G. (2002). Armed conflict as a public health problem. *British Medical Journal* 324: 346–9.

Murray C.J., Lopez A.D. (1997). The utility of DALYs for public health policy and research: a reply. *Bulletin of the World Health Organization* 75: 377–81.

Mushkacheva G., Rabinovich E., Privalov V., Povolotskaya S, Shorokhova V., Sokolova S., Turdakova V., Ryzhova E., Hall P, Schneider A.B., Preston D.L., Ron E. (2006). Thyroid abnormalities associated with protracted childhood exposure to 131I from atmospheric emissions from the Mayak weapons facility in Russia. *Radiation Research* 166: 715–22.

Napier B.A. (2002). A re-evaluation of the 131I atmospheric releases from the Hanford site. *Health Physics* 83: 204–26.

NAS (1984). (Oversight Committee on Radioepidemiologic Tables). Assigned share for radiation as a cause of cancer: review of radioepidemiological tables assigning probability of causation. Washington, D.C., National Academy Press, 210.

NAS (1988). (Committee on the Biological Effects of Ionizing Radiations). Health risks of radon and other internally deposited alpha-emitters (BEIR IV). Washington, National Academy Press.

NAS (1990). (Committee on Health Risks of Exposure to Ionizing Radiation) Health effects of exposure to low levels of ionizing radiation: BEIR V. Washington, DC, National Academy Press: 500.

NAS (1999). (Committee on Health Risks of Exposure to Radon) Health effects of exposure to radon: BEIR VI. Washington, DC, National Academy Press: 500.

NAS (2002). Estimating the public health benefit of proposed air pollution regulations. Washington, DC, National Academies Press: 182.

NAS (2006). (Committee to Assess Health Risks from Exposure to Low Levels of Ionizing Radiation). Health risks from exposure to low levels of ionizing radiation: BEIR VII—Phase 2. Washington, DC, National Academies Press.

Navidi W. (1998). Bidirectional case-crossover designs for exposures with time trends. *Biometrics* 54: 596–605.

Navidi W., Lurman F. (1995). Measurement error in air pollution exposure assessment. *Journal of the Exposure Science and Environmental Epidemiology* 5: 111–24.

NCI (1997). Estimated exposures and thyroid doses received by the American people from iodine-131 in fallout following Nevada atmospheric nuclear bomb tests, a report from the National Cancer Institute. Washington, D.C., U.S. Department of Health and Human Services.

NCRP (2001). Evaluation of the linear-nonthreshold dose–response model for ionizing radiation (Report No. 136). Bethesda, National Council on Radiation Protection and Measurements.

Nelder J.A., Wedderburn RWM (1972). Generalized linear models. *Journal of the Royal Statistical Society, Series A* 135: 370–84.

Nelson C.M., Bunge R.G. (1974). Semen analysis: evidence for changing parameters of male fertility potential. *Fertility and Sterility* 25: 503–7.

Nelson L.M., Longstreth WTJ, Koepsell T.D., et al. (1990). Proxy respondends in epidemiologic research. *Epidemiologic Reviews* 12: 71–86.

Nemery B., Hoet P.H., Nemmar A. (2001). The Meuse Valley fog of 1930: an air pollution disaster. *Lancet* 357: 704–8.

Neriishi K., Stram D.O., Vaeth M., Mizuno S., Akiba S. (1991). The observed relationship between the occurrence of acute radiation effects and leukemia mortality among A-bomb survivors. *Radiation Research* 125: 206–13.

Neuhaus J.M., Kalbfleisch J.D. (1998). Between- and within-cluster covariate effects in the analysis of clustered data. *Biometrics* 54: 638–45.

Neyman J. (1938). Contribution to the theory of sampling human populations. *Journal of the American Statistical Association* 33: 101–16.

Neyman J., Scott E.L. (1948). Consistent estimates based on partially consistent observations. *Econometrica* 16: 1–32.

Nicolas P., Kim K.M., Shibata D., Tavare S. (2007). The stem cell population of the human colon crypt: analysis via methylation patterns. *Public Library of Science Computational Biology* 3: e28.

Nieuwenhuijsen M. (2003). *Exposure Assessment in Occupational and Environmental Epidemiology*. Oxford, Oxford University Press.

Nieuwenhuijsen M.J., Toledano M.B., Bennett J., Best N., Hambly P, de Hoogh C., Wellesley D., Boyd P.A., Abramsky L., Dattani N., Fawell J, Briggs D., Jarup L., Elliott P. (2008). Chlorination disinfection by-products and risk of congenital anomalies in England and Wales. *Environmental Health Perspectives* 116: 216–22.

NIH (1985). Report of the National Institutes of Health ad hoc working group to develop radioepidemiologic tables. Washington, D.C., U.S. Department of Health and Human Services, 355.

Nikolov M.C., Coull B.A., Catalano P.J., Godleski J.J. (2006). An informative Bayesian structural equation model to assess source-specific health effects of air pollution. *Biostatistics*.

Noji E.J. (ed.) (1997). *The public health consequences of disasters*. Oxford, Oxford University Press.

Noji E.K. (2005). Disasters: introduction and state of the art. *Epidemiologic Reviews* 27: 3–8.

Nowak M.A., Komarova N.L., Sengupta A., Jallepalli P.V., Shih Ie M, Vogelstein B., Lengauer C. (2002). The role of chromosomal instability in tumor initiation. *Proceedings of the National Academy of Sciences* 99: 16226–31.

NRC (2000). A review of the draft report of the NCI-CDC working group to revise the 1985 radioepidemiological tables. Washington, DC, National Academy Press.

Nuckols J.R., Ward M.H., Jarup L. (2004). Using geographic information systems for exposure assessment in environmental epidemiology studies. *Environmental Health Perspectives* 112: 1007–15.

OEHHA (2002). Technical support document for describing available cancer potency factors. Sacramento, CA, California Environmental Protection Agency, Office of Environmental Health Hazard Assessment, Air Toxicology and Epidemiology Section.

Ohtaki K., Kodama Y., Nakano M., Itoh M., Awa A.A., Cologne J, Nakamura N. (2004). Human fetuses do not register chromosome damage inflicted by radiation exposure in lymphoid precursor cells except for a small but significant effect at low doses. *Radiation Research* 161: 373–9.

Openshaw S. (1984). *The modifiable areal unit problem: concepts and techniques in modern geography.* Norwich, UK, Geo Books.

Otake M., Prentice R.L. (1984). The analysis of chromosomally aberrant cells based on beta-binomial distribution. *Radiation Research* 98: 456–70.

Ozkaynak H., Xue J., Spengler J., Wallace L., Pellizzari E, Jenkins P. (1996). Personal exposure to airborne particles and metals: results from the Particle TEAM study in Riverside, California. *Journal of Exposure Analysis Environmental Epidemiology* 6: 57–78.

Paik M.C. (2004). Nonignorable missingness in matched case-control data analysis. *Biometrics* 60: 306–14.

Pan X., Yue W., He K., Tong S. (2007). Health benefit evaluation of the energy use scenarios in Beijing, China. *Science of the Total Environment* 374: 242–51.

Parascandola M., Weed D.L. (2001). Causation in epidemiology. *Journal of Epidemiology and Community Health* 55: 905–12.

Parl F., Crooke P., Conti D.V., Thomas D.C. (2008). Pathway-based methods in molecular cancer epidemiology. In: *Fundamentals of Molecular Epidemiology.* Rebbeck TR, Ambrosone C.B., Shields P.G. (eds.). New York, Informa Healthcare: 189–204.

Pascutto C., Wakefield J.C., Best N.G., Richardson S, Bernardinelli L., Staines A., Elliott P. (2000). Statistical issues in the analysis of disease mapping data. *Statistics in Medicine* 19: 2493–519.

Patz J., Campbell-Lendrum D., Gibbs H., Woodruff R. (2008). Health impact assessment of global climate change: expanding on comparative risk assessment approaches for policy making. *Annual Review of Public Health*, 29: 27–39.

Patz J.A., Campbell-Lendrum D., Holloway T., Foley J.A. (2005). Impact of regional climate change on human health. *Nature* 438: 310–17.

Pearce M.S., Cardis E. (2001). Depleted uranium—cause for concern or just a good story? *Pediatric Hematology and Oncology* 18: 367–70.

Pearce N. (2000). The ecological fallacy strikes back. *Journal of Epidemiology and Community Health* 54: 326–7.

Pearl J. (2000). *Causality: models, reasoning, and inference.* Cambridge, Cambridge University Press.

Pepe M.S., Flemming T.R. (1991). A nonparametric method for dealing with mismeasured covariate data. *Journal of the American Statistical Association* 86: 108–13.

Peters A., von Klot S., Heier M., Trentinaglia I., Hormann A, Wichmann H.E., Lowel H. (2004). Exposure to traffic and the onset of myocardial infarction. *New England Journal of Medicine* 351: 1721–30.

Peters J., Hedley A.J., Wong C.M., Lam T.H., Ong S.G., Liu J, Spiegelhalter D.J. (1996). Effects of an ambient air pollution intervention and environmental tobacco smoke on children's respiratory health in Hong Kong. *International Journal of Epidemiology* 25: 821–8.

Peters J.L., Rushton L., Sutton A.J., Jones D.R., Abrams KR, Mugglestone M.A. (2005). Bayesian methods for the cross-design synthesis of epidemiological and toxicological evidence. *Applied Statistics* 54: 159–72.

Peterson A.V., Jr, Prentice R.L., Ishimaru T., Kato H., Mason M (1983). Investigation of circular asymmetry in cancer mortality of Hiroshima and Nagasaki A-bomb survivors. *Radiation Research* 93: 184–99.

Peto R. (1977). Epidemiology, multistage models, and short-term mutagenicity tests. In: *Origins of Human Cancer*. Hiatt H.H., Winsten J.A. (eds.). Cold Spring Harbor, NY, Cold Spring Harbor Laboratory, pp. 1403–28.

Peto R. (1986). Influence of dose and duration of smoking on lung cancer rates. *IARC Scientific Publications*, No. 74: pp. 23–33.

Peto R., Darby S., Deo H., Silcocks P., Whitley E., Doll R (2000). Smoking, smoking cessation, and lung cancer in the UK since 1950: combination of national statistics with two case-control studies. *British Medical Journal* 321: 323–9.

Peto R., Parish S.E., Gray R.G. (1985). There is no such thing as ageing, and cancer is not related to it. *IARC Scientific Publications*, No. 58: 43–53.

Piantadosi S. (1994). Invited commentary: ecologic biases. *American Journal of Epidemiology* 139: 761–4; discussion 769–71.

Piantadosi S., Byar D.P., Green S.B. (1988). The ecological fallacy. *American Journal of Epidemiology* 127: 893–904.

Pickle L.W., Mason T.J., Howard N., Hoover R., Fraumeni JFJ (1987). *Atlas of cancer mortality among whites, 1950–1980*. Washington, D.C., U.S. Government Printing Office.

Pickle L.W., Szczur M., Lewis D.R., Stinchcomb D.G. (2006). The crossroads of GIS and health information: a workshop on developing a research agenda to improve cancer control. *International Journal of Health Geography* 5: 51.

Piegorsch W., Weinberg C., Taylor J. (1994). Non-hierarchical logistic models and case-only designs for assessing susceptibility in population-based case-control studies. *Statistics in Medicine* 13: 153–62.

Pierce D.A. (2003). Mechanistic models for radiation carcinogenesis and the atomic bomb survivor data. *Radiation Research* 160: 718–23.

Pierce D.A., Kellerer A.M. (2004). Adjusting for covariate errors with nonparametric assessment of the true covariate distribution. *Biometrika* 91: 863–76.

Pierce D.A., Mendelsohn M.L. (1999). A model for radiation-related cancer suggested by atomic bomb survivor data. *Radiation Research* 152: 642–54.

Pierce D.A., Preston D.L., Stram D.O., Vaeth M. (1991). Allowing for dose-estimation errors for the A-bomb survivor data. *Journal of Radiation Research (Tokyo)* 32(Suppl.): 108–21.

Pierce D.A., Stram D.O., Vaeth M. (1990). Allowing for random errors in radiation dose estimates for the atomic bomb survivor data. *Radiation Research* 123: 275–84.

Pierce D.A., Stram D.O., Vaeth M., Schafer D.W. (1992). The errors-in-variables problem: considerations provided by radiation dose-response analyses of the a-bomb survivor data. *Journal of the American Statistical Association* 87: 351–9.

Pierce D.A., Vaeth M. (2003). Age-time patterns of cancer to be anticipated from exposure to general mutagens. *Biostatistics* 4: 231–48.

Pierce D.A., Vaeth M., Cologne J. (2007). Allowance for random dose-estimation errors in atomic-bomb survivor studies: a revision.

Pike M.C., Casagrande J., Smith P.G. (1975). Statistical analysis of individually matched case-control studies in epidemiology: factor under study a discrete variable taking multiple values. *British Journal of Preventive and Social Medicine* 29: 196–201.

Pike M.C., Smith P.G. (1974). Case-control approach to examine diseases for evidence of contagion, including diseases with long latent periods. *Biometrics* 30: 263–79.

Plummer M., Clayton D. (1996). Estimation of population exposure in ecological studies. *Journal of the Royal Statistical Society, Series B* 58: 113–26.

Poole C. (1986). Exposure opportunity in case-control studies. *American Journal of Epidemiology* 123: 352–8.

Pope C.A., 3rd (1989). Respiratory disease associated with community air pollution and a steel mill, Utah Valley. *American Journal of Public Health* 79: 623–8.

Pope C.A., 3rd, Burnett R.T. (2007). Confounding in air pollution epidemiology: the broader context. *Epidemiology* 18: 424–6; discussion 427–8.

Pope C.A., 3rd, Burnett R.T., Thun M.J., Calle E.E., Krewski D, Ito K., Thurston G.D. (2002). Lung cancer, cardiopulmonary mortality, and long-term exposure to fine particulate air pollution. *Journal of the American Medical Association* 287: 1132–41.

Pope C.A., 3rd, Kalkstein L.S. (1996). Synoptic weather modeling and estimates of the exposure-response relationship between daily mortality and particulate air pollution. *Environmental Health Perspectives* 104: 414–20.

Pope C.A., 3rd, Rodermund D.L., Gee M.M. (2007). Mortality effects of a copper smelter strike and reduced ambient sulfate particulate matter air pollution. *Environmental Health Perspectives* 115: 679–83.

Pope C.I., Thun M., Namboodiri M., et-al (1995). Particulate are pollution as a predictor of mortality in a prospective study of U.S. adults. *American Journal of Respiratory and Critical Care Medicine* 151: 669–74.

Portier C., Hoel D. (1983). Low-dose-rate extrapolation using the multistage model. *Biometrics* 39: 897–906.

Prentice R., Breslow N. (1978). Retrospective studies and failure time models. *Biometrika* 65: 153–58.

Prentice R., Pyke R. (1979). Logistic disease incidence models and case-control studies. *Biometrika* 86: 403–11.

Prentice R.L. (1982). Covariate measurement errors and parameter estimation in Cox's regression model. *Biometrika* 69: 331–42.

Prentice R.L. (1986). A case-cohort design for epidemiologic studies and disease prevention trials. *Biometrika* 73: 1–11.

Prentice R.L., Pettinger M., Anderson G.L. (2005). Statistical issues arising in the Women's Health Initiative. *Biometrics* 61: 899–911; discussion 911–41.

Prentice R.L., Sheppard L. (1989). Validity of international, time trend, and migrant studies of dietary factors and disease risk. *Preventive Medicine* 18: 167–79.

Prentice R.L., Sheppard L. (1990). Dietary fat and cancer: consistency of the epidemiologic data, and disease prevention that may follow from a practical reduction in fat consumption [erratum appears in Cancer Causes Control 1990 Nov.; 1(3): 253]. *Cancer Causes and Control* 1: 81–97; discussion 99–109.

Prentice R.L., Sheppard L. (1991). Dietary fat and cancer: rejoinder and discussion of research strategies. *Cancer Causes and Control* 2: 53–8.

Prentice R.L., Sheppard L. (1995). Aggregate data studies of disease risk factors. *Biometrika* 82: 113–25.

Prentice R.L., Yoshimoto Y., Mason M.W. (1983). Relationship of cigarette smoking and radiation exposure to cancer mortality in Hiroshima and Nagasaki. *Journal of the National Cancer Institute* 70: 611–22.

Preston D.L., Kusumi S., Tomonaga M., Izumi S., Ron E., Kuramoto A, Kamada N., Dohy H., Matsuo T., Matsui T., et al. (1994). Cancer incidence in atomic bomb survivors. Part III. Leukemia, lymphoma and multiple myeloma, 1950–1987. *Radiation Research* 137: S68–97.

Preston D.L., Lubin J.H., Pierce D.A., McConney M.E. (1993). *Epicure user's guide*. Seattle, WA, Hirosoft International Corporation.

Preston D.L., Pierce D.A., Shimizu Y., Cullings H.M., Fujita S, Funamoto S., Kodama K. (2004). Effect of recent changes in atomic bomb survivor dosimetry on cancer mortality risk estimates. *Radiation Research* 162: 377–89.

Preston D.L., Shimizu Y., Pierce D.A., Suyama A., Mabuchi K (2003). Studies of mortality of atomic bomb survivors. Report 13: Solid cancer and noncancer disease mortality: 1950–1997. *Radiation Research* 160: 381–407.

Prince-Embury S., Rooney J.F. (1988). Psychological symptoms of residents in the aftermath of the Three Mile Island nuclear accident and restart. *Journal of Social Psychology* 128: 779–90.

Pruss A., Corvalan C.F., Pastides H., De Hollander A.E. (2001). Methodologic considerations in estimating burden of disease from environmental risk factors at national and global levels. *International Journal of Occupational and Environmental Health* 7: 58–67.

Pruss A., Kay D., Fewtrell L., Bartram J. (2002). Estimating the burden of disease from water, sanitation, and hygiene at a global level. *Environmental Health Perspectives* 110: 537–42.

Puskin J.S. (2003). Smoking as a confounder in ecologic correlations of cancer mortality rates with average county radon levels. *Health Physics* 84: 526–32.

Rabl A. (2003). Interpretation of air pollution mortality: number of deaths or years of life lost? *Journal of Air and Waste Management Associations* 53: 41–50.

Rabl A. (2005). Air pollution mortality: harvesting and loss of life expectancy. *Journal of Toxicology and Environmental Health, Part A* 68: 1175–80.

Rabl A. (2006). Analysis of air pollution mortality in terms of life expectancy changes: relation between time series, intervention and cohort studies. *Environmental Health* 5: 1.

Rachet B., Siemiatycki J., Abrahamowicz M., Leffondre K (2004). A flexible modeling approach to estimating the component effects of smoking behavior on lung cancer. *Journal of Clinical Epidemiology* 57: 1076–85.

Radiation Science and Health (1998). *Low Level Radiation Health Effects: Compiling the Data*. Needham, MA, Radiation, Science and Health, Inc.

Raftery A.E., Madigan D., Hoeting J.A. (1997). Bayesian model averaging for linear regression models. *Journal of the American Statistical Association* 92: 179–91.

Ramsay T., Burnett R., Krewski D. (2003a). Exploring bias in a generalized additive model for spatial air pollution data. *Environmental Health Perspectives* 111: 1283–8.

Ramsay T.O., Burnett R.T., Krewski D. (2003b). The effect of concurvity in generalized additive models linking mortality to ambient particulate matter. *Epidemiology* 14: 18–23.

Rathouz P.J., Satten G.A., Carroll R.J. (2002). Semiparametric inference in matched case-control studies with missing covariate data. *Biometrika* 89: 905–16.

Reeves G.K., Cox D.R., Darby S.C., Whitley E. (1998). Some aspects of measurement error in explanatory variables for continuous and binary regression models. *Statistics in Medicine* 17: 2157–77.

Reilly M. (1996). Optimal sampling strategies for two-stage studies. *American Journal of Epidemiology* 143: 92–100.

Reilly M., Pepe M.S. (1995). A mean score method for missing and auxiliary covariate data in regression models. *Biometrika* 82: 299–314.

Renn O. (2007). Precaution and analysis: two sides of the same coin? *EMBO Reports* 8: 303–4.

Richardson S., Gilks W.R. (1993a). A Bayesian approach to measurement error problems in epidemiology using conditional independence models. *American Journal of Epidemiology* 138: 430–42.

Richardson S., Gilks W.R. (1993b). Conditional independence models for epidemiological studies with covariate measurement error. *Statistics in Medicine* 12: 1703–22.

Richardson S., Monfort C., Green M., Draper G., Muirhead C (1995). Spatial variation of natural radiation and childhood leukaemia incidence in Great Britain. *Statistics in Medicine* 14: 2487–501.

Richardson S., Stucker I., Hemon D. (1987). Comparison of relative risks obtained in ecological and individual studies: some methodological considerations. *International Journal of Epidemiology* 16: 111–20.

Richardson S., Thomson A., Best N., Elliott P. (2004). Interpreting posterior relative risk estimates in disease-mapping studies. *Environmental Health Perspectives* 112: 1016–25.

Roberts L., Burnham G. (2007). Authors defend study that shows high Iraqi death toll. *Nature* 446: 611.

Roberts L., Lafta R., Garfield R., Khudhairi J., Burnham G (2004). Mortality before and after the 2003 invasion of Iraq: cluster sample survey. *Lancet* 364: 1857–64.

Roberts R.A., Crump K.S., Lutz W.K., Wiegand H.J., Williams GM, Harrison P.T., Purchase I.F. (2001). Scientific analysis of the proposed uses of the T25 dose descriptor in chemical carcinogen regulation. *Archives of Toxicology* 75: 507–12.

Robins J. (1987). A graphical approach to the identification and estimation of causal parameters in mortality studies with sustained exposure periods. *Journal of Chronic Diseases* 40 (suppl. 2): 139S–161S.

Robins J., Greenland S. (1989a). Estimability and estimation of excess and etiologic fractions. *Statistics in Medicine* 8: 845–59.

Robins J., Greenland S. (1989b). The probability of causation under a stochastic model for individual risk. *Biometrics* 45: 1125–38.

Robins J., Greenland S. (1991). Estimability and estimation of expected years of life lost due to a hazardous exposure. *Statistics in Medicine* 10: 79–93.

Robins J., Greenland S., Breslow N.E. (1986). A general estimator for the variance of the Mantel–Haenszel odds ratio. *American Journal of Epidemiology* 124: 719–23.

Robins J., Li L., Tchetgen E., Van Der Vaart AAD (2007). A new approach to semiparametric regression in the analysis of the health effects of pollutants. Health Effects Institute, Annual Meeting.

Robins J., Pike M. (1990). The validity of case-control studies with nonrandom selection of controls. *Epidemiology* 1: 273–84.

Robins JM, Hernan M.A., Brumback B. (2000). Marginal structural models and causal inference in epidemiology. *Epidemiology* 11: 550–60.

Robins J.M., Hsieh F., Newey W. (1995a). Semiparametric efficient estimation of a conditional density with missing or mismeasured covariates. *Journal of the Royal Statistical Society. Series B* 57: 409–24.

Robins J.M., Rotnitzky A.G., Zhao L.P. (1995b). Analysis of semiparametric regression models for repeated outcomes in the presence of missing data. *Journal of the American Statistical Association* 90: 106–20.

Robinson L.D., Jewell N.P. (1991). Some surprising results about covariate adjustment in logistic regression. *International Statistical Review* 59: 227–40.

Robinson W.S. (1950). Ecological correlations and the behavior of individuals. *American Sociological Review* 15: 371–57.

Rodgers A., Ezzati M., Vander Hoorn S., Lopez A.D., Lin RB, Murray C.J. (2004). Distribution of major health risks: findings from the Global Burden of Disease study. *Public Library of Science Medicine* 1: e27.

Roeder K., Carroll R.J., Lindsay B.G. (1996). A semiparametric mixture approach to case-control studies with errors in covariables. *Journal of the American Statistical Association* 91: 722–32.

Roesch W.C. (1977). Microdosimetry of internal sources. *Radiation Research* 70: 494–510.

Rogerson P.A. (2001). Monitoring point patterns for the development of space-time clusters. *Journal of the Royal Statistical Society. Series A* 164: 87–96.

Ron E., Lubin J.H., Shore R.E., Mabuchi K., Modan B., Pottern LM, Schneider A.B., Tucker M.A., Boice J.D., Jr. (1995). Thyroid cancer after exposure to external radiation: a pooled analysis of seven studies. *Radiation Research* 141: 259–77.

Rondeau V., Berhane K., Thomas D.C. (2005). A three-level model for binary time-series data: the effects of air pollution on school absences in the Southern California Children's Health Study. *Statistics in Medicine* 24: 1103–15.

Roosli M., Kunzli N., Braun-Fahrlander C., Egger M. (2005). Years of life lost attributable to air pollution in Switzerland: dynamic exposure–response model. *International Journal of Epidemiology* 34: 1029–35.

Rosenbaum P.R. (1989). Optimal matching for observational studies. *Journal of the American Statistical Association* 84: 1024–32.

Rosenbaum P.R., Rubin D.B. (1983). The central role of the propensity score in observational studies for causal effects. *Biometrika* 70: 41–55.

Rosenbaum P.R., Rubin D.B. (1984). Difficulties with regression analyses of age-adjusted rates. *Biometrics* 40: 437–43.

Rosenbaum P.R., Rubin D.B. (1985). Constructing a control group using multivariate matched sampling methods that incorporate the propensity score. *American Statistician* 39: 33–38.

Rosner B., Spiegelman D., Willett W.C. (1990). Correction of logistic regression relative risk estimates and confidence intervals for measurement error: the case of multiple covariates measured with error. *American Journal of Epidemiology* 132: 734–45.

Rosner B., Spiegelman D., Willett W.C. (1992). Correction of logistic regression relative risk estimates and confidence intervals for random within-person measurement error. *American Journal of Epidemiology* 136: 1400–13.

Rosner B., Willett W.C., Spiegelman D. (1989). Correction of logistic regression relative risk estimates and confidence intervals for systematic within-person measurement error. *Statistics in Medicine* 8: 1051–69; discussion 1071–3.

Rossi H.H. (1959). Specification of radiation quality. *Radiation Research* 10: 522–31.

Rothman K.J. (1974). Synergy and antagonism in cause–effect relationships. *American Journal of Epidemiology* 99: 385–8.

Rothman K.J. (1976a). Causes. *American Journal of Epidemiology* 104: 587–92.

Rothman K.J. (1976b). The estimation of synergy or antagonism. *American Journal of Epidemiology* 103: 506–11.

Rothman K.J. (1990). A sobering start for the cluster busters conference. *American Journal of Epidemiology* 132(Suppl.): S6–13.

Rothman K.J., Greenland S. (1998). *Modern epidemiology*. Philadelphia, Lippincott-Raven.

Rothman N., Wacholder S., Caporaso N.E., Garcia-Closas M, Buetow K., Fraumeni J.F., Jr. (2001). The use of common genetic polymorphisms to enhance the epidemiologic study of environmental carcinogens. *Biochimica et Biophysica Acta* 1471: C1-10.

Royston P., Ambler G., Sauerbrei W. (1999). The use of fractional polynomials to model continuous risk variables in epidemiology. *International Journal of Epidemiology* 28: 964–74.

Royston P., Sauerbrei W. (2005). Building multivariable regression models with continuous covariates in clinical epidemiology with an emphasis on fractional polynomials. *Methods of Inference in Medicine* 44: 561–71.

Rubin D. (1987). *Multiple imputation for nonresponse in surveys*. New York, Wiley.

Rull R.P., Ritz B. (2003). Historical pesticide exposure in California using pesticide use reports and land-use surveys: an assessment of misclassification error and bias. *Environmental Health Perspectives* 111: 1582–9.

Sackett D.L. (1979). Bias in analytic research. *Journal of Chronic Diseases* 32: 51–63.

Salam M.T., Li Y.F., Langholz B., Gilliland F.D., Children's Health Study (2004). Early-life environmental risk factors for asthma: findings from the Children's Health Study. *Environmental Health Perspectives* 112: 760–5.

Sali D., Cardis E., Sztanyik L., Auvinen A., Bairakova A, Dontas N., Grosche B., Kerekes A., Kusic Z., Kusoglu C., et al. (1996). Cancer consequences of the Chernobyl accident in Europe outside the former USSR: a review. *International Journal of Cancer* 67: 343–52.

Salway R., Wakefield J. (2008). A hybrid model for reducing ecological bias. *Biostatistics* 9: 1–17.

Samet J., Dominici F., Curriero F., Coursac I., Zeger S (2000a). Fine particulate air pollution and mortality in 20 U.S. cities, 1987–1994. *New England Journal of Medicine* 343: 1742–9.

Samet J.M. (2002). Air pollution and Epidemiology: "deja vu all over again?" *Epidemiology* 13: 118–19.

Samet J.M., Geyh A.S., Utell M.J. (2007). The legacy of World Trade Center dust. *New England Journal of Medicine* 356: 2233–6.

Samet J.M., Zeger S.L., Dominici F., Curriero F., Coursac I, Dockery D.W., Schwartz J., Zanobetti A. (2000b). The National Morbidity, Mortality, and Air Pollution Study. Part II: Morbidity and mortality from air pollution in the United States. *Research Reports of the Health Effects Institute* 94: 5–70; discussion 71–9.

Samet J.M., Zeger S.L., Kelsall J.E., Xu J. (1997). Particulate Air Pollution and Daily Mortality: Analysis of the Effects of Weather and Multiple Air Pollutants: The Phase I.B Report of the Particle Epidemiology Evaluation Project. Cambridge, MA, Health Effects Institute.

Samoli E., Analitis A., Touloumi G., Schwartz J., Anderson HR, Sunyer J., Bisanti L., Zmirou D., Vonk J.M., Pekkanen J., et al. (2005). Estimating the exposure-response relationships between particulate matter and mortality within the APHEA multicity project. *Environmental Health Perspectives* 113: 88–95.

Samoli E., Schwartz J., Wojtyniak B., Touloumi G., Spix C, Balducci F., Medina S., Rossi G., Sunyer J., Bacharova L., Anderson HR, Katsouyanni K. (2001). Investigating regional differences in short-term effects of air pollution on daily mortality in the APHEA project: a sensitivity analysis for controlling long-term trends and seasonality. *Environmental Health Perspectives* 109: 349–53.

Samoli E., Touloumi G., Zanobetti A., Le Tertre A., Schindler C, Atkinson R., Vonk J, Rossi G., Saez M., Rabczenko D., Schwartz J, Katsouyanni K. (2003). Investigating the dose–response relation between air pollution and total mortality in the APHEA-2 multicity project. *Occupational and Environmental Medicine* 60: 977–82.

Sarnat J.A., Schwartz J., Catalano P.J., Suh H.H. (2001). Gaseous pollutants in particulate matter epidemiology: confounders or surrogates? *Environmental Health Perspectives* 109: 1053–61.

Sasaki M.S., Endo S., Ejima Y., Saito I., Okamura K., Oka Y, Hoshi M. (2006). Effective dose of A-bomb radiation in Hiroshima and Nagasaki as assessed by chromosomal effectiveness of spectrum energy photons and neutrons. *Radiation and Environmental Biophysics* 45: 79–91.

Sastry N. (2002). Forest fires, air pollution, and mortality in southeast Asia. *Demography* 39: 1–23.

Satten G.A., Carroll R.J. (2000). Conditional and unconditional categorical regression models with missing covariates. *Biometrics* 56: 384–8.

Sauerbrei W., Royston P., Binder H. (2007). Selection of important variables and determination of functional form for continuous predictors in multivariable model building. *Statistics in Medicine* 26: 5512–28.

Schafer D.W. (1987). Covariate measurement error in generalized linear models. *Biometrika* 74: 385–91.

Schafer D.W. (2001). Semiparametric maximum likelihood for measurement error model regression. *Biometrics* 57: 53–61.

Schafer D.W., Gilbert E.S. (2006). Some statistical implications of dose uncertainty in radiation dose–response analyses. *Radiation Research* 166: 303–12.

Schafer J.L. (1997). *Analysis of incomplete multivariate data*. London, Chapman & Hall.

Schafer J.L., Schenker N. (2000). Inference with Imputed Conditional Means. *Journal of the American Statistical Association* 95: 144–54.

Schaid D. (1999). Case-parents design for gene–environment interaction. *Genetic Epidemiology* 16: 261–73.

Schaubel D., Hanley J., Collet J.P., Bolvin J.F., Sharpe C, Morrison H.I., Mao Y. (1997). Two-stage sampling for etiologic studies. Sample size and power. *American Journal of Epidemiology* 146: 450–8.

Schauer J.J., Rogge W.F., Hildemann L.M., Mazurek M.A., Cass GR, Simoneit BRT (1996). Source apportionment of airborne particulate matter using organic compounds as tracers. *Atmospheric Environment* 30: 3837–55.

Scherr P.A., Gutensohn N., Cole P. (1984). School contact among persons with Hodgkin's disease. *American Journal of Epidemiology* 120: 29–38.

Schill W., Jockel K.H., Drescher K., Timm J. (1993). Logistic analysis in case-control studies under validation sampling. *Biometrika* 80: 339–52.

Schimmel H., Murawski T.J. (1976). Proceedings: The relation of air pollution to mortality. *Journal of Occupational Medicine* 18: 316–33.

Schmid C.H., Rosner B. (1993). A Bayesian approach to logistic regression models having measurement error following a mixture distribution. *Statistics in Medicine* 12: 1141–53.

Schwartz G. (1978). Estimating the dimension of a model. *Annals of Statistics* 6: 461–4.

Schwartz J. (1993). Air pollution and daily mortality in Birmingham, Alabama. *American Journal of Epidemiology* 137: 1136–47.

Schwartz J. (1994a). Air pollution and daily mortality: a review and meta-analysis. *Environmental Research* 64: 36–52.

Schwartz J. (1994b). Nonparametric smoothing in the analysis of air pollution and respiratory illness. *Canadian Journal of Statististics* 22: 471–88.

Schwartz J. (2000a). Assessing confounding, effect modification, and thresholds in the association between ambient particles and daily deaths. *Environmental Health Perspectives* 108: 563–8.

Schwartz J. (2000b). The distributed lag between air pollution and daily deaths. *Epidemiology* 11: 320–6.

Schwartz J. (2000c). Harvesting and long term exposure effects in the relation between air pollution and mortality [see comments]. *American Journal of Epidemiology* 151: 440–8.

Schwartz J. (2001). Is there harvesting in the association of airborne particles with daily deaths and hospital admissions? *Epidemiology* 12: 55–61.

Schwartz J. (2005). How sensitive is the association between ozone and daily deaths to control for temperature? *American Journal of Respiratory and Critical Care Medicine* 171: 627–31.

Schwartz J., Ballester F., Saez M., Perez-Hoyos S., Bellido J, Cambra K., Arribas F., Canada A, Perez-Boillos M.J., Sunyer J. (2001). The concentration-response relation between air pollution and daily deaths. *Environmental Health Perspectives* 109: 1001–6.

Schwartz J., Coull B.A. (2003). Control for confounding in the presence of measurement error in hierarchical models. *Biostatistics* 4: 539–53.

Schwartz J., Zanobetti A. (2000). Using meta-smoothing to estimate dose-response trends across multiple studies, with application to air pollution and daily death. *Epidemiology* 11: 666–72.

Schwartz S. (1994c). The fallacy of the ecological fallacy: the potential misuse of a concept and the consequences. *American Journal of Public Health* 84: 819–24.

Scott A.J., Lee A.J., Wild C.J. (2007). On the Breslow-Holubkov estimator. *Lifetime Data Analysis* 13: 545–63.

Scott A.J., Wild C.J. (1991). Fitting Logistic Regression Models in Stratified Case-Control Studies. *Biometrics* 47: 497–510.

Scott B.R. (2004). A biological-based model that links genomic instability, bystander effects, and adaptive response. *Mutation Research* 568: 129–43.

Scott R., McPhillips M., Meldrum C., Fitzgerald P., Adams K, Spigelman A.D., du-Sart D., Tucker K., Kirk J, Hunder=Family-Cancer-Service (2001). Hereditary nonpolyposis colorectal cancr in 95 families: differences and similarities between mutation-positive and mutation-negative kindreds. *American Journal of Human Genetics* 68: 118–27.

Seaman J. (1984). *Epidemiology of Natural Disasters: Contributions to Epidemiology and Biostatistics*. Basel, Karger.

See S.W., Balasubramanian R., Rianawati E., Karthikeyan S, Streets D.G. (2007). Characterization and source apportionment of particulate matter < or = 2.5 micrometer in Sumatra, Indonesia, during a recent peat fire episode. *Environmental Science and Technology* 41: 3488–94.

Self S.G., Liang K-Y (1987). Asymptotic properties of maximum likelihood estimators and likelihood ratio tests under nonstandard conditions. *Journal of the American Statistical Association* 82: 605–10.

Selikoff I.J., Hammond E.C., Churg J. (1968). Asbestos exposure, smoking, and neoplasia. *Journal of the American Medical Association* 204: 106–12.

Selvin H.C. (1958). Durkheim's suicide and problems of empirical research. *American Journal of Sociology* 63: 607–19.

Senn S.J. (1989). Covariate imbalance and random allocation in clinical trials. *Statistics in Medicine* 8: 467–75.

Shaw G.L., Tucker M.A., Kase R.G., Hoover R.N. (1991). Problems ascertaining friend controls in a case-control study of lung cancer. *American Journal of Epidemiology* 133: 63–6.

Sheppard L. (2005). Acute air pollution effects: consequences of exposure distribution and measurements. *Journal of Toxicology and Environmental Health, Part A* 68: 1127–35.

Sheppard L., Damian D. (2000). Estimating short-term PM effects accounting for surrogate exposure measurements from ambient monitors. *Environmetrics* 11: 675–87.

Sheppard L., Prentice R.L., Rossing M.A. (1996). Design considerations for estimation of exposure effects on disease risk, using aggregate data studies. *Statistics in Medicine* 15: 1849–58.

Shipler D.B., Napier B.A., Farris W.T., Freshley M.D. (1996). Hanford Environmental Dose Reconstruction Project—an overview. *Health Physics* 71: 532–44.

Shultz J.M., Russell J., Espinel Z. (2005). Epidemiology of tropical cyclones: the dynamics of disaster, disease, and development. *Epidemiologic Reviews* 27: 21–35.

Siemiatycki J. (1978). Mantel's space-time clustering statistic: computing higher moments and a comparison of various data transformations. *Statistical Computation and Simulation* 7: 13–31.

Siemiatycki J. (1989). Friendly control bias. *Journal of Clinical Epidemiology* 42: 687–8.

Siemiatycki J., Thomas D.C. (1981). Biological models and statistical interactions: an example from multistage carcinogenesis. *International Journal of Epidemiology* 10: 383–87.

Simon S.L., Till J.E., Lloyd R.D., Kerber R.L., Thomas DC, Preston-Martin S., Lyon J.L., Stevens W. (1995). The Utah Leukemia Case-Control Study: dosimetry methodology and results. *Health Physics* 68: 460–71.

Singh M.P., Ghosh S. (1987). Bhopal gas tragedy: Model simulation of the dispersal scenario. *Journal of Hazardous Materials* 17: 1–22.

Sinha S., Mukherjee B., Ghosh M., Mallick B.K., Carroll RJ (2005). Semiparametric Bayesian analysis of matched case-control studies with missing exposure. *Journal of the American Statistical Association* 100: 591–601.

Smith A.H., Kark J.D., Cassel J.C., Spears GFS (1977). Analysis of prospective epidemiologic studies by minimum distance case-control matching. *American Journal of Epidemiology* 105: 567–74.

Smith B.J., Field R.W., Lynch C.F. (1998). Residential 222Rn exposure and lung cancer: testing the linear no-threshold theory with ecologic data. *Health Physics* 75: 11–17.

Smith R.L. (2003). Invited commentary: Timescale-dependent mortality effects of air pollution. *American Journal of Epidemiology* 157: 1066–70; discussion 1071–3.

Snow J. (1855). On the mode of communication of cholera. *Edinburgh Medical Journal* 1: 668–70.

Sokol R.Z., Kraft P., Fowler I.M., Mamet R., Kim E., Berhane KT (2006). Exposure to environmental ozone alters semen quality. *Environmental Health Perspectives* 114: 360–5.

Sommer A., Mosley W.H. (1972). East Bengal cyclone of November, 1970. Epidemiological approach to disaster assessment. *Lancet* 1: 1029–36.

Spiegelhalter D.J., Thomas A., Best N.G., Lund D. (2003). *WinBUGS Version 1.4 User Manual*. Cambridge, Medical Research Council Biostatistics Unit.

Spiegelman D., Carroll R.J., Kipnis V. (2001). Efficient regression calibration for logistic regression in main study/internal validation study designs with an imperfect reference instrument. *Statistics in Medicine* 20: 139–60.

Spiegelman D., Casella M. (1997). Fully parametric and semi-parametric regression models for common events with covariate measurement error in main study/validation study designs. *Biometrics* 53: 395–409.

Spiegelman D., Gray R. (1991). Cost-efficient study designs for binary response data with Gaussian covariate measurement error. *Biometrics* 47: 851–69.

Sposto R., Stram D.O., Awa A.A. (1991). An estimate of the magnitude of random errors in the DS86 dosimetry from data on chromosome aberrations and severe epilation. *Radiation Research* 128: 157–69.

Stallones R.A. (1980). To advance epidemiology. *Annual Review of Public Health* 1: 69–82.

Stayner L., Smith R., Bailey A.J. (1995). Modeling epidemiologic studies of occupational cohorts for the qualitative assessment of carcinogenic hazardes. *American Journal of Industrial Medicine* 27: 155–70.

Steenland K., Armstrong B. (2006). An overview of methods for calculating the burden of disease due to specific risk factors. *Epidemiology* 17: 512–19.

Steenland K., Bray I., Greenland S., Boffetta P. (2000). Empirical Bayes adjustments for multiple results in hypothesis-generating or surveillance studies. *Cancer Epidemiology, Biomarkers, and Prevention* 9: 895–903.

Steenland K., Deddens J.A. (1997). Increased precision using countermatching in nested case-control studies [comment]. *Epidemiology* 8: 238–42.

Steenland K., Savitz D.A., Eds. (1997). *Topics in Environmental Epidemiology*. Oxford, Oxford University Press.

Steinbuch M., Weinberg C.R., Buckley J.D., Robison L.L., Sandler DP (1999). Indoor residential radon exposure and risk of childhood acute myeloid leukaemia. *British Journal of Cancer* 81: 900–6.

Stevens W., Thomas D.C., Lyon J.L., Till J.E., Kerber R.A., Simon SL, Lloyd R.D., Elghany N.A., Preston-Martin S. (1990). Leukemia in Utah and radioactive fallout from the Nevada test site. A case-control study. *Journal of the American Medical Association* 264: 585–91.

Stidley C.A., Samet J.M. (1993). A review of ecologic studies of lung cancer and indoor radon. *Health Physics* 65: 234–51.

Stidley C.A., Samet J.M. (1994). Assessment of ecologic regression in the study of lung cancer and indoor radon. *American Journal of Epidemiology* 139: 312–22.

Stolley P.D. (1991). When genius errs: R.A. Fisher and the lung cancer controversy. *American Journal of Epidemiology* 133: 416–25; discussion 426–8.

Stone R.A. (1988). Investigations of excess environmental risks around putative sources: statistical problems and a proposed test. *Statistics in Medicine* 7: 649–60.

Stott R. (2007). Climate change, poverty and war. *Journal of the Royal Society of Medicine* 100: 399–402.

Stram D., Longnecker M., Shames L., Kolonel L., Wilkens L, Pike M., Henderson B. (1995). Cost-efficient design of a diet validation study. *American Journal of Epidemiology* 142: 353–62.

Stram D.O. (1996). Meta-analysis of published data using a linear mixed-effects model. *Biometrics* 52: 536–44.

Stram D.O. (2005). Designs for studies of personal exposure to air pollution and the impact of measurement error. *Journal of Toxicology and Environmental Health, Part A* 68: 1181–7.

Stram D.O., Huberman M., Langholz B. (2000). Correcting for exposure measurement error in uranium miners studies: impact on inverse dose-rate effects. *Radiation Research* 154: 738–9; discussion 739–40.

Stram D.O., Kopecky K.J. (2003). Power and uncertainty analysis of epidemiological studies of radiation-related disease risk in which dose estimates are based on a complex dosimetry system: some observations. *Radiation Research* 160: 408–17.

Stram D.O., Langholz B., Huberman M., Thomas D.C. (1999). Correcting for exposure measurement error in a reanalysis of lung cancer mortality for the Colorado Plateau Uranium Miners cohort. *Health Physics* 77: 265–75.

Stram D.O., Mizuno S. (1989). Analysis of the DS86 atomic bomb radiation dosimetry methods using data on severe epilation. *Radiation Research* 117: 93–113.

Stram D.O., Sposto R. (1991). Recent uses of biological data for the evaluation of A-bomb radiation dosimetry. *Journal of Radiation Research (Tokyo)* 32(Suppl.): 122–35.

Stram D.O., Sposto R., Preston D., Abrahamson S., Honda T., Awa AA (1993). Stable chromosome aberrations among A-bomb survivors: an update. *Radiation Research* 136: 29–36.

Stroup DF, Berlin J.A., Morton S.C., Olkin I, Williamson G.D., Rennie D., Moher D., Becker B.J., Sipe T.A., Thacker SB (2000). Meta-analysis of observational studies in epidemiology: a proposal for reporting. Meta-analysis Of Observational Studies in Epidemiology (MOOSE) group. *Journal of the American Medical Association* 283: 2008–12.

Sturmer T., Schneeweiss S., Avorn J., Glynn R.J. (2005). Adjusting effect estimates for unmeasured confounding with validation data using propensity score calibration. *American Journal of Epidemiology* 162: 279–89.

Sturmer T., Thurigen D., Spiegelman D., Blettner M., Brenner H (2002). The performance of methods for correcting measurement error in case-control studies. *Epidemiology* 13: 507–16.

Sun W. (1998). Comparison of a cokriging method with a Bayesian alternative. *Environmentrics* 9: 445–57.

Swan S.H., Elkin E.P., Fenster L. (2000). The question of declining sperm density revisited: an analysis of 101 studies published 1934–1996. *Environmental Health Perspectives* 108: 961–6.

Talbott E.O., Youk A.O., McHugh-Pemu K.P., Zborowski J.V. (2003). Long-term follow-up of the residents of the Three Mile Island accident area: 1979–1998. *Environmental Health Perspectives* 111: 341–8.

Tango T. (2007). A class of multiplicity adjusted tests for spatial clustering based on case-control point data. *Biometrics* 63: 119–127.

Tao X.G., Massa J., Ashwell L., Davis K., Schwab M., Geyh A (2007). The world trade center clean up and recovery worker cohort study: respiratory health amongst cleanup workers approximately 20 months after initial exposure at the disaster site. *Journal of Occupational and Environmental Medicine* 49: 1063–72.

Tapio S., Jacob V. (2007). Radioadaptive response revisited. *Radiation and Environmental Biophysics* 46: 1–12.

Tatham L.M., Bove F.J., Kaye W.E., Spengler R.F. (2002). Population exposures to I-131 releases from Hanford Nuclear Reservation and preterm birth, infant mortality, and fetal deaths. *International Journal of Hygiene and Environmental Health* 205: 41–8.

Taubes G. (1997). Apocalypse not. *Science* 278: 1004–6.

Thieren M. (2005). Asian tsunami: death-toll addiction and its downside. *Bulletin of the World Health Organization* 83: 82.

Thomas D., Darby S., Fagnani F., Hubert P., Vaeth M., Weiss K (1992a). Definition and estimation of lifetime detriment from radiation exposures: principles and methods. *Health Physics* 63: 259–72.

Thomas D., Pogoda J., Langholz B., Mack W. (1994). Temporal modifiers of the radon-smoking interaction. *Health Physics* 66: 257–62.

Thomas D.C. (1981). General relative risk models for failure time and matched case-control studies. *Biometrics* 37: 673–86.

Thomas D.C. (1982). Temporal effects and interactions in cancer: implications of carcinogenic models. In: *Environmental Epidemiology: Risk Assessment*. Prentice R, Whittemore A.S. (eds.). Philadelphia, Society for Industrial and Applied Mathematics: pp. 107–121.

Thomas D.C. (1983a). Nonparametric estimation and tests of fit for dose–response relations. *Biometrics* 39: 263–8.

Thomas D.C. (1983b). Statistical methods for measuring risk: relevance of epidemiology to environmental standards, compensation and individual behavior (Chapter 16). In: *Methods and issues in occupational and environmental epidemiology*. Chiazze L., Lundiin F.E., Watkins D. (eds.). Ann Arbor, MI, Ann Arbor Science: 149–64.

Thomas D.C. (1987). Pitfalls in the analysis of exposure–time–response relationships. *Journal of Chronic Diseases* 40(Suppl. 2): 71S–8S.

Thomas D.C. (1988). Exposure–time–response relationships with applications to cancer epidemiology. *Annual Review of Public Health* 9: 451–82.

Thomas D.C. (1990). A model for dose rate and duration of exposure effects in radiation carcinogenesis. *Environmental Health Perspectives* 87: 163–71.

Thomas D.C. (1995). Re: "When will nondifferential misclassification of an exposure preserve the direction of a trend?" *American Journal of Epidemiology* 142: 782–4.

Thomas D.C. (1999). The Utah fallout study: How uncertainty has affected estimates of dose-response. In: *Uncertainties in Radiation Dosimetry and Their Impact on*

Dose-Response Analyses. Ron E., Hoffman F.O. (eds.). Bethesda, M.D., National Cancer Institute Monograph 99-4541: pp. 217–24.

Thomas D.C. (2000). Resolved: the probability of causation can be used in an equitable manner to resolve radiation tort claims and design compensation schemes. Pro. *Radiation Research* 154: 717–18.

Thomas D.C. (2004). *Statistical methods in genetic epidemiology.* Oxford, Oxford University Press.

Thomas D.C. (2005a). Why do estimates of the acute and chronic effects of air pollution on mortality differ? *Journal of Toxicology and Environmental Health Part A* 68: 1167–74.

Thomas D.C. (2005b). The need for a comprehensive approach to complex pathways in molecular epidemiology. *Cancer Epidemiology, Biomarkers, and Prevention* 14: 557–9.

Thomas D.C. (2007a). Multistage sampling for latent variable models. *Lifetime Data Analysis* 13: 565–81.

Thomas D.C. (2007b). Using gene–environment interactions to dissect the effects of complex mixtures. *Journal of Exposure Science and Environmental Epidemiology* 17 (suppl. 2: S71–4.

Thomas D.C., Baurley J.W., Brown E.E., Figueiredo J., Goldstein A, Hazra A., Hunter D., Sellers T, Wilson R., Rothman N (2008). Approaches to complex pathways in molecular epidemiology: summary of an AACR special conference. *Cancer Research* (in press).

Thomas D.C., Blettner M., Day N.E. (1992b). Use of external rates in nested case-control studies with application to the international radiation study of cervical cancer patients. *Biometrics* 48: 781–94.

Thomas D.C., Jerrett M., Kuenzli N., Louis T.A., Dominici F, Zeger S., Schwarz J., Burnett R.T., Krewski D., Bates D. (2007a). Bayesian model averaging in time-series studies of air pollution and mortality. *Journal of Toxicology and Environmental Health, Part A* 70: 311–15.

Thomas D.C., Petitti D.B., Goldhaber M., Swan S.H., Rappaport EB, Hertz-Picciotto I. (1992c). Reproductive outcomes in relation to malathion spraying in the San Francisco Bay Area, 1981–1982. *Epidemiology* 3: 32–9.

Thomas D.C., Siemiatycki J., Dewar R., Robins J., Goldberg M, Armstrong B.G. (1985). The problem of multiple inference in studies designed to generate hypotheses. *American Journal of Epidemiology* 122: 1080–95.

Thomas D.C., Stram D., Dwyer J. (1993). Exposure measurement error: Influence on exposure–disease relationships and methods of correction. *Annual Review of Public Health* 14: 69–93.

Thomas D.C., Witte J.S., Greenland S. (2007b). Dissecting effects of complex mixtures: who's afraid of informative priors? *Epidemiology* 18: 186–90.

Thomas D.I., Salmon L., Antell B.A. (1991). Revised Technical Basis for the BNFL/UKAEA Compensation Agreement for radiation linked diseases. *Journal of Radiological Protection*: 111.

Thompson J.R., Carter R.L. (2007). An overview of normal theory structural measurement error models. *International Statistical Review* 75: 183–98.

Thompson W.D. (1990). Nonrandom yet unbiased. *Epidemiology* 1: 262–5.

Thomsen S.R. (1987). *Danish elections 1920-1979: a logit approach to ecological analysis and inference.* Arhus, University of Arhus.

Thun M.J., Henley S.J., Burns D., Jemal A., Shanks T.G., Calle EE (2006). Lung cancer death rates in lifelong nonsmokers. *Journal of the National Cancer Institute* 98: 691–9.

Thurston S.W., Liu G., Miller D.P., Christiani D.C. (2005). Modeling lung cancer risk in case-control studies using a new dose metric of smoking. *Cancer Epidemiology, Biomarkers, and Prevention* 14: 2296–302.

Till JE, Meyer H.R. (1983). *Radiological assessment.* Washington, D.C., U.S. Nuclear Regulatory Commission.

Till J.E., Simon S.L., Kerber R., Lloyd R.D., Stevens W., Thomas DC, Lyon J.L., Preston-Martin S. (1995). The Utah Thyroid Cohort Study: analysis of the dosimetry results. *Health Physics* 68: 472–83.

Tobler W.R. (1970). A computer movie simulating urban growth in the Detroit region. *Economic Geography* 46: 234–40.

Toledano M.B., Nieuwenhuijsen M.J., Best N., Whitaker H., Hambly P, de Hoogh C., Fawell J., Jarup L., Elliott P. (2005). Relation of trihalomethane concentrations in public water supplies to stillbirth and birth weight in three water regions in England. *Environmental Health Perspectives* 113: 225–32.

Tosteson T.D., Ware J.H. (1990). Designing a logistic regression study using surrogate measures for exposure and outcome. *Biometrika* 77: 11–21.

Touloumi G., Katsouyanni K., Zmirou D., Schwartz J., Spix C, de Leon A.P., Tobias A., Quennel P., Rabczenko D., Bacharova L., Bisanti L, Vonk J.M., Ponka A. (1997). Short-term effects of ambient oxidant exposure on mortality: a combined analysis within the APHEA project. Air Pollution and Health: a European Approach. *American Journal of Epidemiology* 146: 177–85.

Truett J., Cornfield J., Kannel W. (1967). A multivariate analysis of the risk of coronary heart disease in Framingham. *Journal of Chronic Diseases* 20: 511–24.

Tsao J.L., Tavare S., Salovaara R., Jass J.R., Aaltonen LA, Shibata D. (1999). Colorectal adenoma and cancer divergence. Evidence of multilineage progression. *American Journal of Pathology* 154: 1815–24.

Tu Y-K, Gilthorpe M.S. (2007). Revisiting the relation between change and initial value: a review and evaluation. *Statistics in Medicine* 26: 443–57.

Tukey J.W. (1949). One degree of freedom for non-additivity. *Biometrics* 5: 232–42.

Ulm K. (1999). Nonparametric analysis of dose–response relationships. *Annals of the New York Academy of Sciences* 895: 223–31.

Umbach D., Weinberg C. (1997). Designing and analysing case-control studies to exploit independence of genotype and exposure. *Statistics in Medicine* 16: 1731–43.

UNSCEAR (2000). *(United Nations Scientific Committee on the Effects of Atomic Radiation) UNSCEAR 2000 report Vol. II: Sources and effects of ionizing radiation.* New York, NY, United Nations.

UNSCEAR (2006). Non-targeted and delayed effects of exposure to ionizing radiation. New York, United Nations Scientific Committee on Effects of Ionizing Radiation: 101.

Ury H. (1975). Efficiency of case-control studies with multiple controls per case: continuous or dichotomous data. *Biometrics* 31: 643–9.

Van Atten C., Brauer M., Funk T., Gilbert N.L., Graham L., Kaden D, Miller P.J., Bracho L.R., Wheeler A., White R.H. (2005). Assessing population exposures to motor vehicle exhaust. *Reviews of Environmental Health* 20: 195–214.

Van Pelt W.R. (2003). Epidemiological associations among lung cancer, radon exposure and elevation above sea level–a reassessment of Cohen's county level radon study. *Health Physics* 85: 397–403.

Varma D.R. (1987). Epidemiological and experimental studies on the effects of methyl isocyanate on the course of pregnancy. *Environmental Health Perspectives* 72: 153 7.

Viallefont V., Raftery A.E., Richardson S. (2001). Variable selection and Bayesian model averaging in case-control studies. *Statistics in Medicine* 20: 3215–30.

Vianna N.J., Polan A.K. (1973). Epidemiologic evidence for transmission of Hodgkin's disease. *New England Journal of Medicine* 289: 499–502.

Vijayan V.K., Sankaran K. (1996). Relationship between lung inflammation, changes in lung function and severity of exposure in victims of the Bhopal tragedy. *Eur Respir J*, 9: 1977–82.

Vijayan V.K., Sankaran K., Sharma S.K., Misra N.P. (1995). Chronic lung inflammation in victims of toxic gas leak at Bhopal. *Respir Med.* 89: 105–11.

Vine M.F., Degnan D., Hanchette C. (1997). Geographic information systems: their use in environmental epidemiologic research. *Environmental Health Perspectives* 105: 598–605.

Vineis P., Alavanja M., Garte S. (2004). Dose–response relationship in tobacco-related cancers of bladder and lung: a biochemical interpretation. *International Journal of Cancer* 108: 2–7.

Vogel C., Brenner H., Pfahlberg A., Gefeller O. (2005). The effects of joint misclassification of exposure and disease on the attributable risk. *Statistics in Medicine* 24: 1881–96.

von Schreeb J., Rosling H., Garfield R. (2007). Mortality in Iraq. *Lancet* 369: 101; author reply 103–4.

Wacholder S. (1995). When measurement errors correlate with truth: surprising effects of nondifferential misclassification. *Epidemiology* 6: 157–61.

Wacholder S., Armstrong B., Hartge P. (1993). Validation studies using an alloyed gold standard. *American Journal of Epidemiology* 137: 1251–8.

Wacholder S., Carroll R.J., Pee D., Gail M.H. (1994). The partial questionnaire design for case-control studies. *Statistics in Medicine* 13: 623–34.

Wacholder S., Dosemeci M., Lubin J.H. (1991). Blind assignment of exposure does not always prevent differential misclassification. *American Journal of Epidemiology* 134: 433–7.

Wacholder S., McLaughlin J.K., Silverman D.T., Mandel JS (1992a). Selection of controls in case-control studies. I. Principles. *American Journal of Epidemiology* 135: 1019–28.

Wacholder S., Silverman D.T., McLaughlin J.K., Mandel JS (1992b). Selection of controls in case–control studies. II. Types of controls. *American Journal of Epidemiology* 135: 1029–41.

Wakefield J. (2003). Sensitivity analyses for ecological regression. *Biometrics* 59: 9–17.

Wakefield J. (2004). Ecologic inference for 2×2 tables. *Journal of the Royal Statistical Society, Series A* 167: 1–42.

Wakefield J., Elliott P. (1999). Issues in the statistical analysis of small area health data. *Statistics in Medicine* 18: 2377–99.

Wakefield J., Quinn M., Raab G. (2001). Editorial: Disease Clusters and Ecological Studies. *Journal of the Royal Statistical Society, Series A* 164: 1–2.

Wakefield J., Salway R. (2001). A statistical framework for ecological and aggregate studies. *Journal of the Royal Statistical Society, Series A* 164: 119–37.

Walker A.M. (1982). Anamorphic analysis: sampling and estimation for covariate effects when both exposure and disease are known. *Biometrics* 38: 1025–32.

Wall M.M. (2004). A close look at the spatial structure implied by the CAR and SAR models. *Journal of Statistical Planning and Inference* 121: 311–24.

Waller L.A., Gotway C.A. (2004). *Applied Spatial Statistics for Public Health*. New York, Wiley-Interscience.

Waller R.E., Brooks A.G., Adler M.W. (1973). The 1952 fog cohort study. *British Journal of Preventive and Social Medicine* 27: 68–9.

Wartenberg D. (2001). Investigating Disease Clusters: Why, When and How? *Journal of the Royal Statistical Society, Series A* 164: 13–22.

Wartenberg D., Greenberg M., Lathrop R. (1993). Identification and characterization of populations living near high-voltage transmission lines: a pilot study. *Environmental Health Perspectives* 101: 626–32.

Webster D.W., Vernick J.S., Hepburn L.M. (2002). Effects of Maryland's law banning "Saturday night special" handguns on homicides. *American Journal of Epidemiology* 155: 406–12.

Weinberg C.R., Umbach D.M. (2000). Choosing a retrospective design to assess joint genetic and environmental contributions to risk. *American Journal of Epidemiology* 152: 197–203.

Weinberg C.R., Wacholder S. (1990). The design and analysis of case-control studies with biased sampling. *Biometrics* 46: 963–75.

Weiss C.H. (1972). *Evaluation research*. Englewood Cliffs, N.J., Prentice-Hall.

Wheeler K., McKelvey W., Thorpe L., Perrin M., Cone J., Kass D, Farfel M., Thomas P., Brackbill R. (2007). Asthma diagnosed after 11 September 2001 among rescue and recovery workers: findings from the World Trade Center Health Registry. *Environmental Health Perspectives* 115: 1584–90.

Whitaker H., Best N., Nieuwenhuijsen M.J., Wakefield J., Fawell J, Elliott P. (2005). Modelling exposure to disinfection by-products in drinking water for an epidemiological study of adverse birth outcomes. *Journal of Exposure Analysis Environmental Epidemiology* 15: 138–46.

White J.E. (1982). A two stage design for the study of the relationship between a rare exposure and a rare disease. *American Journal of Epidemiology* 115: 119–28.

Whittemore A.S. (1977). The age distribution of human cancer for carcinogenic exposures of varying intensity. *American Journal of Epidemiology* 106: 418–32.

Whittemore A.S. (1988). Effect of cigarette smoking in epidemiological studies of lung cancer. *Statistics in Medicine* 7: 223–38.

Whittemore A.S. (1997). Multistage sampling designs and estimating equations. *Journal of the Royal Statistical Society, Series B* 59: 589–602.

Whittemore A.S., Keller J.B. (1978). Quantitative theories of carcinogenesis. *SIAM Review* 20: 1–30.

Whittemore A.S., Keller J.B. (1988). Approximations for regression with covariate measurement error. *Journal of the American Statistical Association* 83: 1057–66.

Whittemore A.S., McMillan A. (1983). Lung cancer mortality among U.S. uranium miners: a reappraisal. *Journal of the National Cancer Institute* 71: 489–99.

WHO (2004). Global burden of disease. http://www.who.int/evidence/bod.

Williams F.L., Ogston S.A. (2002). Identifying populations at risk from environmental contamination from point sources. *Occupational and Environmental Medicine* 59: 2–8.

Williams J.R., Alexander F.E., Cartwright R.A., McNally RJQ (2001). Methods for eliciting aetiological clues from geographically clustered cases of disease, with application to leukaemia lymphoma data. *Journal of the Royal Statistical Society, Series A* 164: 49–60.

Willis A., Jerrett M., Burnett R., Krewski D. (2003a). The association between sulfate air pollution and mortality at the county scale: An exploration of the impact of scale on a long-term exposure study. *Journal of Toxicology and Environmental Health, Part A* 66: 1605–24.

Willis A., Krewski D., Jerrett M., Goldberg M., Burnett R (2003b). Selection of ecologic covariates in the American Cancer Society study. *Journal of Toxicology and Environmental Health, Part A* 66: 1563–90.

Wing S., Richardson D., Armstrong D., Crawford-Brown D (1997). A reevaluation of cancer incidence near the Three Mile Island nuclear plant: the collision of evidence and assumptions. *Environmental Health Perspectives* 105: 52–7.

Wong M.Y., Day N.E., Bashir S.A., Duffy S.W. (1999). Measurement error in epidemiology: the design of validation studies I: univariate situation. *Statistics in Medicine* 18: 2815–29.

Woolf B. (1955). On estimating the relation between blood group and disease. *Annals of Human Genetics* 19: 251–3.

Wraith D., Mengersen K. (2007). Assessing the combined effect of asbestos exposure and smoking on lung cancer: a Bayesian approach. *Statistics in Medicine* 26: 1150–69.

Wulfsohn M.S., Tsiatis A.A. (1997). A joint model for survival and longitudinal data measured with error. *Biometrics* 53: 330–9.

Xu X., Wang L. (1998). Synergistic effects of air pollution and personal smoking on adult pulmonary function. *Archives of Environmental Health* 53: 44–53.

Yi N., George V., Allison D.B. (2003). Stochastic search variable selection for identifying multiple quantitative trait loci. *Genetics* 164: 1129–38.

Zaffanella L., Savitz D.A., Greenland S., Ebi K.L. (1998). The residential case-specular method to study wire codes, magnetic fields, and disease. *Epidemiology* 9: 16–20.

Zanobetti A., Schwartz J., Samoli E., Gryparis A., Touloumi G, Atkinson R., Le Tertre A., Bobros J., Celko M., Goren A., Forsberg B, Michelozzi P., Rabczenko D., Aranguez Ruiz E., Katsouyanni K. (2002). The temporal pattern of mortality responses to air pollution: a multicity assessment of mortality displacement. *Epidemiology* 13: 87–93.

Zanobetti A., Schwartz J., Samoli E., Gryparis A., Touloumi G, Peacock J., Anderson R.H., Le Tertre A., Bobros J., Celko M., et al. (2003). The temporal pattern of respiratory and heart disease mortality in response to air pollution. *Environmental Health Perspectives* 111: 1188–93.

Zanobetti A., Wand M.P., Schwartz J., Ryan L.M. (2000). Generalized additive distributed lag models: quantifying mortality displacement. *Biostatistics* 1: 279–92.

Zeger S., Dominici F., Samet J. (1999). Harvesting-resistant estimates of air pollution effects on mortality. *Epidemiology* 10: 171–5.

Zeger S.L. (1988). A regression model for time series of counts. *Biometrika* 75: 621–9.

Zeger S.L., Qaqish B. (1988). Markov regression models for time series: a quasi-likelihood approach. *Biometrics* 44: 1019–31.

Zeger S.L., Thomas D., Dominici F., Samet J.M., Schwartz J, Dockery D., Cohen A. (2000). Exposure measurement error in time-series studies of air pollution: concepts and consequences. *Environmental Health Perspectives* 108: 419–26.

Zeise L., Hattis D., Andersen M., Bailer A., Bayard S., Chen C, Clewell H., Conolly R., Crump K., Dunson D., Finkel A., Haber L, Jarabek A., Kodell R., Krewski D., Thomas D., Thorslund T, Wassell JT (2002). Research opportunities in dose response modeling to improve risk assessment. *Human and Ecological Risk Assessment* 8: 1421–44.

Zhang Y., Bi P., Hiller J.E. (2007). Climate change and disability-adjusted life years. *Journal of Environmental Health* 70: 32–6.

Zhao L.P., Lipsitz S. (1992). Designs and analysis of two-stage studies. *Statistics in Medicine* 11: 769–82.

Zheng M., Cass G.R., Schauer J.J., Edgerton E.S. (2002). Source apportionment of PM2.5 in the Southeastern United States using solvent-extractable organic compounds as tracers. *Environmental Science and Technology* 36: 2361–71.

Index

Printed in Japan
落丁、乱丁本のお問い合わせは
Amazon.co.jp カスタマーサービスへ

2303548R00240